LECTURES ON
FINSLER GEOMETRY

LECTURES ON FINSLER GEOMETRY

Zhongmin Shen
Indiana University–Purdue University Indianapodis, USA

World Scientific
Singapore • New Jersey • London • Hong Kong

Published by

World Scientific Publishing Co. Pte. Ltd.
P O Box 128, Farrer Road, Singapore 912805
USA office: Suite 1B, 1060 Main Street, River Edge, NJ 07661
UK office: 57 Shelton Street, Covent Garden, London WC2H 9HE

British Library Cataloguing-in-Publication Data
A catalogue record for this book is available from the British Library.

LECTURES ON FINSLER GEOMETRY

ISBN 981-02-4530-0
ISBN 981-02-4531-9 (pbk)

Printed in Singapore by World Scientific Printers

to my daughter, Alice

Preface

In 1854, B. Riemann introduced the notion of curvature for spaces with a family of inner products (Riemannian metrics). There was no significant progress in the general case until 1918, when P. Finsler studied the variation problem for spaces with a family of norms (Finsler metrics). Meanwhile, A. Einstein used Riemannian geometry to present his general relativity. By that times, however, the geometry of Finsler spaces was still at its infant stage. Until 1926, L. Berwald extended Riemann's notion of curvature to Finsler spaces and discovered a new non-Riemannian quantity using his connection. In his Paris address in 1900, D. Hilbert formulated 23 problems, the 4th and 23rd problems being in Finsler's category. Finsler geometry has broader applications in many areas of natural science [AIM] and will continue to develop through the efforts of many geometers around the world [BCS3].

This book comes out of a series of lecture notes based on my work at IHES (*Curvature, distance and volume in Finsler geometry*, preprint, 1997). Viewing Finsler spaces as regular metric spaces, we discuss the problems from the modern geometric point of view. The book is intended to provide basic materials for Finsler geometry including Riemannian geometry. The presentation is aimed at a reader who has completed a one-year graduate course in differential geometry, and who knows the elements of homotopy theory in topology.

The first four chapters cover some basic theories on regular metrics (Finsler metrics) and regular measures (volume forms) on manifolds.

In Chapter 5, we introduce the notion of geodesics via the calculus of variation. Using the geodesics, we define the Chern connection. The Chern connection is an important tool to investigate the geometric structure of Finsler spaces.

In Chapter 6, we introduce the notion of Riemann curvature via geodesic variation. This is different from Riemann's and Berwald's approaches. We discuss the dependence of the Riemann curvature on geodesics. In Chapter 7, we introduce various non-Riemannian quantities (e.g., the Chern curvature, the Landsberg curvature and the S-curvature, etc) and their geometric meanings.

In Chapters 8 and 9, we use Cartan's exterior differential method to discuss various curvatures and relationships. In particular, we discuss Finsler spaces of constant curvature. Several important examples are studied. These examples make Finsler geometry more fruitful.

Starting with Chapter 10, we study the metric properties of Finsler spaces and metric-measure properties of Finsler m spaces. We first derive the second variation formula of length and show the geometric meaning of the T-curvature. Then in Chapters 11 and 12, we discuss the index form, Jacobi fields, exponential maps and their relationship. The discussion leads to some basic comparison theorems that are discussed in Chapter 13.

In Chapters 14, 15 and 16, we use a distance function to study the global geometric structure of Finsler spaces and Finsler m spaces. We obtain several estimates on the geometry of level sets of distance functions. The geometric meaning of the S-curvature lies in the volume comparison theorem.

In Chapters 17 and 18, we apply the Morse theory to the canonical energy functional on the loop space. In particular, we prove a vanishing theorem for the homotopy groups of Finsler spaces with pinched curvature.

In the last chapter, we view a Finsler space as a point of a space of Finsler spaces equipped with the Gromov-Hausdorff distance. We briefly discuss the precompactness and finiteness of certain classes of Finsler spaces.

I would like to take this opportunity to thank several people in my personal life and academic life. First, I would like to thank my thesis advisor Detlef Gromoll for his help and advice in Riemannian geometry during my graduate study in SUNY at Stony Brook, and thank another advisor after the thesis, S.S. Chern for bringing me to a wider field — Finsler geometry. I thank Bart Ng for providing me with a good research environment at my current institution. Finally, I thank my wife, Tianping, for her consistent support and understanding. Without them, I would not have written this book, for sure.

Zhongmin Shen
Indianapolis, USA

Contents

Preface vii

Chapter 1 Finsler Spaces **1**
1.1 Metric Spaces 1
1.2 Minkowski Spaces 6
1.3 Finsler Spaces 12

Chapter 2 Finsler m Spaces **19**
2.1 Measure Spaces 19
2.2 Volume on a Finsler Space 21
2.3 Hyperplanes in a Minkowski m Space 27
2.4 Hypersurfaces in a Finsler m Space 30

Chapter 3 Co-Area Formula **35**
3.1 Legendre Transformations 35
3.2 Gradients of Functions 41
3.3 Co-Area Formula 46

Chapter 4 Isoperimetric Inequalities **55**
4.1 Isoperimetric Profiles 56
4.2 Sobolev Constants and First Eigenvalue 58
4.3 Concentration of Finsler m Spaces 64
4.4 Observable Diameter 69

Chapter 5 Geodesics and Connection **75**
5.1 Geodesics 75

5.2 Chern Connection 80
5.3 Covariant Derivatives 88
5.4 Geodesic Flow 90

Chapter 6 Riemann Curvature **95**
6.1 Birth of the Riemann Curvature 95
6.2 Geodesic Fields 99
6.3 Projectively Related Finsler Metrics 101

Chapter 7 Non-Riemannian Curvatures **107**
7.1 Cartan Torsion 107
7.2 Chern Curvature 112
7.3 S-Curvature 117

Chapter 8 Structure Equations **125**
8.1 Structure Equations of Finsler Spaces 125
8.2 Structure Equations of Riemannian Metrics 130
8.3 Riemann Curvature of Randers Metrics 134

Chapter 9 Finsler Spaces of Constant Curvature **137**
9.1 Finsler Metrics of Constant Curvature 137
9.2 Examples 141
9.3 Randers Metrics of Constant Curvature 145

Chapter 10 Second Variation Formula **153**
10.1 T-Curvature 153
10.2 Second Variation of Length 156
10.3 Synge Theorem 162

Chapter 11 Geodesics and Exponential Map **165**
11.1 Exponential Map 165
11.2 Jacobi Fields 167
11.3 Minimality of Geodesics 171
11.4 Completeness of Finsler Spaces 175

Chapter 12 Conjugate Radius and Injectivity Radius **179**
12.1 Conjugate Radius 179
12.2 Injectivity Radius 183
12.3 Geodesic Loops and Closed Geodesics 189

Chapter 13 Basic Comparison Theorems **193**
13.1 Flag Curvature Bounded Above 193
13.2 Positive Flag Curvature 198
13.3 Ricci Curvature Bounded Below 201
13.4 Green-Dazord Theorem 203

Chapter 14 Geometry of Hypersurfaces **207**
14.1 Hessian and Laplacian 207
14.2 Normal Curvature 213
14.3 Mean Curvature 218
14.4 Shape Operator 221

Chapter 15 Geometry of Metric Spheres **229**
15.1 Estimates on the Normal Curvature 229
15.2 Convexity of Metric Balls 234
15.3 Estimates on the Mean Curvature 236
15.4 Metric Spheres in a Convex Domain 241

Chapter 16 Volume Comparison Theorems **245**
16.1 Volume of Metric Balls 245
16.2 Volume of Tubular Neighborhoods 252
16.3 Gromov Simplicial Norms 256
16.4 Estimates on the Expansion Distance 259

Chapter 17 Morse Theory of Loop Spaces **265**
17.1 A Review on the Morse Theory 265
17.2 Indexes of Geodesic Loops 268
17.3 Energy Functional on a Loop Space 271
17.4 Approximation of Loop Spaces 273

Chapter 18 Vanishing Theorems for Homotopy Groups **279**
18.1 Intermediate Curvatures 279
18.2 Vanishing Theorem for Homotopy Groups 280
18.3 Finsler Spaces of Positive Constant Curvature 286

Chapter 19 Spaces of Finsler Spaces **291**
19.1 Gromov-Hausdorff Distance 291
19.2 Precompactness Theorem 293

Bibliography 299

Index 305

Chapter 1

Finsler Spaces

1.1 Metric Spaces

Metric spaces exist everywhere in our life. A metric space is a set of points equipped with a metric (distance function). With a metric, we measure the distance from a point to another point in the set. More precisely, a metric on a set M is a function $d : M \times M \to \mathrm{R}$ which has the following properties

(i) for any $p, q \in M$,

$$d(p, q) \geq 0,$$

equality holds if and only if $p = q$;

(ii) for any $p, q, r \in M$,

$$d(p, q) \leq d(p, r) + d(r, q).$$

In addition, if d has the following reversibility,

(iii) for any two points $p, q \in M$,

$$d(p, q) = d(q, p),$$

then d is called a *reversible metric.*

Let

$$\mathrm{R}^n := \Big\{ (x^i) = (x^1, \cdots, x^n) \;\Big|\; x^i \in \mathrm{R},\ i = 1, \cdots, n \Big\}$$

denote the canonical n-dimensional real vector space. The *canonical Euclidean norm* $|\cdot|$ on R^n is defined by

$$|y| := \sqrt{\sum_{i=1}^{n}(y^i)^2}, \qquad y = (y^i) \in \mathrm{R}^n.$$

Define $d_E : \mathrm{R}^n \times \mathrm{R}^n \to [0, \infty)$ by

$$d_E(u, v) := |v - u|.$$

It is easy to see that d_E is a reversible metric on R^n. We call d_E the *canonical Euclidean metric.* For simplicity, we shall denote the pair $(\mathrm{R}^n, |\cdot|)$ or (R^n, d_E) by $\mathbb{R}^n$ and call it the *canonical Euclidean space.* Euclidean spaces are the simplest metric spaces.

Let V be a finite dimensional real vector space and $\langle\cdot,\cdot\rangle$ an inner product on V. Define

$$\|y\| := \sqrt{\langle y, y\rangle}, \qquad y \in \mathrm{V}.$$

We obtain a metric d on V,

$$d(u, v) := \|v - u\|, \qquad u, v \in \mathrm{V}.$$

We call $\|\cdot\|$ and d the *Euclidean norm* and the *Euclidean metric* on V, respectively.

For any n-dimensional Euclidean space $(V, \|\cdot\|)$, there is a linear transformation $\varphi : \mathrm{V} \to \mathrm{R}^n$ such that

$$|\varphi(y)| = \|y\|, \qquad y \in \mathrm{V}.$$

In this sense, there is a unique Euclidean space in each dimension up to a linear isometry. Euclidean metrics can be generalized to more "flat" metrics.

Example 1.1.1 Let V be a finite-dimensional vector space and $\|\cdot\| : \mathrm{V} \to \mathrm{R}$ be a function with the following properties

(i) $\|u\| \geq 0$, equality holds if and only if $u = 0$;
(ii) $\|\lambda u\| = \lambda\|u\|,\ \lambda > 0$;
(iii) $\|u + v\| \leq \|u\| + \|v\|,\ u, v \in \mathrm{V}$.

Define $d : \mathrm{V} \times \mathrm{V} \to [0, \infty)$ by

$$d(u, v) := \|v - u\|.$$

d is a "flat" metric on V. ♯

Example 1.1.2 (Funk) Let Ω be a bounded domain in $\mathbb{R}^n = (\mathrm{R}^n, |\cdot|)$. Assume that Ω is strictly convex, i.e., for any line segment L in $\mathbb{R}^n$, if the endpoints of L are contained in Ω, then the whole L is contained in Ω. For an ordered pair of points $p, q \in \Omega$, let L_{pq} denote the ray issuing from p and passing through q. Since Ω is strictly convex, there is a unique intersection point $z_{pq} := L_{pq} \cap \partial\Omega$.

Define

$$d(p, q) := \ln \frac{|z_{pq} - p|}{|z_{pq} - q|}. \tag{1.1}$$

We will show that d is a non-reversible metric on Ω. d is called the *Funk metric*.

It suffices to prove the triangle inequality:

$$d(p, q) \le d(p, r) + d(r, q).$$

Assume that r is not on the line passing through p and q. Let

$$\theta := \angle p x z_{pr}, \qquad \phi := \angle p z_{pr} z_{rq}, \qquad \psi := \angle r z_{rq} z_{pr}.$$

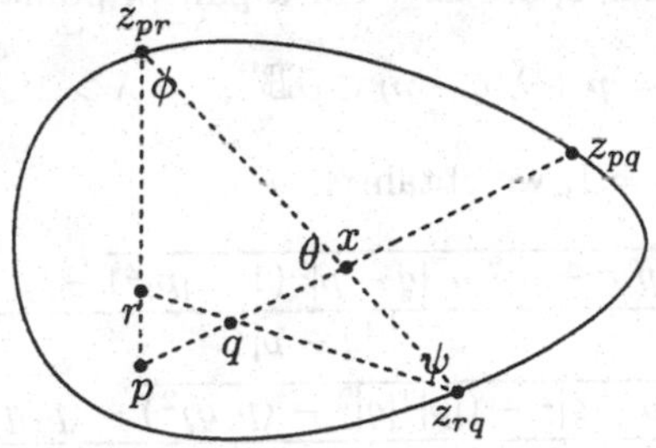

By the law of sines, we have the following identities.

$$\frac{|z_{pr} - p|}{\sin\theta} = \frac{|x - p|}{\sin\phi}, \qquad \frac{|z_{rq} - q|}{\sin\theta} = \frac{|x - q|}{\sin\psi}, \qquad \frac{|z_{rq} - r|}{\sin\phi} = \frac{|z_{pr} - r|}{\sin\psi}.$$

We obtain

$$\begin{aligned}\frac{|x-p|}{|x-q|} &= \frac{|z_{pr}-p|\sin\phi/\sin\theta}{|z_{rq}-q|\sin\psi/\sin\theta}\\ &= \frac{|z_{pr}-p|\sin\phi}{|z_{rq}-q|\sin\psi}\\ &= \frac{|z_{pr}-p|}{|z_{rq}-q|}\frac{\sin\phi}{\sin\psi}\\ &= \frac{|z_{pr}-p|}{|z_{rq}-q|}\frac{|z_{rq}-r|}{|z_{pr}-r|}.\end{aligned}$$

On the other hand,

$$\frac{|z_{pq}-p|}{|z_{pq}-q|} = \frac{|z_{pq}-x|+|x-p|}{|z_{pq}-x|+|x-q|} < \frac{|x-p|}{|x-q|}.$$

Combining them yields

$$\frac{|z_{pq}-p|}{|z_{pq}-q|} < \frac{|z_{pr}-p|}{|z_{pr}-r|}\frac{|z_{rq}-r|}{|z_{rq}-q|}.$$

This gives

$$d(p,q) < \ln\frac{|z_{pq}-p|}{|z_{pq}-q|} = \ln\frac{|z_{pq}-p|}{|z_{pq}-r|} + \ln\frac{|z_{pq}-r|}{|z_{pq}-q|} = d(p,r)+d(r,q).$$

Thus d is a metric on Ω.

Now we take a look at the special case when $\Omega = \mathbb{B}^n$ is the standard unit ball in the Euclidean space $\mathbb{R}^n$. For a pair of points $p, q \in \mathbb{B}^n$, let

$$z_{pq} = p + \lambda(q-p) \in \partial\mathbb{B}^n, \qquad \lambda > 1.$$

From the equation $|z|^2 = 1$, we obtain

$$\begin{aligned}\lambda &= \frac{\sqrt{\langle p, q-p\rangle^2 + |q-p|^2(1-|p|^2)} - \langle p, q-p\rangle}{|q-p|^2}\\ &= \frac{\sqrt{|p-q|^2 - (|p|^2|q|^2 - \langle p,q\rangle^2)} - \langle p, q-p\rangle}{|q-p|^2}.\end{aligned}$$

By (1.1), we obtain a formula for the Funk metric on $\mathbb{B}^n$

$$d(p,q) = \ln\frac{\lambda}{\lambda-1} = \ln\frac{\sqrt{|p-q|^2 - (|p|^2|q|^2-\langle p,q\rangle^2)} - \langle p, q-p\rangle}{\sqrt{|p-q|^2 - (|p|^2|q|^2-\langle p,q\rangle^2)} - \langle q, q-p\rangle}.$$

Note that

$$\lim_{q\to\partial\mathrm{B}^n} d(0,q) = \infty, \qquad \lim_{p\to\partial\mathrm{B}^n} d(p,0) = \ln 2.$$

♯

Example 1.1.3 (Klein) Let d denote the Funk metric on a strongly convex domain $\Omega \subset \mathbb{R}^n$. Define $d_K : \Omega \times \Omega \to [0,\infty)$ by

$$d_K(p,q) := \frac{1}{2}\Big\{d(p,q) + d(q,p)\Big\}, \tag{1.2}$$

Observe that for $p, q, r \in \Omega$,

$$\begin{aligned} d_K(p,q) &= \frac{1}{2}\Big\{d(p,q) + d(q,p)\Big\} \\ &\leq \frac{1}{2}\Big\{d(p,r) + d(r,q)\Big\} + \frac{1}{2}\Big\{d(q,r) + d(r,p)\Big\} \\ &= \frac{1}{2}\Big\{d(p,r) + d(r,p)\Big\} + \frac{1}{2}\Big\{d(q,r) + d(r,q)\Big\} \\ &= d_K(p,r) + d_K(r,q). \end{aligned}$$

Thus d_K is a metric too. We call d_K the *Klein metric.* ♯

Let (M,d) be a metric space. For a Lipschitz continuous curve $c : [a,b] \to M$, define

$$L_d(c) := \sup \sum_{i=0}^{k-1} d\Big(c(t_i), c(t_{i+1})\Big), \tag{1.3}$$

where the supremum is taken over all partitions $a = t_0 < \cdots < t_k = b$. L_d is called the *length structure* induced by d. The length structure $L = L_d$ induces a metric d_L on M by

$$d_L(p,q) := \inf\Big\{L(c),\ c \text{ is Lipschitz}, c(a) = p, c(b) = q\Big\}.$$

From the definition,

$$d \leq d_L.$$

d is called a *path metric* if $d = d_L$.

Example 1.1.4 Let $\mathbb{S}^n$ denote the unit sphere in the Euclidean space $\mathbb{R}^{n+1}$. Define two metrics d_1 and d_2 on $\mathbb{S}^n$ by

$$\begin{aligned} d_1(u,v) : &= \theta(u,v), \qquad u,v \in \mathbb{S}^n, \\ d_2(u,v) : &= |u-v|, \qquad u,v \in \mathbb{S}^n, \end{aligned}$$

where $\theta(u,v)$ denotes the arc-length of the smaller portion of the great circle passing through u and v. Both d_1 and d_2 define the same length structure $L_{d_1} = L_{d_2}$ on $\mathbb{S}^n$. It is easy to verify that d_1 is a path metric, but d_2 is not a path metric. ♯

1.2 Minkowski Spaces

In this section, we will discuss the simplest regular "flat" metric spaces.

Definition 1.2.1 A *Minkowski norm* on V is a nonnegative function $F : \mathrm{V} \to [0,\infty)$ which has the following properties

(M1) F is C^∞ on $\mathrm{V} \setminus \{0\}$;
(M2) $F(\lambda y) = \lambda F(y)$, for all $\lambda > 0$ and $y \in \mathrm{V}$;
(M3) For any $y \in \mathrm{V}\setminus\{0\}$, the symmetric bilinear form $\mathbf{g}_y$ on V is positive definite, where

$$\mathbf{g}_y(u,v) := \frac{1}{2}\frac{\partial^2}{\partial s \partial t}\Big[F^2(y+su+tv)\Big]|_{s=t=0}.$$

The pair (V, F) is called a *Minkowski space*

We will show that any Minkowski norm F on a vector space V satisfies

$$F(u+v) \le F(u) + F(v), \qquad u,v \in \mathrm{V}. \tag{1.4}$$

See Lemma 1.2.2 below. Hence it induces a metric d on V by

$$d(u,v) := F(v-u), \qquad u,v \in \mathrm{V}.$$

We call a metric d on V a *Minkowski metric* if it is induced by a Minkowski norm F on V.

There are many Minkowski norms on a vector space.

Example 1.2.1 (Randers norm) Let α an Euclidean norm and β a linear form on an n-dimensional vector space V. Define

$$F(y) := \alpha(y) + \beta(y). \tag{1.5}$$

Clearly, F satisfies (M1) and (M2) in Definition 1.2.1. We are going to show that (M3) holds if and only if $\|\beta\| < 1$.

Fix a basis $\{\mathbf{b}_i\}_{i=1}^n$ for V and express

$$\alpha(y) = \sqrt{a_{ij}y^i y^j} \quad \text{and} \quad \beta(y) = b_i y^i, \qquad y = y^i \mathbf{b}_i \in \mathrm{V},$$

where $\left(a_{ij}\right)$ is a positive definite symmetric matrix. We have

$$\|\beta\| := \sup_{\alpha(y)=1} \beta(y) = \sqrt{a^{ij}b_i b_j},$$

where $\left(a^{ij}\right) = \left(a_{ij}\right)^{-1}$. Let

$$g_{ij}(y) := \mathbf{g}_y(\mathbf{b}_i, \mathbf{b}_j) = \frac{1}{2}[F^2]_{y^i y^j}(y).$$

An easy computation yields

$$g_{ij}(y) = \frac{F}{\alpha}\Big(a_{ij} - \frac{y_i}{\alpha}\frac{y_j}{\alpha}\Big) + \Big(\frac{y_i}{\alpha} + b_i\Big)\Big(\frac{y_j}{\alpha} + b_j\Big), \tag{1.6}$$

where $y_i := a_{is}y^s$. From (1.6), we can show that $\mathbf{g}_y$ is positive definite for all $y \neq 0$ if and only if $\|\beta\| < 1$. Thus $F = \alpha + \beta$ is a Minkowski norm if and only if $\|\beta\| < 1$. A function $F = \alpha + \beta$ on V with $\|\beta\| < 1$ is called a *Randers norm.*

Now we verify (1.4) in a direct way. Observe that

$$\begin{aligned} F(u+v) &= \alpha(u+v) + \beta(u) + \beta(v) \\ &\leq \alpha(u) + \alpha(v) + \beta(u) + \beta(v) \\ &= F(u) + F(v). \end{aligned}$$

By an elementary argument in linear algebra, we obtain

$$\det\left(g_{ij}\right) = \left(\frac{F}{\alpha}\right)^{n+1} \det\left(a_{ij}\right). \tag{1.7}$$

The inverse $\left(g^{ij}\right) := \left(g_{ij}\right)^{-1}$ can be expressed in the following form

$$g^{ij}(y) = \frac{\alpha}{F}a^{ij} + \left(\frac{\alpha}{F}\right)^2 \frac{\beta + \alpha b^2}{F}\frac{y^i}{\alpha}\frac{y^j}{\alpha} - \left(\frac{\alpha}{F}\right)^2 \left(b^j \frac{y^i}{\alpha} + b^i \frac{y^j}{\alpha}\right), \qquad (1.8)$$

where $b^i := a^{ij}b_j$. Both (1.7) and (1.8) are useful in the study on Randers norms. ♯

Take an open domain Ω of the origin in V with regular boundary $\mathrm{S} = \partial\Omega$. Suppose that Ω is convex, i.e., any line segment is contained in the closure $\bar{\Omega}$ if its endpoints lie inside $\bar{\Omega}$. Define $F : \mathrm{V} \to [0, \infty)$ by

$$F(\lambda y) := \lambda, \qquad \lambda > 0,\ y \in \mathrm{S}.$$

Then $\mathrm{S} = F^{-1}(1)$. We call F the *defining function* of S. Clearly, F satisfies (M1) and (M2) in Definition 1.2.1. Further, the convexity of Ω implies that

$$F(u + v) \le F(u) + F(v), \qquad u, v \in \mathrm{V}.$$

Thus $d(u, v) := F(v - u)$ is a metric on V. The convexity of Ω also implies that for any $y \neq 0$, the induced bilinear form $\mathbf{g}_y$ on V is positive semi-definite. A convex domain Ω is said to be *strongly convex* if the defining function F satisfies (M3) in Definition 1.2.1.

Example 1.2.2 Consider the following domain in R^n

$$\Omega := \left\{ y \,\middle|\, \sum_{i=1}^{n} (y^i)^4 < 1 \right\}.$$

Ω is strictly convex. The defining function of $\mathrm{S} = \partial\Omega$ is given by

$$F(y) := \left\{ \sum_{i=1}^{n} (y^i)^4 \right\}^{\frac{1}{4}}. \qquad (1.9)$$

The induced bilinear form $\mathbf{g}_y$ is positive semi-definite for all $y \in \mathrm{S}$, but it is not positive definite at certain points $y \in \mathrm{S}$. Thus Ω is not strongly convex. ♯

Example 1.2.3 *Consider the following function on* R^2

$$F(u, v) := \left\{ u^4 + 3c\, u^2 v^2 + v^4 \right\}^{\frac{1}{4}}, \qquad (u, v) \in \mathrm{R}^2.$$

Let

$$g_{11} := \frac{1}{2}[F^2]_{uu}, \qquad g_{12} := \frac{1}{2}[F^2]_{uv} =: g_{21}, \qquad g_{22} := \frac{1}{2}[F^2]_{vv}.$$

A direct computation gives

$$\begin{aligned} g_{11} &= \frac{2u^6 + 9cu^4v^2 + 6u^2v^4 + 3cv^6}{2F^6(u,v)} \\ g_{12} &= \frac{(9c^2-4)u^3v^3}{2F^6(u,v)} = g_{21} \\ g_{22} &= \frac{3cu^6 + 6u^4v^2 + 9cu^2v^4 + 2v^6}{2F^6(u,v)}, \end{aligned}$$

and

$$\det(g_{ij}) = g_{11}g_{22} - g_{12}g_{21} = \frac{6cu^4 + 3(4-3c^2)u^2v^2 + 6cv^4}{4F^4(u,v)}.$$

Clearly, $g_{11} > 0$ *and* $g_{22} > 0$ *if and only if* $c > 0$. *Assume that* $c > 0$. *Let us study the sign of* $\det(g_{ij})$. *Write*

$$6cu^4 + 3(4-3c^2)u^2v^2 + 6cv^4 = 6c\Big(u^4 + 2\delta u^2v^2 + v^4\Big),$$

where $\delta := (4-3c^2)/(4c)$. *The above polynomial is positive for all* $(u,v) \neq (0,0)$ *if and only if* $\delta > -1$. *Clearly,* $\delta > -1$ *if and only if* $c < 2$ *(under the assumption* $c > 0$*). Thus* $\mathbf{g}_y$ *is positive definite for all* $y = (u,v) \neq 0$ *if and only if* $0 < c < 2$.

♯

Let (V, F) be a Minkowski space. It follows from (M2) in Definition 1.2.1 that

$$\mathbf{g}_y(y,u) = \frac{1}{2}\frac{\partial}{\partial s}\Big[F^2(y+su)\Big]|_{s=0}, \tag{1.10}$$

$$\mathbf{g}_y(y,y) = F^2(y). \tag{1.11}$$

For a vector $y \in \mathrm{S} = F^{-1}(1)$, the tangent space $T_y\mathrm{S}$ can be naturally identified with the following hyperplane

$$\mathrm{W}_y := \Big\{ w \in \mathrm{V},\ \mathbf{g}_y(y,w) = 0 \Big\} \subset \mathrm{V}.$$

We obtain a decomposition for V

$$\mathrm{V} = \mathrm{R}\cdot y \oplus \mathrm{W}_y.$$

Let

$$\mathbf{h}_y(u,v) := \mathbf{g}_y(u,v) - \frac{1}{F^2(y)}\mathbf{g}_y(y,u)\mathbf{g}_y(y,v).$$

$\mathbf{h}_y$ is called the *angular form.* Note that

$$\mathbf{h}_y(y,y) = 0, \qquad \mathbf{h}_y(y,w) = 0,\ \forall w \in \mathrm{W}_y.$$

Thus for any $v = w + \lambda y \in \mathrm{V}$, where $w \in \mathrm{W}_y$,

$$\mathbf{h}_y(v,v) = \mathbf{g}_y(w,w) \geq 0.$$

The equality holds if and only if $v = \lambda y$. Thus $\mathbf{h}_y$ is positive semi-definite on V and positive definite on W_y.

Lemma 1.2.2 *F satisfies*

$$F(y+v) \leq F(y) + F(v), \qquad y, v \in \mathrm{V}. \tag{1.12}$$

Equality holds if and only if $y = \lambda v$ for some $\lambda \geq 0$.

Proof. Fix a vector $v \in \mathrm{V}$ and let $y(t) := (1-t)y + tv \neq 0$ and

$$\phi(t) := F(y(t)).$$

Suppose that $y(t) \neq 0$ for any $0 \leq t \leq 1$. Then $\phi(t)$ is C^∞ on $(0,1)$. Observe that

$$\phi''(t) = \frac{1}{F(y(t))}\mathbf{h}_{y(t)}(v-y, v-y) \geq 0.$$

This implies

$$2\phi\Big(\frac{1}{2}\Big) \leq \phi(0) + \phi(1).$$

Namely,

$$F(y+v) \leq F(y) + F(v).$$

Suppose that $y(t_o) = 0$ for some $0 < t_o < 1$. Without loss of generality, we may assume that $t_o \geq 1/2$. Then $v = -\frac{1-t_o}{t_o}y$ and $y + v = \frac{2t_o - 1}{t_o}y$. Thus

$$F(y+v) = \frac{2t_o-1}{t_o}F(y) = F(y) - \frac{1-t_o}{t_o}F(y) \leq F(y) + F(v).$$

This proves (1.12). Q.E.D.

By a similar argument, one obtains that

$$F(y) \leq F(y + w), \qquad w \in \mathrm{W}_y. \tag{1.13}$$

Equality holds if and only if $w = 0$.

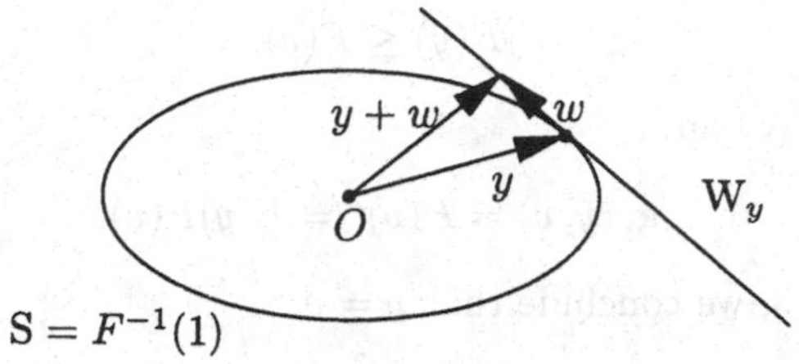

Lemma 1.2.3 (Cauchy-Schwarz inequality) *Let* (V, F) *be a Minkowski space. For any* $y \neq 0$,

$$\mathbf{g}_y(y, v) \leq F(y)F(v), \qquad \forall v \in \mathrm{V}. \tag{1.14}$$

Equality holds if and only if $v = \lambda y$ *for some* $\lambda \geq 0$.

Proof. Let $v = \lambda y + w \in \mathrm{V}$, where $\lambda \in \mathrm{R}$ and $w \in \mathrm{W}_y$. Observe that

$$\mathbf{g}_y(y, v) = \lambda \mathbf{g}_y(y, y) = \lambda F^2(y). \tag{1.15}$$

If $\lambda \leq 0$, then (1.15) implies (1.14), and the equality in (1.14) holds if and only if $\lambda = 0$ and $v = 0$. Assume that $\lambda > 0$. Then (1.13) implies

$$\mathbf{g}_y(y, v) = \lambda F(y)F(y) \leq \lambda F(y + \frac{1}{\lambda}w)F(y) = F(v)F(y).$$

Equality holds if and only if $w = 0$, i.e., $v = \lambda y$ for some $\lambda \geq 0$. Q.E.D.

Lemma 1.2.3 has the following application.

Lemma 1.2.4 *Let* (V, F) *be a Minkowski space. Suppose that* $y, v \in \mathrm{V} \setminus \{0\}$ *satisfies the following equation*

$$\mathbf{g}_y(y, w) = \mathbf{g}_v(v, w), \qquad w \in \mathrm{V}.$$

Then $y = v$.

Proof. Taking $w = v$ yields

$$F(v)^2 = \mathbf{g}_v(v, v) = \mathbf{g}_y(y, v) \le F(y)F(v).$$

This implies

$$F(v) \le F(y).$$

By taking $w = y$, we obtain

$$F(y) \le F(v).$$

Thus $F(y) = F(v)$ and

$$\mathbf{g}_y(y, v) = F(v)^2 = F(y)F(v).$$

By Lemma 1.2.3, we conclude that $y = v$. Q.E.D.

1.3 Finsler Spaces

In this section, we will introduce Finsler spaces. Finsler spaces are viewed as regular metric spaces.

Definition 1.3.1 A function $F : TM \to [0, \infty)$ is called a *Finsler metric* if it has the following properties:

(F1) F is C^∞ on $TM \setminus \{0\}$;
(F2) for each $x \in M$, $F_x := F|_{T_xM}$ is a Minkowski norm on T_xM.

The pair (M, F) is called a *Finsler space*. A Finsler metric F is said to be *reversible* if $F(-v) = F(v)$ for all $v \in TM$.

Given a Finsler metric F on a manifold M. For a Lipschitz continuous curve $c : [a, b] \to M$, the function $t \to F(\dot{c}(t))$ is measurable. Define

$$L_F(c) := \int_a^b F(\dot{c}(t))dt,$$

We obtain a length structure L_F. This length structure L_F induces a function d_F on $M \times M$ by

$$d_F(p, q) := \inf_c L_F(c),$$

where the infimum is taken over all Lipschitz continuous curves $c : [0,1] \to M$ with $c(0) = p$ and $c(1) = q$.

It is easy to verify that

$$d_F(p,q) \leq d_F(p,r) + d(r,q).$$

At any point $x \in M$, there are an open neighborhood $\mathcal{U}_x$ of x, a constant $C \geq 1$ and a diffeomorphism $\varphi : \mathcal{U}_x \to \mathbb{B}^n \subset \mathbb{R}^n$ such that

$$C^{-1}|u-v| \leq d_F\Big(\varphi^{-1}(u), \varphi^{-1}(v)\Big) \leq C|u-v|, \qquad u, v \in \mathbb{B}^n. \tag{1.16}$$

Thus, $d_F(p,q) = 0$ if and only if $p = q$. We conclude that d_F is a metric on M and the manifold topology coincides with the metric topology.

The induced metric d_F also defines a length structure L_{d_F} by (1.3). It has been proved that

$$L_F = L_{d_F}$$

(cf. [BuMa]). This implies

$$d_F(p,q) = \inf_c L_{d_F}(c),$$

where the infimum is taken over all Lipschitz curves $c : [0,1] \to M$ with $c(0) = p$ and $c(1) = q$. Thus d_F is a path metric. Further, this path metric is related to F by the following

$$F(y) = \limsup_{t_1 < t_2, \max|t_i| \to 0} \frac{d\Big(c(t_1), c(t_2)\Big)}{t_2 - t_1}, \qquad y \in T_xM. \tag{1.17}$$

where $c : (-\varepsilon, \varepsilon) \to M$ is an arbitrary C^1 curve with $\dot{c}(0) = y$.

The study of Finsler metrics began in P. Finsler's dissertation [Fi] in 1918, which was published in 1951. Thereafter geometers call these metrics the *Finsler metrics*. An important class of Finsler metrics is that of Riemannian metrics.

Example 1.3.1 (Riemannian metric) Let $g = \{g_x\}_{x \in M}$, where g_x is a positive definite symmetric bilinear form in T_xM such that in local coordinates (x^i),

$$g_{ij}(x) = g_x\Big(\frac{\partial}{\partial x^i}|_x, \frac{\partial}{\partial x^j}|_x\Big)$$

are C^∞ functions. g is called a *Riemannian metric* . Let

$$F_x(y) = \sqrt{g_x(y,y)}, \qquad y \in T_xM. \tag{1.18}$$

From the definition, we see that F_x is an Euclidean norm. The family of Euclidean norms $F = \{F_x\}_{x\in M}$ is a Finsler metric on M. A Finsler metric F is said to be *Riemannian* if it can be expressed by (1.18) for some Riemannian metric g. Such a Finsler metric is usually called a Riemannian metric in this book. ♯

Example 1.3.2 (Randers metric) Let $\alpha(y) = \sqrt{a_{ij}(x)y^iy^j}$ and $\beta(y) = b_i(x)y^i$ be a Riemannian metric and 1-form on a manifold M. Assume that

$$\|\beta\|_x = \sup_{y\in T_xM} \frac{\beta(y)}{\alpha(y) = 1} < 1, \qquad x \in M.$$

By Example 1.2.1,

$$F(y) := \alpha(y) + \beta(y)$$

is a Finsler metric on M. We call $F = \alpha + \beta$ a *Randers metric.* ♯

Example 1.3.3 (Psychometric Function [DzCo]) Let Ω be a domain in R^n. A *psychometric function* on Ω is a function $\Psi : \Omega \times \Omega \to \mathrm{R}$ with the following properties:

(a) For each $x \in \Omega$, the function $\Psi(x,\cdot)$ attains the absolute minimum $\Psi(x,x)$ at x;

(b) There exists an increasing continuous function $\Phi : [0,\varepsilon) \to [0,\infty)$ with $\Phi(0) = 0$ such that the following limit exists for any $x \in \Omega$ and $y \in T_x\Omega = \mathrm{R}^n$

$$F(x,y) = \lim_{s\to 0^+} \frac{\Phi[\Psi(x, x+sy) - \Psi(x,x)]}{s}.$$

We obtain a nonnegative function $F = F(x,y)$ on $T\Omega = \Omega \times \mathrm{R}^n$. We call F the *Fechner function* and Φ a *psychometric transformation* associated with Ψ. Clearly, F satisfies

$$F(x,\lambda y) = \lambda F(x,y), \qquad \lambda > 0.$$

A psychometric function is said to be *Finslerian* if the associated Fechner function is a Finsler metric. From the definition, given any Finsler metric F on a domain $\Omega \subset \mathrm{R}^n$ and any $\mu > 0$, the function

$$\Psi(x,y) := [F(x,y)]^{\mu}$$

is a psychometric function on Ω. ♯

Example 1.3.4 Let d be the Funk metric on a strictly convex domain Ω in R^n (see Exercise 1.1.2). Assume that Ω is strongly convex. By definition, there is a Minkowski norm φ on R^n and a point $p \in \Omega$,

$$\Omega - \{p\} = \Big\{ y \in \mathrm{R}^n \;\Big|\; \varphi(y) < 1 \Big\}.$$

The Funk metric d on Ω is actually induced by a Finsler metric. To find the candidate, we take a point $x \in \Omega$ and $y \in T_x\Omega \setminus \{0\}$. Let L_y denote the ray issuing from x in the direction y and $z = L_y \cap \partial\Omega$. From (1.1), the candidate F must satisfy

$$\begin{aligned} F(y) &= \lim_{\varepsilon \to 0+} \frac{d(x, x+\varepsilon y)}{\varepsilon} \\ &= \lim_{\varepsilon \to 0+} \frac{1}{\varepsilon} \ln \frac{|z-x|}{|z-x-\varepsilon y|} \\ &= \frac{\langle y, z-x \rangle}{|z-x|^2}, \qquad y \in T_x\Omega = \mathrm{R}^n. \end{aligned}$$

Clearly, the above equation is equivalent to the following equation:

$$x + \frac{y}{F(y)} = z \in \partial\Omega. \tag{1.19}$$

Thus if d is induced by a Finsler metric, then F is determined by (1.19). We can easily show that F determined by (1.19) is indeed a Finsler metric and this Finsler metric F induces the Funk metric d. We shall also call F the *Funk metric* on Ω.

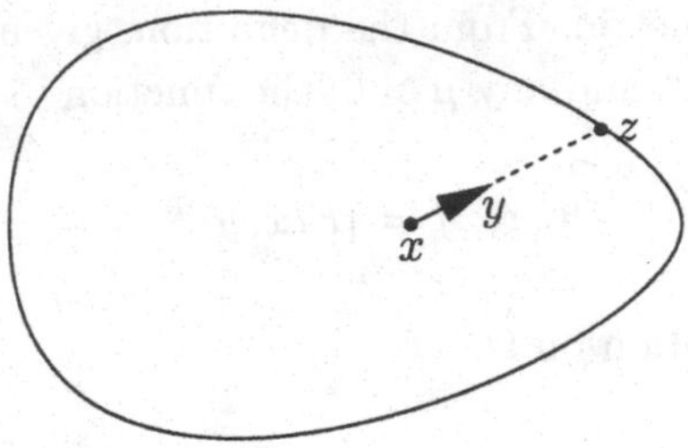

The Klein metric d_K on Ω is defined by

$$d_K(p,q) := \frac{1}{2}\Big(d(p,q) + d(q,p)\Big).$$

We can show that d_K is induced by the following Finsler metric

$$F_K(y) := \frac{1}{2}\Big(F(-y) + F(y)\Big), \tag{1.20}$$

where F is the Funk metric on Ω. We shall also call F_K the *Klein metric* on Ω.

Take a look at the special case when $\Omega = \mathbb{B}^n$ is the unit ball in the Euclidean space $\mathbb{R}^n$. A direct computation gives an explicit formula for the Funk metric,

$$F(y) := \frac{\sqrt{|y|^2 - (|x|^2|y|^2 - \langle x, y\rangle^2)} + \langle x, y\rangle}{1 - |x|^2}, \qquad y \in T_x\mathbb{B}^n = \mathrm{R}^n,$$

where $|\cdot|$ and $\langle , \rangle$ denote the Euclidean norm and inner product in $\mathbb{R}^n$, respectively. The Funk metric on $\mathbb{B}^n$ is a Randers metric. The Klein metric F_K is given by

$$F_K(y) = \frac{\sqrt{|y|^2 - (|x|^2|y|^2 - \langle x, y\rangle^2)}}{1 - |x|^2}, \qquad y \in T_x\mathbb{B}^n = \mathrm{R}^n,$$

Thus the Klein metric on $\mathbb{B}^n$ is a Riemannian metric. ♯

The following lemma is due to Okada [Ok].

Lemma 1.3.2 (Okada) *The Funk metric $F : T\Omega = \Omega \times \mathrm{R}^n \to [0, \infty)$ satisfies*

$$F_{x^k} = F F_{y^k}. \tag{1.21}$$

Proof. By assumption, Ω is strongly convex, i.e., there is a Minkowski norm φ on R^n and a point $p \in \Omega$ such that $\varphi(y-p)=1$, $y \in \partial\Omega$. Thus

$$\varphi\Big(x+\frac{y}{F(y)}-p\Big)=1. \tag{1.22}$$

By differentiating (1.22) with respect to x^j and y^j respectively, we obtain

$$\Big(\delta^i_j - F^{-2}F_{x^j}y^i\Big)\varphi_{z^i}(z)=0 \tag{1.23}$$

$$\Big(\delta^i_j - F^{-1}F_{y^j}y^i\Big)F^{-1}\varphi_{z^i}(z)=0, \tag{1.24}$$

where $z := x + y/F(y) - p$. It follows from (1.23) and (1.24) that

$$\Big(F_{x^j} - FF_{y^j}\Big)\varphi_{z^i}(z)y^i = 0. \tag{1.25}$$

Observe that $v=(v^i) \in T_z\mathrm{R}^n$ is tangent to $\partial\Omega$ if and only if

$$\varphi_{z^i}(z)v^i = 0.$$

Thus $\varphi_{z^i}(z)y^i \neq 0$. It follows from (1.25) that

$$F_{x^j} - FF_{y^j} = 0.$$

This implies (1.22). Q.E.D.

Chapter 2

Finsler m Spaces

In his recent book [Gr4], M. Gromov has set the foundation of the theory of metric measure spaces and shown some potential applications to various problems, in particular, the Levy concentration problem. A *metric measure space* is a triple $M = (M, d, \mu)$, where (M, d) is a metric space with a countable base and μ is a σ-finite Borel measure. Here a measure μ on a metric space (M, d) always means a Borel measure, i.e., all Borel subsets are measurable. In some cases, the underlying spaces are not manifolds. Here we are concerned with *regular* metric measure spaces, so that we can employ calculus to study the geometric properties. Finsler spaces with a volume form are viewed as regular metric measure spaces. They are simply called Finsler m spaces. In this chapter, we will discuss some basics on Finsler m spaces.

2.1 Measure Spaces

Let M be a C^∞ n-manifold. A *volume form* $d\mu$ on M is a collection of non-degenerate n-forms $d\mu_i = \sigma_i(x)dx^1 \cdots dx^n$ on coordinate neighborhoods $\{\varphi_i, U_i\}$ such that

(V1) $M = \bigcup_i U_i$;

(V2) if $U_i \bigcap U_j \neq \emptyset$, then $d\mu_i = d\mu_j$ on $U_i \cap U_j$, namely, if $d\mu_i = \sigma_i(x)dx^1 \cdots dx^n$ and $d\mu_j = \sigma_j(u)du^1 \cdots du^n$, then

$$\sigma_i(x) = \left| \det\Big(\frac{\partial u^i}{\partial x^j}\Big)\right| \sigma_j(u).$$

By (V2), we have

$$\int_{U_i \cap U_j} f \, d\mu_j = \int_{U_i \cap U_j} f \, d\mu_i, \qquad f \in C_c^\infty(U_i \cap U_j).$$

Take a partition of a unity $\{\psi_i\}$ for the covering $\{U_i\}$, that is a set of nonnegative C^∞ functions $\psi_i \in C_c^\infty(U_i)$ with the following properties:

(a) for each $x \in M$, there are only finitely many U_i's such that $\psi_i(x) \neq 0$;
(b) $\sum_i \psi_i(x) = 1$ for all $x \in M$.

Every volume form $d\mu = \{d\mu_i\}$ defines a measure μ on M by

$$\mu(f) := \sum_{i=1} \int_{U_i} \psi_i f \, d\mu_i, \qquad f \in C_c^0(M).$$

For any two volume forms $d\mu_1$ and $d\mu_2$ on M, there is a positive C^∞ function φ such that

$$d\mu_2 = \varphi \, d\mu_1.$$

Thus if we fix a volume form $d\mu_o$ on M, volume forms $d\mu$ one-to-one correspond to positive C^∞ functions φ by $d\mu = \varphi \, d\mu_o$.

The simplest volume form on R^n is the Euclidean volume form

$$dV := dx^1 \cdots dx^n. \tag{2.1}$$

The Euclidean volume of a bounded open subset $\Omega \subset \mathrm{R}^n$ is given by

$$\mathrm{Vol}(\Omega) = \int_\Omega dV = \int_\Omega dx^1 \cdots dx^n.$$

More general, on a Riemannian manifold (M, F), where $F(y) = \sqrt{g_{ij}(x) y^i j y^j}$, the Riemannian metric F determines a canonical volume form

$$dV_g := \sqrt{\det\Big(g_{ij}(x)\Big)} \, dx^1 \cdots dx^n. \tag{2.2}$$

We call dV_g the *Riemannian volume form* of F.

We consider an oriented manifold M equipped with a volume form $d\mu$. We can view $d\mu$ as an n-form on M. Let X a vector field on M. Define an $(n-1)$-form $X\rfloor d\mu$ on M by

$$X\rfloor d\mu(X_2, \cdots, X_n) := d\mu(X, X_2, \cdots, X_n).$$

Define

$$d(X\rfloor d\mu) = \mathrm{div}(X) d\mu. \tag{2.3}$$

We call $\mathrm{div}(X)$ the *divergence* of X. Clearly, $\mathrm{div}(X)$ depends only on the volume form $d\mu$. In a local coordinate system (x^i), express $d\mu = \sigma(x) dx^1 \cdots dx^n$. Then for a vector field $X = X^i \frac{\partial}{\partial x^i}$ on M,

$$\mathrm{div}(X) = \frac{1}{\sigma} \frac{\partial}{\partial x^i}\Big(\sigma X^i\Big) = \frac{\partial X^i}{\partial x^i} + \frac{X^i}{\sigma}\frac{\partial \sigma}{\partial x^i}. \tag{2.4}$$

Applying the Stokes theorem to $\eta = X\rfloor d\mu$, we obtain

$$\int_M \mathrm{div}(X) d\mu = \int_M d(X\rfloor d\mu) = 0, \qquad \text{if } \partial M = \emptyset \tag{2.5}$$

$$\int_M \mathrm{div}(X) d\mu = \int_M d(X\rfloor d\mu) = \int_{\partial M} X\rfloor d\mu, \qquad \text{if } \partial M \neq \emptyset. \tag{2.6}$$

2.2 Volume on a Finsler Space

There are two canonical volume forms on a Finsler space. Both reduce to the Riemannian volume form when the Finsler metric becomes Riemannian.

Let (M, F) be an n-dimensional Finsler space. Let $\{\mathbf{b}_i\}_{i=1}^n$ be an arbitrary basis for T_xM and $\{\theta^i\}_{i=1}^n$ the dual basis for T_x^*M. The set

$$\mathrm{B}_x^n := \Big\{(y^i) \in \mathrm{R}^n, \ F(y^i\mathbf{b}_i) < 1\Big\} \tag{2.7}$$

is a bounded open strongly convex open subset in R^n. Define

$$dV_F := \sigma_F(x)\theta^1 \wedge \cdots \wedge \theta^n,$$

where

$$\sigma_F(x) := \frac{\mathrm{Vol}(\mathbb{B}^n)}{\mathrm{Vol}(\mathrm{B}^n_x)}. \tag{2.8}$$

The volume form dV_F determines a regular measure Vol_F. H. Busemann [Bu1][Bu2] proved that if F is reversible, then Vol_F is the Hausdorff measure of the induced metric d_F. dV_F is called the *Busemann-Hausdorff volume form* .

In general, the Euclidean volume of B^n_x in (2.8) can not be expressed by F in an explicit form. Below we will discuss some special metrics for which we can compute dV_F.

Example 2.2.1 Let (V, F) be an n-dimensional Minkowski space and

$$\mathrm{B} := \Big\{ y \in \mathrm{V},\ F(y) < 1 \Big\}$$

denote the unit ball of F. Let (x^i) denote the global coordinate system on V determined by a basis $\{\mathbf{b}_i\}_{i=1}^n$. Let

$$\mathrm{B}^n := \Big\{ (y^i) \in \mathrm{R}^n,\ F(y^i \mathbf{b}_i) < 1 \Big\}.$$

Then

$$\sigma_F(x) = \frac{\mathrm{Vol}(\mathbb{B}^n)}{\mathrm{Vol}(\mathrm{B}^n)} = constant.$$

We obtain

$$\mathrm{Vol}_F(\mathrm{B}) = \int_{\mathrm{B}} dV_F = \int_{\mathrm{B}^n} \sigma_F(x) dx^1 \cdots dx^n = \mathrm{Vol}(\mathbb{B}^n).$$

♮

Example 2.2.2 Consider a Randers metric $F = \alpha + \beta$ on a manifold M, where $\alpha = \sqrt{a_{ij}(x) y^i y^j}$ is a Riemannian metric and $\beta = b_i(x) y^i$ is a 1-form satisfying

$$\|\beta\|_x := \sup_{\alpha_x(y)=1} \beta(y) = \sqrt{a^{ij}(x) b_i(x) b_j(x)} < 1,$$

where $(a^{ij}(x)) = (a_{ij}(x))^{-1}$. Let dV_F and dV_α denote the Busemann-Hausdorff volume form of F and α, respectively. By a simple computation,

we obtain

$$dV_\alpha = \sqrt{\det(a_{ij}(x))}\, dx^1 \cdots dx^n.$$

This is just the Riemannian volume form of α defined in (2.2). To find dV_F, we take an orthonormal basis $\{\mathbf{b}_i\}_{i=1}^n$ for (T_xM, α_x) such that $\beta_x(y) = \|\beta\|_x y^1$, where $y = y^i\mathbf{b}_i$. Then the open subset B_x^n in (2.7) is a convex body in R^n given by

$$\left(1 - \|\beta\|_x^2\right)^2 \left(y^1 + \frac{\|\beta\|_x}{1 - \|\beta\|_x^2}\right)^2 + \left(1 - \|\beta\|_x^2\right) \sum_{a=2}^n (y^a)^2 < 1.$$

The Euclidean volume of B_x^n is given by

$$\mathrm{Vol}(\mathrm{B}_x^n) = \frac{\mathrm{Vol}(\mathbb{B}^n)}{\left(1 - \|\beta\|_x^2\right)^{\frac{n+1}{2}}}. \tag{2.9}$$

Thus

$$dV_F = \left(1 - \|\beta\|_x^2\right)^{\frac{n+1}{2}} dV_\alpha. \tag{2.10}$$

This implies

$$\mathrm{Vol}_F \le \mathrm{Vol}_\alpha. \tag{2.11}$$

Assume that M is closed. Then

$$\mathrm{Vol}_F(M) = \int_M \left(1 - \|\beta\|_x^2\right)^{\frac{n+1}{2}} dV_\alpha \le \mathrm{Vol}_\alpha(M)$$

and equality holds if and only if $\beta = 0$. ♯

Example 2.2.3 Let

$$\|u\| := \sqrt{\sum_{\mu=1}^m (u^\mu)^2} + \sum_{\mu=1}^m B_\mu u^\mu, \qquad u = (u^\mu) \in \mathrm{R}^m,$$

with $\sqrt{\sum_{\mu=1}^m (B_\mu)^2} < 1$. Consider an immersion $\varphi = (\varphi^\mu) : M \to (\mathrm{R}^{m+1}, \|\cdot\|)$. φ induces a Finsler metric on M,

$$F(y) := \sqrt{\frac{\partial \varphi^\mu}{\partial x^i}(x) \frac{\partial \varphi^\mu}{\partial x^j}(x) y^i y^j} + \sum_{\mu=1}^m B_\mu \frac{\partial \varphi^\mu}{\partial x^i}(x) y^i, \qquad y = y^i \frac{\partial}{\partial x^i}|_x.$$

Let $\alpha := \sqrt{a_{ij}(x)y^iy^j}$ and $\beta := b_i(x)y^i$, where

$$a_{ij}(x) := \frac{\partial\varphi^\mu}{\partial x^i}(x)\frac{\partial\varphi^\mu}{\partial x^j}(x), \qquad b_i(x) := \sum_{\mu=1}^{m} B_\mu \frac{\partial\varphi^\mu}{\partial x^i}(x).$$

Then the norm of β with respect to α is given by

$$\|\beta\| = \sqrt{a^{ij}(x)b_i(x)b_j(x)} \leq \sqrt{\sum_{\mu=1}^{m}(B_\mu)^2} < 1,$$

where $(a^{ij}(x)) = (a_{ij}(x))^{-1}$. Thus $F = \alpha + \beta$ is a Randers metric on M.

Let b with $|b| < 1$ and

$$\|u\|_b := \sqrt{\sum_{\mu=1}^{n+1}(u^\mu)^2} + bu^{n+1}, \qquad u = (u^i) \in \mathrm{R}^{n+1}.$$

Consider a graph in a Randers space $(\mathrm{R}^{n+1}, \|\cdot\|_b)$,

$$u^{n+1} = f(u), \qquad u = (u^i) \in \Omega \subset \mathrm{R}^n.$$

The standard immersion $\varphi : \Omega \to \mathrm{R}^{n+1}$ is given by

$$\varphi\Big(x^1, \cdots, x^n\Big) = \Big(x^1, \cdots, x^n, f(x^1, \cdots, x^n)\Big).$$

We obtain

$$a_{ij} = \delta_{ij} + \frac{\partial f}{\partial x^i}(x)\frac{\partial f}{\partial x^j}(x), \qquad b_i = b\frac{\partial f}{\partial x^i}(x).$$

This gives

$$\det(a_{ij}) = 1 + |df|^2,$$

where

$$|df| = \sqrt{\Big(\frac{\partial f}{\partial x^1}\Big)^2 + \cdots + \Big(\frac{\partial f}{\partial x^n}\Big)^2}.$$

The inverse of (a_{ij}) is given by

$$a^{ij} = \delta_{ij} - \frac{1}{1+|df|^2}\frac{\partial f}{\partial x^i}(x)\frac{\partial f}{\partial x^j}(x).$$

This gives

$$\|\beta\|^2 := b^2 \frac{|df|^2}{1+|df|^2}.$$

We obtain

$$\begin{aligned} dV_F &= \Big(1 - b^2\frac{|df|^2}{1+|df|^2}\Big)^{\frac{n+1}{2}} dV_\alpha \\ &= \Big(1 - b^2\frac{|df|^2}{1+|df|^2}\Big)^{\frac{n+1}{2}} \sqrt{1+|df|^2}\, dx^1\cdots dx^n. \end{aligned}$$

♯

Example 2.2.4 Let Ω be a strongly convex domain in R^n. The Funk metric F is defined by

$$z = x + \frac{y}{F(y)}, \qquad y \in T_x\Omega \approx \mathrm{R}^n. \tag{2.12}$$

where $z \in \partial\Omega$ (cf. Example 1.3.4). Note

$$\mathrm{B}^n_x := \Big\{(y^i) \in \mathrm{R}^n,\ \ F(y^i\mathbf{b}_i) < 1\Big\} = \bar{\Omega} - \{x\}.$$

Thus

$$\sigma_F(x) = \frac{\mathrm{Vol}(\mathbb{B}^n)}{\mathrm{Vol}(\mathrm{B}^n_x)} = \frac{\mathrm{Vol}(\mathbb{B}^n)}{\mathrm{Vol}(\Omega)} = constant. \tag{2.13}$$

This implies that the Busemann-Hausdorff volume of (Ω, F) is constant

$$\mathrm{Vol}_F(\Omega) = \int_\Omega \sigma_F(x)dx^1\cdots dx^n = \mathrm{Vol}(\mathbb{B}^n). \tag{2.14}$$

♯

There is another important volume form on Finsler spaces. Let (M, F) be an n-dimensional Finsler space. At a point $x \in M$, fix a basis $\{\mathbf{b}_i\}_{i=1}^n$ for T_xM and its dual basis $\{\theta^i\}_{i=1}^n$ for T_x^*M. For a vector $y = y^i\mathbf{b}_i \in T_xM\backslash\{0\}$, put

$$g_{ij}(y) := \mathbf{g}_y(\mathbf{b}_i, \mathbf{b}_j).$$

Each $g_{ij}(y)$ is a C^∞ function on $\mathrm{R}^n \setminus \{0\}$. Define

$$\tilde{\sigma}_F(x) := \frac{\int_{\mathrm{B}^n_x} \det\big(g_{ij}(y)\big)\, dy^1 \cdots dy^n}{\mathrm{Vol}(\mathbb{B}^n)},$$

where B^n_x is defined in (2.7). The n-form

$$d\tilde{V}_F := \tilde{\sigma}_F(x)\, \theta^1 \wedge \cdots \wedge \theta^n \tag{2.15}$$

is a well-defined volume form on M.

Let

$$\omega := g_{ij}(y) y^j dx^i.$$

ω is called the *Hilbert form*. Observe that

$$d\omega = \frac{\partial g_{jk}}{\partial x^i} y^k dx^i \wedge dx^j - g_{ij} dx^i \wedge dy^i.$$

Thus

$$(d\omega)^n = d\omega \wedge \cdots \wedge d\omega = (-1)^{\frac{n(n+1)}{2}} n! \det\big(g_{ij}(y)\big) dx^1 \cdots dx^n dy^1 \cdots dy^n.$$

The Hilbert form ω defines a volume form on $TM \setminus \{0\}$ by

$$dV := (-1)^{\frac{n(n+1)}{2}} \frac{1}{n!} (d\omega)^n. \tag{2.16}$$

P. Dazord discussed $(d\omega)^n$ in his Ph.D. thesis [Da1]. He further defined the volume of a compact Finsler space (M, F) by

$$\mathrm{Vol}(M) := \frac{1}{\omega_n} \int_{\mathrm{B}M} dV, \tag{2.17}$$

where $\pi : \mathrm{B}M \to M$ denotes the unit ball bundle of M. Dazord actually defined the volume using the tangent sphere bundle $\mathrm{S}M$. But his definition is essentially same as (2.17) due to the homogeneity of F (cf. [Da2]). Observe that for a function f on M,

$$\int_{\mathrm{B}M} \pi^* f\, dV = \omega_n \int_M f\, d\tilde{V}_F.$$

Thus the volume form dV on $\mathrm{B}M$ gives rise to the volume form $d\tilde{V}_F$ on M.

R.D. Holmes and A.C. Thompson [Th] took a different approach to study Minkowski geometry and discovered this special volume form $d\tilde{V}_F$.

Therefore, $d\tilde{V}_F$ is called the *Holmes-Thompson volume form* in literatures [Al][AlFe].

2.3 Hyperplanes in a Minkowski m Space

Given a volume form $d\mu$ on a manifold M, there is no canonical way to define a volume form on hypersurfaces from $d\mu$. If M is also equipped with a Finsler metric F, then F determines a (local) normal vector field along any hypersurface. Using the normal vector field, one can define a volume form on hypersurfaces from $d\mu$.

Let $(M, F, d\mu)$ be a Finsler m space. Let dV_F denote the Busemann-Hausdorff volume form of F. The volume form $d\mu$ can written

$$d\mu = \varphi dV_F,$$

where φ is a positive C^∞ function on M. Thus it suffices to define the "induced volume form" on a hypersurface for the Busemann-Hausdorff volume form dV_F. Since dV_F at a point $x \in M$ is completely determined by the Minkowski norm F_x on T_xM, we shall first study Minkowski spaces.

Consider a Minkowski space (V, F). Given a hyperplane $\mathrm{W} \subset \mathrm{V}$, we claim that there is a unit vector $\mathbf{n} \in \mathrm{V}$ such that

$$\mathrm{W} = \Big\{ v \in \mathrm{V} \mid \mathbf{g_n}(\mathbf{n}, w) = 0 \Big\}. \tag{2.18}$$

To prove the claim, take a vector $v \notin \mathrm{W}$ and let

$$\phi(w) := F(v - w), \qquad w \in \mathrm{W}.$$

Clearly, ϕ attains its minimum $m := \min \phi$ at a unique point $w_o \in \mathrm{W}$. Let

$$\mathbf{n} := \frac{v - w_o}{m}.$$

If $\bar{v} = \lambda v + \bar{w}$, $\lambda > 0$, is another vector on the same side of W, then $\bar{\phi}(w) := F(\bar{v} - w)$ attains its minimum $\bar{m} = \lambda m$ at $\bar{w}_o = \bar{w} - \lambda w_o$. This implies that

$$\frac{\bar{v} - \bar{w}_o}{\bar{m}} = \frac{(\lambda v + \bar{w}) - (\bar{w} - \lambda w_o)}{\lambda m} = \frac{v - w_o}{m}.$$

Thus $\mathbf{n}$ is independent of the vectors on the same side of W as v. We call $\mathbf{n}$ a *normal vector* to W.

Fix a vector $w \in \mathrm{W}$ and let

$$f(t) := \frac{1}{2m^2}\phi^2(w_o + tm\, w) = \frac{1}{2m^2}F^2(v - w_o - tm\, w) = \frac{1}{2}F^2(\mathbf{n} + t\, w).$$

Differentiating f yields

$$0 = f'(0) = \frac{1}{2}\frac{d}{dt}\Big[F^2(\mathbf{n} + t\, w)\Big]|_{s=0} = \mathbf{g}_{\mathbf{n}}(\mathbf{n}, w).$$

This implies (2.18).

For any hyperplane $\mathrm{W} \subset \mathrm{V}$, there are exactly two normal vectors $\mathbf{n}, \mathbf{n}' \in \mathrm{V}$ to W. In general, $\mathbf{n}$ and $\mathbf{n}'$ are not parallel unless F is reversible.

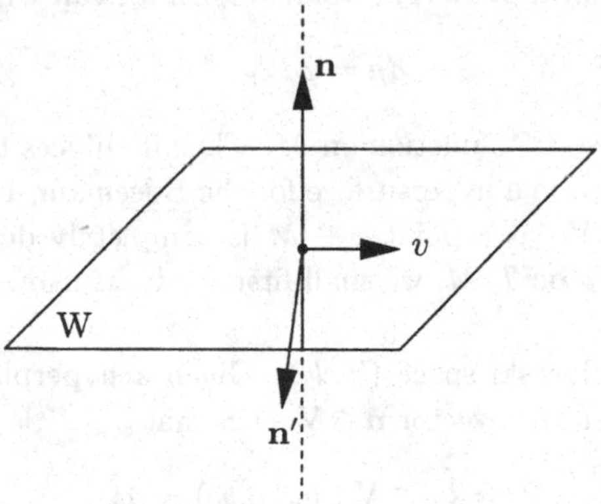

In what follows, we are going to introduce an important function on a Minkowski space. This function will be used to define the volume form on hyperplanes in a Minkowski space.

Let (V, F) be a Minkowski space. For a vector $y \in \mathrm{V} \setminus \{0\}$, we obtain a hyperplane

$$\mathrm{W}_y := \Big\{w \in \mathrm{V}, \mid \mathbf{g}_y(y, v) = 0\Big\}.$$

Take a basis $\{\mathbf{b}_i\}_{i=2}^n$ for W_y and $\mathbf{b}_1 = y$ such that $\{\mathbf{b}_i\}_{i=1}^n$ is a basis for V. Let

$$\mathrm{B}^n := \Big\{(y^i) \in \mathrm{R}^n,\ F\Big(\sum_{i=1}^n y^i \mathbf{b}_i\Big) < 1\Big\},$$

$$\mathrm{B}_y^{n-1} := \Big\{(y^a) \in \mathrm{R}^{n-1},\ F\Big(\sum_{a=2}^{n} y^a \mathbf{b}_a\Big) < 1\Big\}.$$

Both B^n and B_y^{n-1} depend on the choice of $\{\mathbf{b}_a\}_{a=2}^n$. Define

$$\zeta(y) := \frac{\mathrm{Vol}(\mathbb{B}^n)}{\mathrm{Vol}(\mathbb{B}^{n-1})} \cdot \frac{\mathrm{Vol}(\mathrm{B}_y^{n-1})}{F(y)\mathrm{Vol}(\mathrm{B}^n)}. \tag{2.19}$$

The function ζ is independent of the choice of $\{\mathbf{b}_i\}_{i=2}^n$. Moreover, ζ has the following homogeneity property

$$\zeta(\lambda y) = \zeta(y), \qquad \lambda > 0,\ y \neq 0.$$

Note that if F is Euclidean, then $\zeta(y) = 1$. A natural question is whether or not F is Euclidean when $\zeta = 1$. This is not clear yet.

Lemma 2.3.1 *Let* (V, F) *be an n-dimensional Minkowski space. Let* $c(n) := \mathrm{Vol}(\mathbb{B}^n)/\mathrm{Vol}(\mathbb{B}^{n-1})$ *and* $\lambda := \sup_{y \in \mathrm{V}\setminus\{0\}} F(y)/F(-y)$. *Then*

$$(1+\lambda)^{-n} c_n \leq \zeta(y) \leq 2^n c(n), \qquad y \neq 0. \tag{2.20}$$

In particular, if F is reversible ($\lambda = 1$), *then*

$$2^{-n} c_n \leq \zeta(y) \leq 2^n c_n. \tag{2.21}$$

Proof. Fix a unit vector $y = y^i \mathbf{b}_i \in \mathrm{V}$. For any $t \in (0,1)$ and $(v^a) \in \mathrm{B}_y^{n-1}$,

$$F(ty + v^a \mathbf{b}_a) \leq tF(y) + F(v^a \mathbf{b}_a) \leq 2.$$

Thus $[0,1] \times \mathrm{B}_y^{n-1} \subset 2\mathrm{B}^n$ and

$$\mathrm{Vol}(\mathrm{B}_y^{n-1}) \leq 2^n \mathrm{Vol}(\mathrm{B}^n).$$

This gives the right hand side of (2.20).

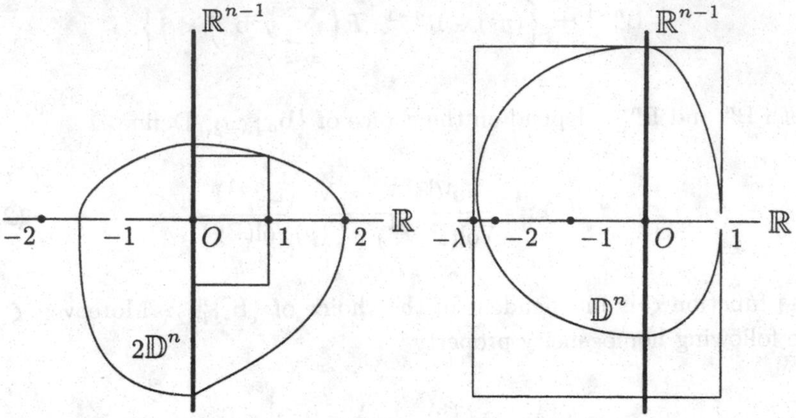

Let $(t, v^a) \in \mathrm{B}^n$, i.e., $F(ty + v^a\mathbf{b}_a) \leq 1$. If $t > 0$, then it follows from (1.13) that

$$t = F(ty) < F(ty + v^a\mathbf{b}_a) \leq 1,$$

$$F(v^a\mathbf{b}_a) \leq F(ty + v^a\mathbf{b}_a) + F(-ty) \leq 1 + \lambda.$$

If $t < 0$, then it follows from (1.13) again that

$$-t = F(-ty) < F(-ty - v^a\mathbf{b}_a) \leq \lambda F(ty + v^a\mathbf{b}_a) \leq \lambda.$$

$$F(v^a\mathbf{b}_a) \leq F(ty + v^a\mathbf{b}_a) + F(-ty) \leq 1 - t \leq 1 + \lambda,$$

Thus

$$\mathrm{B}^n \subset (-\lambda, 1) \times (1 + \lambda)\mathrm{B}_y^{n-1}.$$

This gives the left hand side of (2.20). Q.E.D.

2.4 Hypersurfaces in a Finsler m Space

Now we are going to define the volume form on hypersurfaces in a Finsler m space.

Let $(M, F, d\mu)$ be a Finsler m space and $i : N \to M$ an embedding. Our goal is to find a volume dA_F on hypersurfaces such that the following equality holds,

$$A_F\Big(S(x,\rho)\Big) = \lim_{\varepsilon \to 0^+} \frac{\mathrm{Vol}_F\Big(B(x,\rho+\varepsilon)\Big) - \mathrm{Vol}_F\Big(B(x,\rho)\Big)}{\varepsilon}. \tag{2.22}$$

However, (2.22) does not hold for the Busemann-Hausdorff volume form $dV_{\bar{F}}$ of the induced Finsler metric $\bar{F} := i^*F$ on N.

Recall that for the Minkowski norm $F_x = F|_{T_xM}$, there is an important quantity $\zeta_x : T_xM \setminus \{0\} \to (0, \infty)$ (cf. (2.19)). Thus we obtain a scalar function $\zeta = \{\zeta_x\}_{x \in M}$ on $TM \setminus \{0\}$. With this scalar function, we can define the desired volume form on hypersurfaces.

Definition 2.4.1 Let N be a hypersurface in a Finsler space (M, F) and $\mathbf{n}$ a normal vector field along N. Let $dV_{\bar{F}}$ denote the Finsler volume form of the induced Finsler metric $\bar{F}$ on N. Set

$$dA_F := \zeta(\mathbf{n})dV_{\bar{F}}, \tag{2.23}$$

where ζ is defined in (2.19). dA_F is called the *induced volume form* of dV_F with respect to $\mathbf{n}$. For an arbitrary volume form $d\mu$ on M, write $d\mu := \varphi dV_F$, where $\varphi \in C^\infty(M)$ and define

$$d\nu := \varphi dA_F.$$

$d\nu$ is called the *induced volume form* by $d\mu$ with respect to $\mathbf{n}$.

From the definition of dA_F, we see that dA_F depends on the choice of normal vectors unless F is reversible. By Lemma 2.3.1, we know that if F is reversible, then ζ is bounded by two universal positive numbers $c_1(n)$ and $c_2(n)$. Thus

$$c_1(n)dV_{\bar{F}} \le dA_F \le c_2(n)dV_{\bar{F}}.$$

There is another way to express dA_F on a hypersurface $i : N \to M$. Fix a normal vector field $\mathbf{n}$ along N. For a point $x \in N \subset M$, let $\{\mathbf{b}_i\}_{i=1}^n$ be a basis for T_xM such that $\mathbf{b}_1 = \mathbf{n}_x$ and $\{\mathbf{b}_a\}_{a=2}^n$ is a basis for T_xN. Put

$$dV_F = \sigma_F(x)\theta^1 \wedge \cdots \wedge \theta^n,$$

where σ_F is given in (2.8). Let $\{\theta^i\}_{i=1}^n$ denote the dual basis for T_x^*M and $\{\bar{\theta}^a\}_{a=2}^n$ the pull-back of $\{\theta^a\}_{a=2}^n$ on N. Express

$$dV_{\bar{F}} = \sigma_{\bar{F}}(x)\bar{\theta}^2 \wedge \cdots \wedge \bar{\theta}^n.$$

From the definition of $\zeta(\mathbf{n}_x)$, we have

$$\sigma_F(x) = \zeta(\mathbf{n}_x)\sigma_{\bar{F}}(x)$$

and

$$dA_F = \sigma_F(x)\bar{\theta}^2 \wedge \cdots \wedge \bar{\theta}^n. \tag{2.24}$$

Let X be a vector field on M. Write

$$X|_N = \mathbf{g_n}(\mathbf{n}, X)\mathbf{n} + \bar{X},$$

where $\bar{X} \in TN$. Then for any set of tangent vectors $X_2, \cdots, X_n \in TN$

$$X \rfloor d\mu(X_2, \cdots, X_n) = \mathbf{g_n}(\mathbf{n}, X)\, d\mu(\mathbf{n}, X_2, \cdots, X_n). \tag{2.25}$$

At a point $x \in N$, let $\{\mathbf{b}_i\}_{i=1}^n$ and $\{\theta^i\}_{i=1}^n$ be as above. Put

$$d\mu = \sigma(x)\theta^1 \wedge \cdots \wedge \theta^n.$$

By (2.24), we obtain the induced volume ν on N,

$$d\nu = \sigma(x)\bar{\theta}^2 \wedge \cdots \wedge \bar{\theta}^n.$$

Thus

$$d\mu(\mathbf{n}, X_2, \cdots, X_n) = d\nu(X_2, \cdots, X_n).$$

It follows from (2.25) that

$$i^*(X \rfloor d\mu) = \mathbf{g_n}(\mathbf{n}, X)d\nu. \tag{2.26}$$

Assume that M is a compact oriented manifold with smooth boundary ∂M. By the Stokes theorem, for any $(n-1)$-form η on M,

$$\int_M d\eta = \int_{\partial M} i^*\eta.$$

Applying the Stokes theorem to $\eta = X \rfloor d\mu$, we obtain the following

Theorem 2.4.2 *Let $(M, F, d\mu)$ be a compact oriented Finsler m space and* $\mathbf{n}$ *be the out-ward pointing normal vector to ∂M. Then for any vector field X on M,*

$$\int_M \operatorname{div}(X) d\mu = \int_{\partial M} \mathbf{g}_{\mathbf{n}}(\mathbf{n}, X) d\nu. \tag{2.27}$$

In particular, if M is a closed oriented manifold, then

$$\int_M \operatorname{div}(X) d\mu = 0.$$

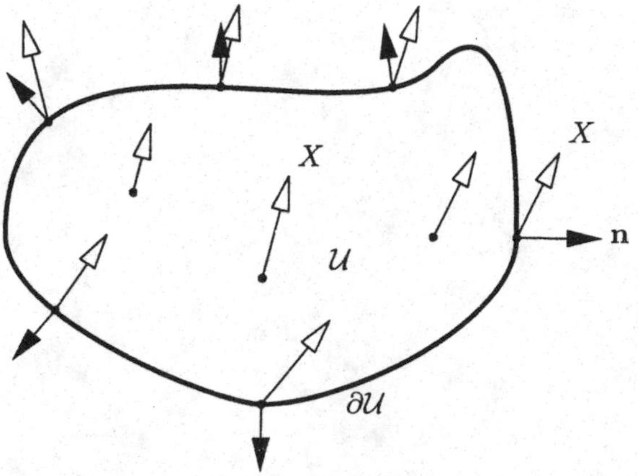

Theorem 2.4.2 is important in many global geometric analysis problems.

Chapter 3

Co-Area Formula

To study Finsler m spaces, we use a family of level hypersurfaces of a distance function. On each level hypersurface, there is an induced volume form. The co-area formula relates the induced volume on level hypersurfaces and the volume on the manifold.

3.1 Legendre Transformations

First, let us study the relationship between a Minkowski norm and its dual norm. Let V be a finite-dimensional vector space and V^* the dual vector space. Given a Minkowski norm F on V. F is a norm in the sense that for any $y, v \in V$ and $\lambda > 0$,

$$F(\lambda y) = \lambda F(y), \qquad F(y+v) \leq F(y) + F(v).$$

Let V^* denote the vector space dual to V. Define

$$F^*(\xi) := \sup_{F(y)=1} \xi(y). \tag{3.1}$$

F^* is a norm on V^* again, i.e., for any $\xi, \eta \in V^*$ and $\lambda > 0$,

$$F^*(\lambda\xi) = \lambda F^*(\xi), \qquad F^*(\xi + \eta) \leq F^*(\xi) + F^*(\eta).$$

We shall show that F^* is a Minkowski norm on V^* via the Legendre transformation between V and V^*.

Lemma 3.1.1 *Let F be a Minkowski norm on* V *and F^* the dual norm* *on* V^*. *For any vector* $y \in V \setminus \{0\}$, *the covector* $\xi = \mathbf{g}_y(y, \cdot) \in V^*$ *satisfies*

$$F(y) = F^*(\xi) = \frac{\xi(y)}{F(y)} \tag{3.2}$$

For any covector $\xi \in V^* \setminus \{0\}$, *there exists a unique vector* $y \in V \setminus \{0\}$ *such that* $\xi = \mathbf{g}_y(y, \cdot)$.

Proof: Note that by (1.15) and (3.1)

$$F(y) = \frac{\mathbf{g}_y(y, y)}{F(y)} = \frac{\xi(y)}{F(y)} \leq F^*(\xi).$$

On the other hand, inequality (1.14) implies

$$\xi(v) = \mathbf{g}_y(y, v) \leq F(y)F(v).$$

Thus

$$F^*(\xi) = \sup_{v \in V - \{0\}} \frac{\xi(v)}{F(v)} \leq F(y).$$

We conclude that $F^*(\xi) = F(y)$. This gives (3.2).

Given an arbitrary covector $\xi \in V^* \setminus \{0\}$. The level sets

$$W^\lambda := \Big\{ v \in V, \ \xi(v) = \lambda \Big\}$$

are hyperplanes in V. Since the indicatrix $S := F^{-1}(1)$ is strongly convex, there are exactly one positive number $\lambda_o > 0$ and a vector $y \in S$ such that W^{λ_o} is tangent to S at y. Thus

$$\xi(v) = 0, \qquad v \in T_y S.$$

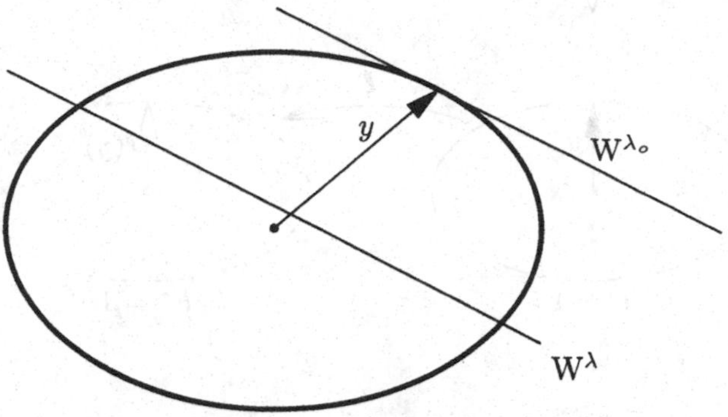

On the other hand,

$$T_y\mathrm{S} = \mathrm{W}_y := \Big\{ v \in \mathrm{V},\ \mathbf{g}_y(y, v) = 0 \Big\}.$$

Thus

$$\xi(v) = \lambda_o\, \mathbf{g}_y(y, v), \qquad v \in \mathrm{V}. \tag{3.3}$$

Letting $\tilde{y} = \lambda_o y$, we obtain the desired vector such that $\xi = \mathbf{g}_{\tilde{y}}(\tilde{y}, \cdot)$. This gives (3.3). Q.E.D.

We define a map $\ell : \mathrm{V} \setminus \{0\} \to \mathrm{V}^* \setminus \{0\}$ by

$$\ell(y) := \mathbf{g}_y(y, \cdot), \qquad y \in \mathrm{V} \setminus \{0\}. \tag{3.4}$$

ℓ is a C^∞ map satisfying

$$\ell(\lambda y) = \lambda \ell(y), \qquad \lambda > 0.$$

Set $\ell(0) = 0$. It follows from (3.2) that

$$F(y) = F^*(\ell(y)). \tag{3.5}$$

Thus ℓ is a norm-preserving map. Moreover, by Lemma 3.1.1, we know that ℓ is an onto map. We call ℓ the *Legendre transformation*. The Legendre transformation is first used by R. Miron in Finsler geometry and Lagrange geometry.

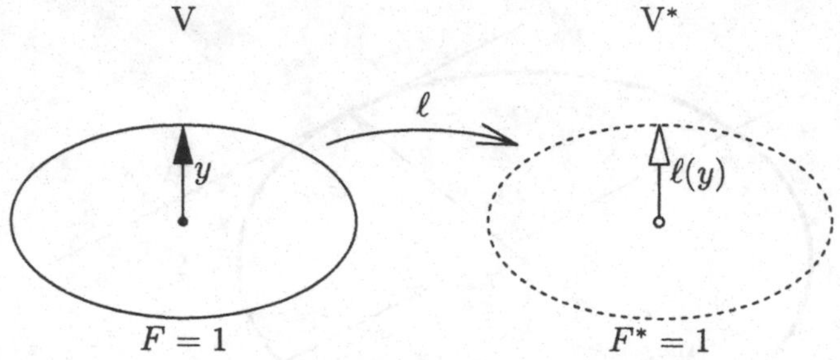

Take a basis $\{\mathbf{b}_i\}_{i=1}^n$ for V and its dual basis $\{\theta^i\}_{i=1}^n$ for V*. Express

$$\xi = \xi_i \theta^i = g_{ij}(y) y^j \, \theta^i = \ell(y), \tag{3.6}$$

where $g_{ij}(y) = \frac{1}{2}[F^2]_{y^i y^j}(y)$. The Jacobian of ℓ is given by

$$\frac{\partial \xi_i}{\partial y^j}(y) = g_{ij}(y).$$

Thus ℓ is a diffeomorphism from $V \setminus \{0\}$ onto $V^* \setminus \{0\}$. From (3.5), we can see that F^* is C^∞ on $V^* \setminus \{0\}$.

Lemma 3.1.2 *For any Minkowski norm F on V, the dual norm F^* on V* is always a Minkowski norm.*

Proof. It suffices to prove that for any $\xi = \ell(y)$,

$$g^{*kl}(\xi) := \frac{1}{2}[F^{*2}]_{\xi^k \xi^l}(\xi) = g^{kl}(y), \tag{3.7}$$

is positive definite, where $\Big(g^{kl}(y)\Big) := \Big(g_{ij}(y)\Big)^{-1}$. Differentiating $F^2(y) = F^{*2}(\ell(y))$ with respect to y^i yields

$$\frac{1}{2}[F^2]_{y^i}(y) = \frac{1}{2}[F^{*2}]_{\xi^k}(\xi) g_{ik}(y). \tag{3.8}$$

This implies

$$g^{*kl}(\xi)\xi_l = \frac{1}{2}[F^{*2}]_{\xi^k}(\xi) = \frac{1}{2} g^{ik}(y)[F^2]_{y^i}(y) = y^k.$$

Note that

$$g^{*kl}\xi_l \frac{\partial g_{ik}}{\partial y^j} = y^k \frac{\partial g_{ik}}{\partial y^j} = y^k \frac{\partial g_{ij}}{\partial y^k} = 0.$$

Differentiating (3.8) with respect to y^j gives

$$g_{ij}(y) = g^{*kl}(\xi)g_{ik}(y)g_{jl}(y) + g^{*kl}(\xi)\xi_l \frac{\partial g_{ik}}{\partial y^j}(y) = g^{*kl}(\xi)g_{ik}(y)g_{jl}(y).$$

Therefore $\left(g^{*kl}(\xi)\right)$ is positive definite. Q.E.D.

Let (V, F) be a Minkowski space and (V^*, F^*) the dual Minkowski space. We can define the Legendre transformation $\ell^* : \mathrm{V}^* \to \mathrm{V}^{**}$ and the Minkowski norm F^{**} on V^{**}. Identifying $\mathrm{V}^{**} = \mathrm{V}$, we have

$$\ell^* = \ell^{-1}, \qquad F^{**} = F. \tag{3.9}$$

For a covector $\xi \in \mathrm{V}^* \setminus \{0\}$, let $\mathbf{g}^{*\xi}$ denote the induced inner product on V^*. It is given by

$$\mathbf{g}^{*\xi}(\zeta, \eta) := g^{*kl}(\xi)\zeta_k\eta_l, \qquad \zeta = \zeta_k\theta^k, \ \eta = \eta_k\theta^k.$$

The vector $y := \ell^{-1}(\xi)$ is determined by

$$\zeta(y) = \mathbf{g}^{*\xi}(\xi, \zeta), \qquad \zeta = \zeta_l\theta^l \in \mathrm{V}^*.$$

Thus y is given by

$$y = y^l \mathbf{b}_l = g^{*kl}(\xi)\xi_k \ \mathbf{b}_l = \ell^{-1}(\xi). \tag{3.10}$$

Example 3.1.1 ([HrSh]) Let $F = \alpha + \beta$ be a Randers norm on a vector space V, where α is an Euclidean norm and β is a linear functional with $\|\beta\| := \sup_{\alpha(y)=1} \beta(y) < 1$. Let $\{\mathbf{b}_i\}_{i=1}^n$ be a basis and $\{\theta^i\}_{i=1}^n$ the dual basis for V. Express α and β by

$$\alpha(y) = \sqrt{a_{ij}y^iy^j}, \qquad \beta(y) = b_iy^i, \qquad y = y^i\mathbf{b}_i.$$

Then $\|\beta\| = \sqrt{a^{ij}b_ib_j}$, where $(a^{ij}) := (a_{ij})^{-1}$. The dual Randers norm F^* on V^* is still of Randers type. More precisely, $F^* = \alpha^* + \beta^*$, where α^* is an Euclidean norm and β^* is a linear functional on V^*. They are expressed by

$$\alpha^*(\xi) = \sqrt{a^{*ij}\xi_i\xi_j}, \qquad \beta^*(\xi) = b^{*i}\xi_i, \qquad \xi = \xi_i\theta^i,$$

where

$$a^{*ij} = \frac{(1-\|\beta\|^2)a^{ij} + b^i b^j}{(1-\|\beta\|^2)^2}$$

$$b^{*i} = -\frac{b^i}{1-\|\beta\|^2},$$

where $b^i := b_j a^{ij}$. Let $(a_{*ij}) := (a^{*ij})^{-1}$. a_{*ij} are given by

$$a_{*ij} = (1-\|\beta\|^2)\Big(a_{ij} - b_i b_j\Big).$$

The norm $\|\beta^*\| := \sup_{\alpha^*(\xi)=1} \beta^*(\xi)$ is given by

$$\|\beta^*\|^2 = a_{*ij} b^{*i} b^{*j} = \frac{1}{1-\|\beta\|^2}\Big(a_{ij} - b_i b_j\Big) b^i b^j = \|\beta\|^2.$$

Thus β and β^* have the same length with respect to α and α^*, respectively. By (1.6) and (3.6), the Legendre transformation $\ell : \mathrm{V} \to \mathrm{V}^*$ is given by

$$\xi = \ell(y) = g_{ij}(y) y^j\, \theta^i = F(y)\Big\{\frac{a_{ij}y^j}{\alpha(y)} + b_i\Big\}\theta^i. \tag{3.11}$$

Similarly, by (1.6) and (3.10), the Legendre transformation $\ell^* = \ell^{-1} : \mathrm{V}^* \to \mathrm{V}$ is given by

$$y = \ell^{-1}(\xi) = g^{*kl}(\xi)\xi_l\, \mathbf{b}_k = F^*(\xi)\Big\{\frac{a^{*kl}\xi_l}{\alpha^*(\xi)} + b^{*i}\Big\}\mathbf{b}_k. \tag{3.12}$$

♯

Let F be a Finsler metric on a manifold M. For a point $x \in M$, $F_x := F|_{T_xM}$ is a Minkowski norm. The dual norm $F_x^* : T_x^*M \to \mathrm{R}$ is defined by

$$F_x^*(\xi) = \sup_{F_x(y)=1} \xi(y).$$

We obtain a family of Minkowski norms $F^* := \{F_x^*\}_{x\in M}$. Let $\ell_x : T_xM \to T_x^*M$ denote the Legendre transformation of F_x. We obtain a family of Legendre transformations $\ell := \{\ell_x\}_{x\in M}$. According to Section 3.1, the map $\ell : TM \setminus \{0\} \to T^*M \setminus \{0\}$ is a C^∞ diffeomorphism satisfying

$$F^*(\ell(y)) = F(y), \qquad y \in TM.$$

3.2 Gradients of Functions

Given a function f on a manifold M, the differential df_x at a point $x \in M$,

$$df_x = \frac{\partial f}{\partial x^i}(x)dx^i,$$

is a linear functional on T_xM. To convert df_x to a vector $\nabla f_x \in T_x^*M$, we need a Minkowski norm on T_xM.

Let F be a Finsler metric on M. By definition, F_x is a Minkowski norm on T_xM. Assume that $df_x \neq 0$. There is a unique unit vector $\mathbf{n}_x \in \mathrm{S}_xM := F_x^{-1}(1)$ and a positive number $\lambda_x > 0$ such that

$$\mathrm{W}^{\lambda_x} := \left\{ df_x(v) = \lambda_x \right\}$$

is tangent to S_xM at $\mathbf{n}_x$. By (3.3),

$$df_x(v) = \lambda_x\, \mathbf{g}_{\mathbf{n}_x}(\mathbf{n}_x, v), \qquad v \in T_xM. \tag{3.13}$$

where

$$F_x^*(df_x) = df_x(\mathbf{n}_x) = \lambda_x\, \mathbf{g}_{\mathbf{n}_x}(\mathbf{n}_x, \mathbf{n}_x) = \lambda_x.$$

Define

$$\nabla f_x := \lambda_x \mathbf{n}_x = F_x^*(df_x)\, \mathbf{n}_x. \tag{3.14}$$

Equation (3.13) can be rewritten as

$$df_x(v) = \mathbf{g}_{\nabla f_x}(\nabla f_x, v), \qquad v \in T_xM. \tag{3.15}$$

If $df_x = 0$, we set $\nabla f_x = 0$. We call ∇f_x the *gradient* of f at x. It follows from (3.2) that

$$F_x(\nabla f_x) = F_x^*(df_x) = \frac{df_x(\nabla f_x)}{F_x(\nabla f_x)}. \tag{3.16}$$

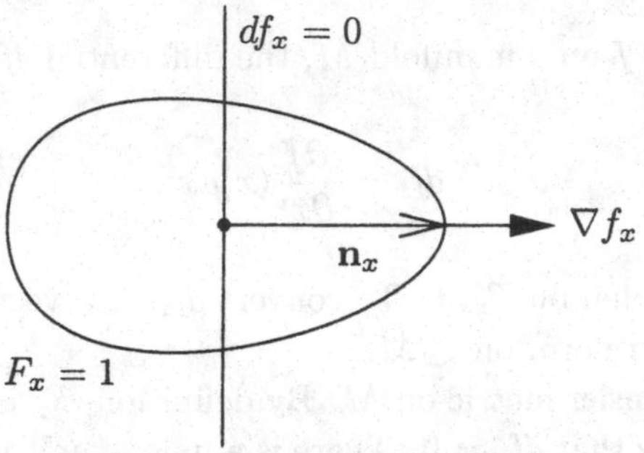

Let $\ell : TM \to T^*M$ and $\ell^* : T^*M \to TM$ denote the Legendre transformations. By the regularity of ℓ, we see that $\nabla f = \ell^{-1}(df)$ is C^∞ on the open subset $\{df \neq 0\}$ and C^0 at $\{df = 0\}$.

Example 3.2.1 Consider a Randers metric $F = \alpha + \beta$ on a manifold M. In local coordinates (x^i) in M, express

$$\alpha(y) = \sqrt{a_{ij}(x)y^iy^j}, \qquad \beta(y) = b_i(x)y^i. \qquad y = y^i \frac{\partial}{\partial x^i}|_x.$$

Then

$$\|\beta\|_x = \sqrt{a^{ij}(x)b_i(x)b_j(x)}.$$

By Example 3.1.1,

$$F(\nabla f) = F^*(df) = \frac{\sqrt{(1-\|\beta\|^2)|df|^2 + \langle\beta, df\rangle^2} - \langle\beta, df\rangle}{1-\|\beta\|^2},$$

where

$$|df|_x := \sqrt{a^{ij}(x)\frac{\partial f}{\partial x^i}(x)\frac{\partial f}{\partial x^j}(x)}, \qquad \langle\beta, df\rangle_x := a^{ij}(x)b_i(x)\frac{\partial f}{\partial x^j}(x).$$

By (3.12), we obtain

$$\nabla f = \frac{F^*(df)}{1-\|\beta\|^2}\left\{\frac{(1-\|\beta\|^2)a^{ij}\frac{\partial f}{\partial x^j} + \langle\beta, df\rangle a^{ij}b_j}{\sqrt{(1-\|\beta\|^2)|df|^2 + \langle\beta, df\rangle^2}} - a^{ij}b_j\right\}\frac{\partial}{\partial x^i}$$

Clearly, ∇f is not C^∞ at points where $df = 0$. ♯

The basic properties of ∇f are shown in the following lemmas.

Lemma 3.2.1 *Suppose that f is C^∞ on an open subset $\mathcal{U}$ with $df \neq 0$. Let $\hat{g} := \mathbf{g}_{\nabla f}$. Then*

$$\nabla f = \hat{\nabla} f, \tag{3.17}$$

where $\hat{\nabla} f$ denotes the gradient of f with respect to $\hat{F} := \sqrt{\hat{g}}$. Further,

$$F(\nabla f) = \hat{F}(\hat{\nabla} f). \tag{3.18}$$

Proof: (3.15) gives

$$\mathbf{g}_{\nabla f}(\nabla f, w) = df(w) = \hat{g}(\hat{\nabla} f, w) = \mathbf{g}_{\nabla f}(\hat{\nabla} f, w), \qquad \forall w \in T\mathcal{U}. \tag{3.19}$$

Thus $\nabla f = \hat{\nabla} f$. Taking $w = \hat{\nabla} f$ in (3.19) yields (3.18). Q.E.D.

Lemma 3.2.2 *Let f be a C^∞ function on an open subset $\mathcal{U}$ with $df \neq 0$. Then $\mathbf{n} := \frac{1}{F(\nabla f)} \nabla f|_{N_t}$ is orthogonal to $N_t := f^{-1}(t)$ with respect to $\mathbf{g}_{\mathbf{n}}$.*

Proof: Since f is a constant on N_t,

$$df(w) = 0, \qquad \forall w \in TN_t.$$

It follows from (3.15) that

$$0 = df(w) = F(\nabla f)\mathbf{g}_{\mathbf{n}}(\mathbf{n}, w), \quad \forall w \in TN_t.$$

By definition, $\mathbf{n}$ is orthogonal to TN_t with respect to $\mathbf{g}_{\mathbf{n}}$. Q.E.D.

Let $d = d_F$ denote the metric induced by F, that is defined by

$$d_F(p, q) = \inf_c \int_0^1 F(\dot{c}(t))dt,$$

where the infimum is taken over all piecewise C^∞ curves $c : [0, 1] \to M$ with $c(0) = p$ and $c(1) = q$. If F is reversible, i.e., $F(-y) = F(y)$, then d_F is reversible.

Given a compact subset $A \subset M$, define

$$\rho_+(x) := d(A, x), \qquad \rho_-(x) := -d(x, A). \tag{3.20}$$

For $x_1, x_2 \in M$ and $z \in A$,

$$d(z, x_1) \leq d(z, x_2) + d(x_2, x_1).$$

This implies that

$$d(A, x_1) \le d(A, x_2) + d(x_2, x_1).$$

and

$$d(A, x_2) \le d(A, x_1) + d(x_1, x_2).$$

Thus

$$-d(x_2, x_1) \le \rho_+(x_2) - \rho_+(x_1) \le d(x_1, x_2). \tag{3.21}$$

By (1.16), we conclude that ρ_+ is locally Lipschitz continuous on M. Similarly, one can show that ρ_- is locally Lipschitz continuous on M. Thus both ρ_+ and ρ_- are differentiable almost everywhere.

Lemma 3.2.3 *Let (M, F) be a Finsler space and ρ_+, ρ_- be the functions defined in (3.20). Suppose that for any points $p, q \in M$, there is a shortest unit speed curve from p to q. Then*

$$F(\nabla\rho_+) = 1, \qquad F(\nabla\rho_-) = 1$$

hold almost everywhere.

Proof. We shall prove $F(\nabla\rho_-) = 1$ only. Suppose that ρ_- is differentiable at $x \in M$. For a vector $v \in T_xM$, let c be a constant speed curve with $\dot{c}(0) = v \in T_xM$. According to Section 1.3, we have

$$F(v) = \lim_{s\to 0^+} \frac{d(x, c(s))}{s}.$$

For $s > 0$,

$$\frac{\rho_-(c(s)) - \rho_-(c(0))}{s} = \frac{d(x, A) - d(c(s), A)}{s} \le \frac{d(x, c(s))}{s}.$$

Letting $s \to 0^+$ yields

$$d\rho_-(v) \le F(v), \qquad \forall v \in T_xM. \tag{3.22}$$

On the other hand, there is a point $p_- \in A$ such that $d(x, p_-) = d(x, A)$. By assumption, there is a unit speed curve $c : [0, a] \to M$ with $a = d(x, A)$ from $x = c(0)$ to $p_- = c(a)$. Thus for any $0 \le s_1 \le s_2 \le a$,

$$d(c(s_1), c(s_2)) = s_2 - s_1.$$

For small $s > 0$,

$$\frac{\rho_-(c(s)) - \rho_-(x)}{s} = \frac{d(x, c(a)) - d(c(s), c(a))}{s} \geq \frac{d(x, c(s))}{s} = 1.$$

This implies

$$d\rho_-(\dot{c}(0)) \geq 1. \tag{3.23}$$

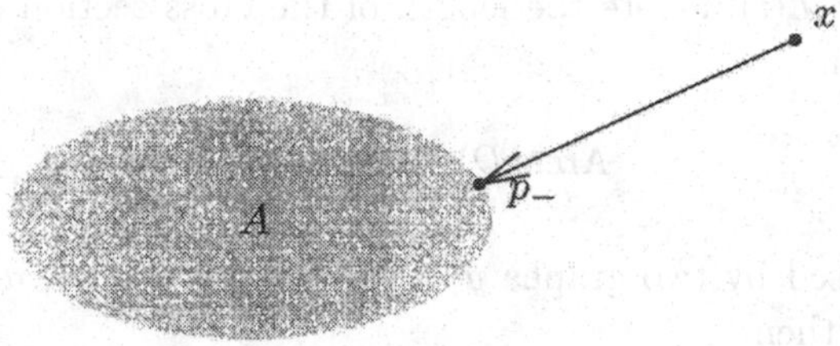

Combining (3.22) and (3.23), one obtains

$$F^*(d\rho_-) = \sup_{F(v)=1} d\rho_-(v) = 1.$$

By (3.16), one concludes that $F(\nabla\rho_-) = F^*(d\rho_-) = 1$. Q.E.D.

Remark 3.2.4 *The condition in Lemma 3.2.3 is satisfied if the Finsler space is positively complete. See Lemma 11.4.1 below.*

Based on Lemma 3.2.3, we make the following

Definition 3.2.5 A locally Lipschitz function ρ on a Finsler space (M, F) is called a *distance function* if

$$F(\nabla\rho) = 1 = F^*(d\rho) \tag{3.24}$$

holds almost everywhere.

It follows from Lemma 3.2.1 that if ρ is a distance function with respect to F, then

$$\hat{F}(\hat{\nabla}\rho) = F(\nabla\rho) = 1,$$

where $\hat{F} = \sqrt{g_{\nabla\rho}}$. Thus ρ is also a distance function with respect to the Riemannian metric $\hat{F}$.

3.3 Co-Area Formula

In calculus, we have several important integral formulas for area and volume. Consider a bounded domain D in the Euclidean plane $\mathbb{R}^2$. For a number t, let $L(t)$ denote the length of the cross section of the line $x = t$ with D. Then

$$\text{Area}(D) = \int_{-\infty}^{\infty} L(t)\, dt.$$

If D is enclosed by two graphs $y = f(x)$ and $y = g(x)$, $a \le x \le b$, where $f(x) \ge g(x)$, then

$$\text{Area}(D) = \int_a^b \Big\{ f(x) - g(x) \Big\} dx.$$

This formula can be generalized to Finsler m spaces.

We have the following co-area formula.

Theorem 3.3.1 *Let $(M, F, d\mu)$ be a Finsler m space. Let φ be a piecewise C^1 function on M such that every $\varphi^{-1}(t)$ is compact. Then for any continuous function f on M,*

$$\int_M f F(\nabla\varphi) d\mu = \int_{-\infty}^{\infty} \Big(\int_{\varphi^{-1}(t)} f d\nu \Big) dt. \tag{3.25}$$

Proof. It suffices to prove (3.25) in a coordinate neighborhood U. For the sake of simplicity, we assume that φ is C^∞ with $d\varphi \ne 0$ on U. Fix a number t_o such that $\varphi^{-1}(t_o) \cap U \ne \emptyset$. Consider the vector field

$$X := \frac{\nabla\varphi}{F(\nabla\varphi)^2}.$$

For a point $x \in \varphi^{-1}(t_o) \cap U$, let $c(t)$ denote the integral curve of X with $c(t_o) = x$. Then

$$\frac{d}{dt}\Big[\varphi \circ c(t)\Big] = d\varphi(X) = \frac{d\varphi(\nabla\varphi)}{F(\nabla\varphi)^2} = 1.$$

Thus

$$\varphi \circ c(t) = t.$$

The integral curves of X give rise to a coordinate map $\psi = (x^1, \cdots, x^n)$: $U \to (-\varepsilon, \varepsilon) \times \mathbb{B}^{n-1}$ such that

$$\varphi \circ \psi^{-1}\Big(x^1, x^a\Big) = x^1.$$

Thus

$$N_t := \varphi^{-1}(t) \cap U = \psi^{-1}\Big(\{t\} \times \mathbb{B}^{n-1}\Big).$$

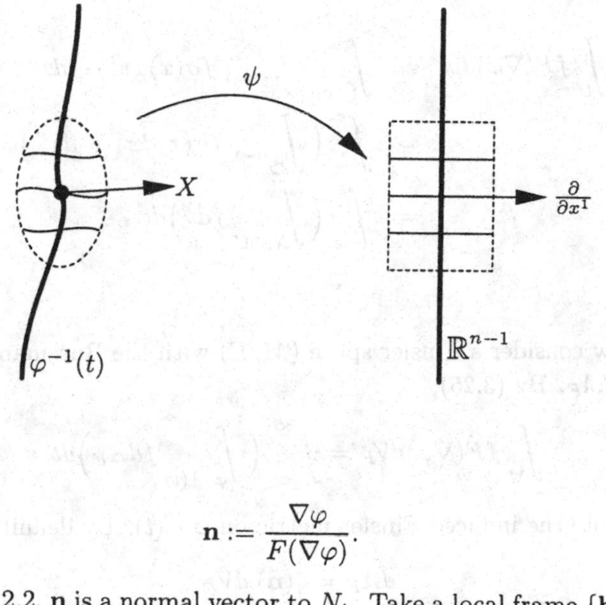

Let

$$\mathbf{n} := \frac{\nabla\varphi}{F(\nabla\varphi)}.$$

By Lemma 3.2.2, $\mathbf{n}$ is a normal vector to N_t. Take a local frame $\{\mathbf{b}_i\}_{i=1}^n$ for TM with $\mathbf{b}_1 = \mathbf{n}$ and $\mathbf{b}_a = \frac{\partial}{\partial x^a}$, $a = 2, \cdots, n$. Let $\{\theta^i\}_{i=1}^n$ be the coframe for T^*M dual to $\{\mathbf{b}_i\}_{i=1}^n$. We have

$$\theta^1 = \lambda\, dx^1 = \lambda\, d\varphi, \qquad \theta^a = dx^a, \quad (a = 2, \cdots, n).$$

Observe that

$$1 = \theta^1(\mathbf{n}) = \lambda\, d\varphi\Big(\frac{\nabla\varphi}{F(\nabla\varphi)}\Big) = \lambda\, \frac{d\varphi(\nabla\varphi)}{F(\nabla\varphi)} = \lambda\, F(\nabla\varphi).$$

We obtain

$$\theta^1 = \frac{1}{F(\nabla\varphi)} dx^1.$$

Put $d\mu = \sigma(x)\theta^1 \wedge \cdots \wedge \theta^n$. Then by (2.24)

$$d\nu = \sigma(x)\bar{\theta}^2 \wedge \cdots \bar{\theta}^n,$$

where $\bar{\theta}^a$ denotes the pull-back of θ^a to N_t, $a = 2, \cdots, n$. We obtain

$$\begin{aligned} d\mu &= \frac{\sigma(x)}{F(\nabla\varphi)} dx^1 \wedge \cdots \wedge dx^n, \\ d\nu &= \sigma(x) dx^2 \wedge \cdots \wedge dx^n. \end{aligned}$$

We obtain

$$\begin{aligned} \int_U fF(\nabla\varphi)d\mu &= \int_{(-\varepsilon,\varepsilon)\times\mathbb{B}^{n-1}} f\sigma(x)dx^1 \cdots dx^n \\ &= \int_{-\varepsilon}^{\varepsilon} \Big(\int_{\mathbb{B}^{n-1}} f\sigma(x)dx^2 \cdots dx^n \Big) dx^1 \\ &= \int_{-\varepsilon}^{\varepsilon} \Big(\int_{N_t\cap U} f d\nu \Big) dt. \end{aligned}$$

Q.E.D.

We now consider a Finsler space (M, F) with the Busemann-Hausdorff measure Vol_F. By (3.25),

$$\int_M fF(\nabla\varphi)dV_F = \int_{-\infty}^{\infty} \Big(\int_{\varphi^{-1}(t)} f dA_F \Big) dt. \tag{3.26}$$

Let $\bar{F}$ denote the induced Finsler metric on $\varphi^{-1}(t)$. By definition,

$$dA_F = \zeta(\mathbf{n})\, dV_{\bar{F}}.$$

where ζ is defined in (2.19). By Lemma 2.3.1, $\zeta(\mathbf{n})$ satisfies

$$(1+\lambda)^{-n} c_n dV_{\bar{F}} \le dA_F \le 2^n c(n) dV_{\bar{F}}, \tag{3.27}$$

where

$$\lambda := \sup_{y\in SM} F(y)/F(-y), \qquad c_n := \mathrm{Vol}(\mathbb{B}^n)/\mathrm{Vol}(\mathbb{B}^{n-1}). \tag{3.28}$$

By (3.25), we obtain the following co-area inequalities for Finsler spaces.

Theorem 3.3.2 *Let (M, F) be a Finsler space. Let φ be a piecewise C^1 function on M such that every $\varphi^{-1}(t)$ is compact. Then for any nonnegative continuous function f on M,*

$$\int_M fF(\nabla\varphi)dV_F \leq 2^n c_n \int_{-\infty}^{\infty}\Big(\int_{\varphi^{-1}(t)} f dV_{\bar{F}}\Big)dt, \tag{3.29}$$

$$\int_M fF(\nabla\varphi)dV_F \geq (1+\lambda)^{-n} c_n \int_{-\infty}^{\infty}\Big(\int_{\varphi^{-1}(t)} f dV_{\bar{F}}\Big)dt. \tag{3.30}$$

There are many important applications of Theorem 3.3.2. For example, one can use it to extend Gromov's estimates on the Filling radius of Riemannian spaces to Finsler spaces [Gr1]. Below is a simple application of Theorem 3.3.2

Proposition 3.3.3 *Let* S *be the indicatrix in a Minkowski space* (V, F). *Let $\bar{F}$ denote the induced Finsler metric on* S. *Then*

$$2^{-n} n\omega_{n-1} \leq \mathrm{Vol}_{\bar{F}}(\mathrm{S}) \leq (1+\lambda)^n n\omega_{n-1}, \tag{3.31}$$

where λ is given in (3.28) and $\omega_{n-1} = \mathrm{Vol}(\mathbb{B}^{n-1})$.

Proof. Let $\rho(x) := F(x)$ denote the distance function from the origin. By definition, the volume of the unit ball B in (V, F) is equal to $\mathrm{Vol}(\mathbb{B}^n)$. Let $dV_{\bar{F}}$ denote the volume form of the induced Finsler metric $\bar{F}$ on $\mathrm{S}(r)$. Note that

$$\mathrm{Vol}_{\bar{F}}(\mathrm{S}(r)) = r^{n-1}\mathrm{Vol}_{\bar{F}}(\mathrm{S}).$$

It follows from (3.29) that

$$\begin{aligned}\mathrm{Vol}(\mathbb{B}^n) = \mathrm{Vol}_F(\mathrm{B}) &\leq 2^n c_n \int_0^1 \mathrm{Vol}_{\bar{F}}(\mathrm{S}(t))dt \\ &= 2^n c_n \int_0^1 t^{n-1}\mathrm{Vol}_{\bar{F}}(\mathrm{S})dt \\ &= \frac{2^n c_n}{n}\mathrm{Vol}_{\bar{F}}(\mathrm{S}).\end{aligned}$$

This gives

$$\mathrm{Vol}_{\bar{F}}(\mathrm{S}) \geq 2^{-n} n \mathrm{Vol}(\mathbb{B}^n)/c_n = 2^{-n} n \mathrm{Vol}(\mathbb{B}^{n-1}).$$

The proof of the inequality on the right hand side of (3.31) is similar, so is omitted. Q.E.D.

When the Minkowski norm is not reversible, the volume $\mathrm{Vol}_{\bar F}(\mathrm{S})$ has no universal upper bound.

Example 3.3.1 *For $0 \le \varepsilon < 1$, consider the following Minkowski norm F_ε on R^2*

$$F_\varepsilon(u,v) := \sqrt{u^2+v^2} - \varepsilon u.$$

The indicatrix $\mathrm{S}_\varepsilon = F_\varepsilon^{-1}(1)$ is an ellipse,

$$\Big(1-\varepsilon^2\Big)^2\Big(u - \frac{\varepsilon}{1-\varepsilon^2}\Big)^2 + \Big(1-\varepsilon^2\Big)v^2 = 1.$$

Parametrize S_ε by

$$u = \frac{1}{1-\varepsilon^2}\cos\theta + \frac{\varepsilon}{1-\varepsilon^2}, \qquad v = \frac{1}{\sqrt{1-\varepsilon^2}}\sin\theta,$$

so that

$$F_\varepsilon(\dot u, \dot v) = \sqrt{\Big(\frac{\varepsilon}{1-\varepsilon^2}\Big)^2 \sin^2\theta + \frac{1}{1-\varepsilon^2}} + \frac{\varepsilon}{1-\varepsilon^2}\sin\theta.$$

This gives

$$\mathrm{Vol}_{\bar F_\varepsilon}(\mathrm{S}_\varepsilon) = \int_0^{2\pi} \sqrt{\Big(\frac{\varepsilon}{1-\varepsilon^2}\Big)^2 \sin^2\theta + \frac{1}{1-\varepsilon^2}}\, d\theta > \frac{2\pi}{\sqrt{1-\varepsilon^2}} \to \infty,$$

as $\varepsilon \to 1^-$. Compare [Ma2]. ♮

Let (V, F) be a Minkowski space. F induces a Riemannian metric $\hat g$ on $\mathrm{V} - \{0\}$ by

$$\hat g(u,v) := \mathbf{g}_y(u,v), \qquad u, v \in T_y\mathrm{V} = \mathrm{V}.$$

Let (y^i) be a global coordinate system in V associated with a basis $\{\mathbf{b}_i\}_{i=1}^n$. The Riemannian volume form of $\hat g$ is given by

$$dV_{\hat g} = \sqrt{\det(g_{ij}(y))}\, dy^1 \cdots dy^n,$$

where $g_{ij}(y) = \mathbf{g}_y(\mathbf{b}_i, \mathbf{b}_j)$.

Let $\dot g$ denote the induced Riemannian metric on $\mathrm{S} := F^{-1}(1)$. Identifying $\mathrm{V}\setminus\{0\}$ with $(0,\infty)\times \mathrm{S}$ in a natural way, i.e., $y \equiv (F(y), y/F(y))$. Since

$\mathbf{g}_y(y, w) = 0$ for any $w \in T_y\mathrm{S} \subset \mathrm{V}$, $\hat{g}$ can be decomposed to the following form.

$$\hat{g} = dt^2 \oplus t^2 \dot{g}. \tag{3.32}$$

Denote by $\mathrm{Vol}_r(\mathrm{B})$ and $\mathrm{Vol}_r(\mathrm{S})$ the Riemannian volume of $(\mathrm{B}, \hat{g})$ and $(\mathrm{S}, \dot{g})$, respectively. By (3.32),

$$\mathrm{Vol}_r(\mathrm{B}) = \int_0^1 t^{n-1}\mathrm{Vol}_r(\mathrm{S})dt = \frac{1}{n}\mathrm{Vol}_r(\mathrm{S}). \tag{3.33}$$

Recall that for the induced Finsler metric $\bar{F}$ on S, the volume $\mathrm{Vol}_{\bar{F}}(\mathrm{S})$ has a universal lower bound, but no universal upper bound. In addition, if F is reversible, then $\mathrm{Vol}_{\bar{F}}(\mathrm{S})$ has a universal upper bound. For the induced Riemannian metric $\dot{g}$ on S, we have the following

Proposition 3.3.4 *Let* (V, F) *be an* n*-dimensional reversible Minkowski space. Then the Riemannian volume of the indicatrix* S *satisfies*

$$\mathrm{Vol}_r(\mathrm{B}) \le \mathrm{Vol}(\mathbb{B}^{n-1}), \qquad \mathrm{Vol}_r(\mathrm{S}) \le \mathrm{Vol}(\mathbb{S}^{n-1}).$$

Equality holds if and only if F *is Euclidean .*

Proof. Let $\{\mathbf{b}_i\}_{i=1}^n$ be a basis for V and $\{\theta^i\}_{i=1}^n$ the dual basis for V^*. Let

$$\mathrm{B}^n := \Big\{(y^i),\ F(y^i\mathbf{b}_i) < 1\Big\}, \qquad \mathrm{B}^{*n} := \Big\{(\xi_i),\ F^*(\xi_i\theta^i) < 1\Big\}.$$

By Santaló inequality [MePa] which requires the symmetry of B^n and B^{*n}.

$$\mathrm{Vol}(\mathrm{B}^n)\mathrm{Vol}(\mathrm{B}^{*n}) \le \mathrm{Vol}(\mathbb{B}^n).$$

Equality holds if and only if B^n is an ellipsoid. The Legendre transformation $\ell : \mathrm{B}^n \to \mathrm{B}^{*n}$ given by

$$y^i \to \xi_i := g_{ij}(y)y^j$$

has Jacobian $\det(g_{ij}(y))$. Thus

$$\int_{\mathrm{B}^n} \det(g_{ij}(y))dy = \int_{\mathrm{B}^{*n}} d\xi = \mathrm{Vol}(\mathrm{B}^{*n}).$$

Observe that

$$\begin{aligned}\mathrm{Vol}_r(\mathrm{B}) &= \int_{\mathrm{B}^n}\sqrt{\det(g_{ij}(y))}dy \\ &\le \Big[\int_{\mathrm{B}^n}\det(g_{ij}(y))dy\Big]^{1/2}\mathrm{Vol}(\mathrm{B}^n)^{1/2} \\ &= \Big[\mathrm{Vol}(\mathrm{B}^{*n})\mathrm{Vol}(\mathrm{B}^n)\Big]^n \\ &\le \mathrm{Vol}(\mathbb{B}^n).\end{aligned}$$

Equality holds if and only if $\det(g_{ij}(y)) = constant$ and B^n is an ellipsoid. This completes the proof. Q.E.D.

The idea of the proof is borrowed from [Du], where C. Duran established a sharp upper bound on the volume of the Sasaki metric on the unit tangent sphere bundle. Here we are only concerned with the volume of the induced Riemannian metric on the indicatrix.

When F is not reversible, then $\mathrm{Vol}_r(\mathrm{S})$ can be greater than $\mathrm{Vol}(\mathbb{S}^{n-1})$. Consider the case when $\dim \mathrm{V} = 2$. Let

$$c(t) = u(t)\mathbf{b}_1 + v(t)\mathbf{b}_2, \qquad a \le t \le b,$$

be an arbitrary parametrization for the indicatrix $\mathrm{S} = F^{-1}(1)$. For any vector $v = \lambda c(t) + \zeta\dot{c}(t) \in T_{c(t)}\mathrm{V}$,

$$\dot{g}_{c(t)}(v,v) = \lambda^2 + \frac{u'(t)v''(t) - u''(t)v'(t)}{u(t)v'(t) - u'(t)v(t)}\,\zeta^2. \tag{3.34}$$

This gives a formula for the Riemannian perimeter $\mathrm{L}_r(\mathrm{S})$ with respect to $\dot{g}$,

$$\mathrm{L}_r(\mathrm{S}) = \int_a^b \sqrt{\frac{u'(t)v''(t) - u''(t)v'(t)}{u(t)v'(t) - u'(t)v(t)}}\,dt.$$

Example 3.3.2 Consider the following Randers norm on R^2,

$$F_\varepsilon(u,v) := \sqrt{u^2+v^2} - \varepsilon u,$$

where $0 < \varepsilon < 1$. Parametrize $\mathrm{S}_\varepsilon = F_\varepsilon^{-1}(1)$ by $c(t) = (u(t), v(t))$,

$$u = \frac{1}{1-\varepsilon^2}\cos(t) + \frac{\varepsilon}{1-\varepsilon^2}, \qquad v = \frac{1}{\sqrt{1-\varepsilon^2}}\sin(t).$$

By (3.34), we obtain

$$\dot{g}_\varepsilon\Big(\dot{c}(t), \dot{c}(t)\Big) = \frac{1}{1 + \varepsilon\cos(t)}.$$

The Riemannian perimeter $\mathrm{L}_r(\mathrm{S}_\varepsilon)$ of S_ε with respect to $\dot{g}_\varepsilon$ is

$$\begin{aligned}
\mathrm{L}_r(\mathrm{S}_\varepsilon) &= \int_0^{2\pi} \frac{1}{\sqrt{1 + \varepsilon\cos(t)}} dt \\
&= 2\int_0^{\pi/2} \Big\{ \frac{1}{\sqrt{1 - \varepsilon\cos(t)}} + \frac{1}{\sqrt{1 + \varepsilon\cos(t)}} \Big\} dt \\
&> 2\pi
\end{aligned}$$

♯

Chapter 4

Isoperimetric Inequalities

The classical isoperimetric inequality for the standard unit sphere $\mathbb{S}^n \subset \mathbb{R}^{n+1}$ states as follows: for any domain $\Omega \subset \mathbb{S}^n$ with regular boundary $\partial\Omega$,

$$\frac{\text{Vol}(\partial\Omega)}{\text{Vol}(\mathbb{S}^n)} \geq h_{\mathbb{S}^n}(s), \qquad s = \frac{\text{Vol}(\Omega)}{\text{Vol}(\mathbb{S}^n)},$$

where $h_{\mathbb{S}^n}(s)$ denotes the ratio $\text{Vol}(\partial \text{B}_s)/\text{Vol}(\mathbb{S}^n)$ for a geodesic ball $\text{B}_s \subset \mathbb{S}^n$ with $\text{Vol}(\text{B}_s) = s \cdot \text{Vol}(\mathbb{S}^n)$. The function $h_{\mathbb{S}^n}$ is called the isoperimetric profile of $\mathbb{S}^n$. The isoperimetric profile can be defined for Finsler m spaces. In this chapter, we will discuss the relationship between the isoperimetric profile and the Sobolev constants (first eigenvalue) of a Finsler m space.

The classical isoperimetric inequality is used by P. Levy to discuss his concentration theory [Le]. Levy's concentration theory is developed further to Riemannian spaces and more general metric measure spaces [GrMi][Gr4]. In his study on the concentration of metric and measure spaces, M. Gromov introduces many new geometric invariants. Among them are the two basic quantities: the expansion distance and the observable diameter [Gr4]. In this chapter, we will discuss the relationship between the isoperimetric profile and these two quantities.

4.1 Isoperimetric Profiles

Let $(M, F, d\mu)$ be a Finsler m space. We define the *isoperimetric profile* $h_M : [0,1] \to [0,\infty)$ by

$$h_M(s) := \inf \frac{\nu(\partial\Omega)}{\mu(M)}, \tag{4.1}$$

where the infimum is taken over all regular domains $\Omega \subset M$ such that $\mu(\Omega) = s \cdot \mu(M)$, and ν denotes the induced measure on $\partial\Omega$ (cf. (2.23)). Note that

$$h_M(s) = h_M(1-s) > 0, \quad \forall s \in (0,1)$$

and

$$h_M(0) = 0 = h_M(1).$$

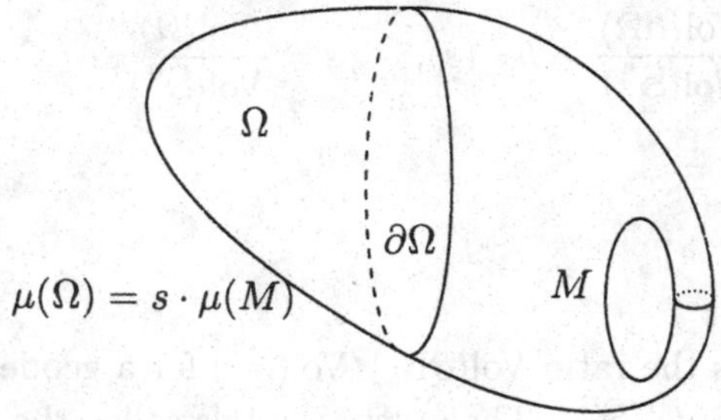

Therefore we make the following

Definition 4.1.1 *A continuous function $h : [0,1] \to \mathrm{R}_+$ is called an isoperimetric function if it satisfies*

$$h(s) = h(1-s) > 0, \ \forall s \in (0,1)$$

and

$$h(0) = 0 = h(1).$$

Given an arbitrary isoperimetric function h, we will construct an open interval $(-L, L)$ and a function $\sigma_h(t)$ on $(-L, L)$ such that

$$\int_{-L}^{t} \sigma_h(u)du = s, \qquad \sigma_h(t) = h(s). \tag{4.2}$$

First, we try to find a solution $\mathbf{s} = \mathbf{s}(t)$ to the following O.D.E.

$$\begin{cases} \dfrac{d\mathbf{s}}{dt} & = h(\mathbf{s}) \\ \mathbf{s}(0) & = \frac{1}{2}. \end{cases} \tag{4.3}$$

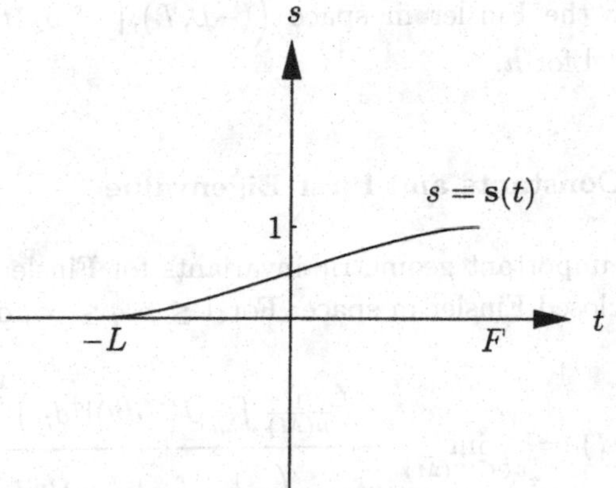

Let

$$\eta(s) := \int_{\frac{1}{2}}^{s} \frac{1}{h(u)} du. \tag{4.4}$$

η is an increasing C^1 function with a symmetric range $(-L, L)$. Let

$$\mathbf{s}(t) := \eta^{-1}(t) \tag{4.5}$$

Then $\mathbf{s}(t)$ is a solution to (4.3).

Let

$$\sigma_h(t) := h(\mathbf{s}(t)) = h(\eta^{-1}(t)).$$

Since the range of $\mathbf{s}(t)$ is the open interval $(0, 1)$, we have

$$\lim_{t \to \pm L} \sigma_h(t) = 0.$$

Further,

$$\mathbf{s}(t) = \int_{-L}^{t} \sigma_h(u)du, \tag{4.6}$$

$$\sigma_h(t) = h\Big(\int_t^L \sigma_h(u)du\Big) = h\Big(\int_{-L}^{t} \sigma_h(u)du\Big). \tag{4.7}$$

This gives (4.2).

We can view the Finsler m space $\Big((-L, L), |\cdot|, \sigma_h(t)dt\Big)$ as a one-dimensional model for h.

4.2 Sobolev Constants and First Eigenvalue

There are many important geometric invariants for Finsler m spaces. Let $(M, F, d\mu)$ be a closed Finsler m space. For $1 \le p, q \le \infty$, define

$$\lambda_{p,q}(M) := \inf_{u \in C^\infty(M)} \frac{\Big(\frac{1}{\mu(M)}\int_M |F^*(du)|^q d\mu\Big)^{\frac{1}{q}}}{\inf_{\lambda \in \mathrm{R}}\Big(\frac{1}{\mu(M)}\int_M |u-\lambda|^p d\mu\Big)^{\frac{1}{p}}}.$$

$\lambda_{p,q}(M)$ is called the (p,q)-*Sobolev constant.*

Another geometric invariant is the the *first eigenvalue* of $(M, F, d\mu)$, defined by

$$\lambda_1(M) := \inf_{u \in C^\infty(M)} \frac{\int_M |F^*(du)|^2 d\mu}{\inf_{\lambda \in \mathrm{R}} \int_M |u-\lambda|^2 d\mu}. \tag{4.8}$$

Clearly,

$$\lambda_1(M) = [\lambda_{2,2}(M)]^2.$$

The Sobolev constants $\lambda_{p,q}(M)$ are closely related to the isoperimetric constants (cf. [Be] and [Ga] in the Riemannian case). In what follows, we shall establish some relationships between the Sobolev constants and the isoperimetric constants for Finsler m spaces.

For an isoperimetric function h and $1 \leq p, q < \infty$, define

$$\lambda_{p,q}(h) := \inf_{\phi} \frac{\left(\int_0^1 |\phi'(s)|^q h(s)^q ds \right)^{1/q}}{\left(\inf_{\lambda \in \mathbf{R}} \int_0^1 |\phi(s) - \lambda|^p ds \right)^{1/p}}, \tag{4.9}$$

where the infimum is taken over all bounded piecewise C^1-functions ϕ on $[0, 1]$. $\lambda_{p,q}(h)$ can be viewed as the (p, q)-Sobolev constant of the Finsler m space $\Big((-L, L), |\cdot|, \sigma_h(t)dt\Big)$.

By symmetrization based on the one-dimensional model, we can prove the following

Theorem 4.2.1 ([GeSh]) *Let (M, d, μ) be a closed Finsler m space. Suppose that $h_M \geq h$ for some isoperimetric function h. Then*

$$\lambda_{p,q}(M) \geq \lambda_{p,q}(h). \tag{4.10}$$

In particular,

$$\lambda_1(M) \geq [\lambda_{2,2}(h)]^2.$$

Proof. Under the assumption $h_M \geq h$, for any regular domain $\Omega \subset M$,

$$\frac{\nu(\partial\Omega)}{\mu(M)} \geq h\Big(\frac{\mu(\Omega)}{\mu(M)}\Big).$$

Now we process to prove (4.10). Take an arbitrary function $u \in C^\infty(M)$. Since u can be approximated by Morse functions, we may assume that u is a Morse function on M. Decompose u into the difference of the positive and negative parts, $u = u_+ - u_-$.

For $a \geq 0$, let

$$\Omega_+(a) := \{x \in M, \quad u_+(x) > a\}.$$

Define

$$t_+ : [0, \sup u_+) \to (-L, L)$$

by

$$\int_{t_+(a)}^L \sigma_h(t)dt = \frac{\mu(\Omega_+(a))}{\mu(M)}. \tag{4.11}$$

The function t_+ is an increasing piecewise C^1 function. By assumption $h_M \geq h$, we have that for $a \geq 0$,

$$\begin{aligned}\frac{\nu(\partial\Omega_+(a))}{\mu(M)} &\geq h\Big(\frac{\mu(\Omega_+(a))}{\mu(M)}\Big)\\ &= h\Big(\int_{t_+(a)}^{L}\sigma_h(t)dt\Big)\\ &= \sigma_h(t_+(a)).\end{aligned} \tag{4.12}$$

The last equality follows from (4.7).

Define $\varphi_+ : (-L, L) \to \mathrm{R}_+$ by setting $\varphi_+(t) = a$ if $t = t_+(a)$ for some $a \geq 0$ and $\varphi_+(t) = 0$ if $t \in (-L, t_+(0))$. We have

$$\varphi_+'(t_+(a)) = \frac{1}{t_+'(a)}.$$

We know that $F(\nabla u_+) = F^*(du_+)$. By the co-area formula (3.25) and (4.11),

$$\int_a^\infty \Big[\int_{\partial\Omega_+(u)} \frac{1}{F^*(du_+)} d\nu\Big] du = \mu(\Omega_+(a)) = \mu(M)\int_{t_+(a)}^{L} \sigma_h(t)dt. \tag{4.13}$$

Differentiating (4.13), one obtains

$$\int_{\partial\Omega_+(a)} \frac{1}{F^*(du_+)} d\nu = \mu(M)\sigma_h(t_+(a))t_+'(a). \tag{4.14}$$

The Hölder inequality yields

$$\begin{aligned}\nu(\partial\Omega_+(a)) &= \int_{\partial\Omega_+(a)} \frac{[F^*(du_+)]^{\frac{q-1}{q}}}{[F^*(du_+)]^{\frac{q-1}{q}}} d\nu\\ &\leq \Big(\int_{\partial\Omega_+(a)} [F^*(du_+)]^{q-1}\Big)^{\frac{1}{q}}\Big(\int_{\partial\Omega_+(a)} [F^*(du_+)]^{-1}d\nu\Big)^{1-\frac{1}{q}}.\end{aligned} \tag{4.15}$$

It follows from (4.12) - (4.15) that

$$\begin{aligned}\int_M [F^*(du_+)]^q d\mu &= \int_0^\infty \Big(\int_{\partial\Omega_+(a)} [F^*(du_+)]^{q-1} d\nu\Big) da\\ &\geq \int_0^\infty \Big[\frac{\nu(\partial\Omega_+(a))}{\Big(\int_{\partial\Omega_+(a)} F^*(du_+)^{-1}d\nu\Big)^{1-1/q}}\Big]^q da\end{aligned}$$

$$
\begin{aligned}
&\geq \mu(M)\int_0^\infty \Big[\frac{\sigma_h(t_+(a))}{\Big(\sigma_h(t_+(a))t'_+(a)\Big)^{1-1/q}}\Big]^q da \\
&= \mu(M)\int_0^\infty \frac{1}{[t'_+(a)]^q}\sigma_h(t_+(a))t'_+(a)da \\
&= \mu(M)\int_0^\infty \Big[\varphi'_+(t_+(a))\Big]^q \sigma_h(t_+(a))t'_+(a)da \\
&= \mu(M)\int_{-L}^{L} |d\varphi_+(t)|^q\sigma_h(t)dt
\end{aligned}
$$

That is,

$$
\|F^*(du_+)\|_q \geq \mu(M)^{\frac{1}{q}}\|d\varphi_+\|_q. \tag{4.16}
$$

Let λ be an arbitrary number. By (4.14) and the co-area formula again, we obtain

$$
\begin{aligned}
\int_{\Omega_+(0)} |u_+ - \lambda|^p d\mu &= \int_0^\infty |a-\lambda|^p\Big[\int_{\partial\Omega_+(a)}\frac{1}{F^*(du_+)}d\nu\Big]da \\
&= \mu(M)\int_0^\infty |a-\lambda|^p\sigma_h(t_+(a))t'_+(a)da \\
&= \mu(M)\int_0^\infty \Big|\varphi_+(t_+(a))-\lambda\Big|^p\sigma_h(t_+(a))t'_+(a)da \\
&= \mu(M)\int_{t_+(0)}^{L} \Big|\varphi_+(t)-\lambda\Big|^p\sigma_h(t)dt
\end{aligned}
$$

That is,

$$
\|u_+ - \lambda\|_{L^p(\Omega_+(0))} = \mu(M)^{\frac{1}{p}}\|\varphi_+ - \lambda\|_{L^p(t_+(0),L)}.
$$

Similarly, for the negative part u_- and $a \geq 0$, let

$$
\Omega_-(a) := \{x \in M,\ u_-(x) > a\}.
$$

Define $t_- : [0,\infty) \to (-L, L)$ by

$$
\int_{-L}^{t_-(a)} \sigma_h(t)dt = \frac{\mu(\Omega_-(a))}{\mu(M)}.
$$

Define $\varphi_- : (-L, L) \to \mathrm{R}_+$ by setting $\varphi_-(t) = a$ if $t = t_-(a)$ for some $a \geq 0$ and $\varphi_-(t) = 0$ if $t > t_-(0)$. By the same argument as above, we obtain

$$
\|u_- + \lambda\|_{L^p(\Omega_+(0))} = \mu(M)^{1/p}\|\varphi_- + \lambda\|_{L^p(-L,t_-(0))},
$$

$$\|F^*(du_-)\|_q \geq \mu(M)^{1/q}\|d\varphi_-\|_q. \tag{4.17}$$

Note that $t_-(a) < t_+(a)$, $\forall a > 0$. Set $\varphi := \varphi_+ - \varphi_-$. It follows from (4.16)-(4.17) that

$$\|u - \lambda\|_p = \mu(M)^{1/p}\|\varphi - \lambda\|_p,$$

$$\|F^*(du)\|_q \geq \mu(M)^{1/q}\|d\varphi\|_q.$$

Therefore

$$\mu(M)^{1/p-1/q}\frac{\|F^*(du)\|_q}{\|u-\lambda\|_p} \geq \frac{\|d\varphi\|_q}{\|\varphi - \lambda\|_p}. \tag{4.18}$$

Take the substitution $s = \mathbf{s}(t)$ and let $\phi(s) := \varphi(t)$. Note that

$$ds = h(s)dt = \sigma_h(t)dt.$$

We have

$$\frac{\left(\int_{-L}^{L}|\varphi'(t)|^q\sigma_h(t)dt\right)^{1/q}}{\left(\inf_{\lambda\in\mathrm{R}}\int_{-L}^{L}|\varphi(t)-\lambda|^p\sigma_h(t)dt\right)^{1/p}} = \frac{\left(\int_{0}^{1}|\phi'(s)|^q h(s)^q ds\right)^{1/q}}{\left(\inf_{\lambda\in\mathrm{R}}\int_0^1|\phi(s)-\lambda|^p ds\right)^{1/p}}. \tag{4.19}$$

(4.18) and (4.19) imply

$$\lambda_{p,q}(M) \geq \lambda_{p,q}(h).$$

Q.E.D.

For an isoperimetric function h, define

$$\mathrm{I}_h(M) := \inf_{\Omega}\frac{\nu(\partial\Omega)}{h\Big(\mu(\Omega)/\mu(M)\Big)\mu(M)}, \tag{4.20}$$

where the infimum is taken over all regular domains Ω in M. $\mathrm{I}_h(M)$ is called the *h-isoperimetric constant* of M.

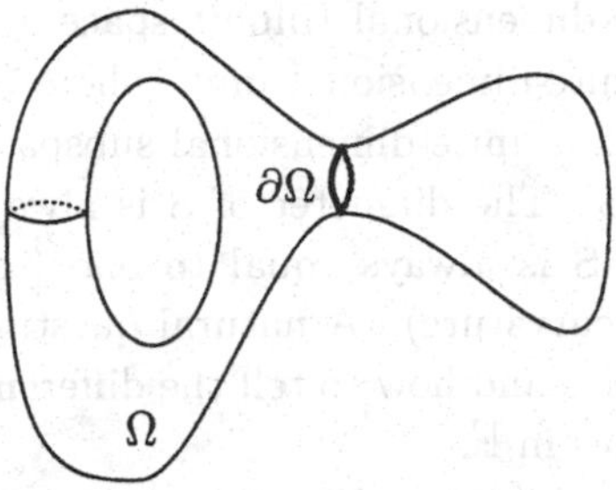

Take an arbitrary domain $\Omega \subset M$ with $\mu(\Omega) = s \cdot \mu(M)$. From (4.20),

$$\frac{\nu(\partial\Omega)}{h(s)\mu(M)} \geq \mathrm{I}_h(M).$$

This gives

$$\frac{\mu(\partial\Omega)}{\mu(M)} \geq \mathrm{I}_h(M)h(s).$$

Thus

$$h_M(s) \geq \mathrm{I}_h(M)h(s). \tag{4.21}$$

Consider the following isoperimetric function

$$h_\alpha(s) := \min(s, 1-s)^{1-\frac{1}{\alpha}}, \qquad 1 < \alpha \leq \infty.$$

Here we assume that $\frac{1}{\alpha} = 0$ if $\alpha = \infty$. It can be shown that if $\frac{1}{\alpha} \geq \frac{1}{q} - \frac{1}{p}$, then $\lambda_{p,q}(h_\alpha) > 0$. In particular, $\lambda_{2,2}(h_\alpha) > 0$ for any $1 < \alpha \leq \infty$. Let

$$\mathrm{I}_\alpha(M) = \mathrm{I}_{h_\alpha}(M).$$

We call $I_\infty(M)$ the *Cheeger constant.*

The isoperimetric profile of a Riemannian manifold can be estimated under a lower Ricci curvature and an upper diameter bound. This is done by M. Gromov using Levy's method. Gromov applies his estimates on the isoperimetric profiles to obtain some estimates for the eigenvalues of the Laplacian on a Riemannian manifold. See [Gr4] [Gr5] for details.

4.3 Concentration of Finsler m Spaces

Let $\mathbb{R}^\infty$ be an infinite-dimensional Hilbert space and S^∞ the unit sphere in $\mathbb{R}^\infty$. Consider a finite-dimensional unit sphere S in $\mathbb{R}^\infty$, which is the intersection of S^∞ with a finite-dimensional subspace in $\mathbb{R}^\infty$. Let us take a look at this sphere S. The diameter of S is always equal to π and the sectional curvature of S is always equal to one (see Section 6.1 for the definition of sectional curvature). A natural question arises: How can we detect the dimension of S and how to tell the difference between two finite-dimensional unit spheres in $\mathbb{R}^\infty$?

We have a satisfied answer to this problem using the Lévy concentration theory of mm spaces [Le]. The Lévy concentration theory of mm spaces has been developed further by M. Gromov [Gr4]. He introduce several geometric quantities for mm spaces. Among them is the so-called *expansion distance.*

Definition 4.3.1 *Let (M, d, μ) be an mm space with $\mu(M) < \infty$. For $0 < \varepsilon < \frac{1}{2}$, the expansion distance* $\mathrm{ExDist}(M; \varepsilon)$ *is defined to be the infimal ρ such that the following statement holds*

If a subset $A \subset X$ satisfies

$$\mu(A) \geq \varepsilon \mu(M),$$

then

$$\mu\Big(U_\rho(A)\Big) \geq (1 - \varepsilon)\mu(M).$$

Here $U_\rho(A)$ denotes the ρ-neighborhood of A in M,

$$U_\rho(A) := \Big\{x \in A,\ d(A, x) < \rho\Big\}.$$

The expansion distance tells us how the mass concentrates, with respect to the underlying metric. We say that a sequence of mm spaces (M_i, d_i, μ_i) *concentrates to a point* if $\mathrm{ExDist}(X_i; \varepsilon) \to 0$ as $i \to \infty$ for any $\varepsilon \in (0, 1/2)$. A trivial inequality is that

$$\mathrm{ExDist}(M; \varepsilon) \leq \mathrm{Diam}(M).$$

Thus if the diameter of (M_i, d_i, μ_i) converges to zero, then (M_i, d_i, μ_i) concentrates to a point. There are sequences of mm spaces which do not collapse to a point, but concentrate to a point.

Example 4.3.1 Let $\mathbb{S}^i$ denote the standard unit spheres in $\mathbb{R}^{i+1} \subset \mathbb{R}^\infty$ with the induced Riemannian metric and Riemannian volume form. We have

$$\mathrm{ExDist}(\mathbb{S}^i; \varepsilon) \approx \frac{C(\varepsilon)}{\sqrt{i}},$$

where $C(\varepsilon)$ depends only on $\varepsilon \in (0, 1/2)$. Thus the sequence $\{\mathbb{S}^i\}$ concentrates to a point as $i \to \infty$, while, $\{\mathbb{S}^i\}$ does not collapse to a point.

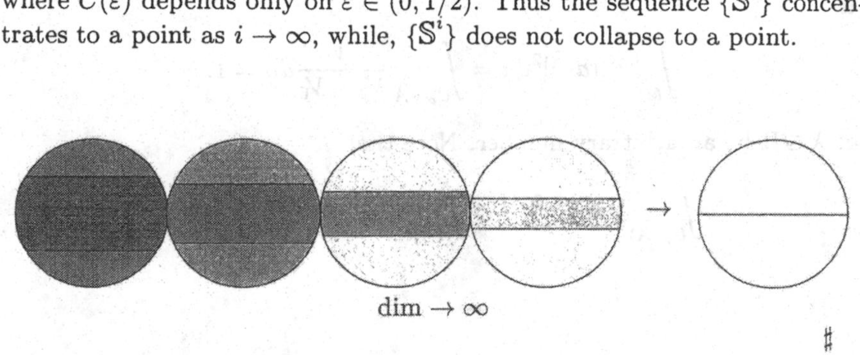

♯

The above concentration phenomenon was first discovered by Paul Lévy [Le]. Lévy explained this by using the sharp isoperimetric inequality for $\mathbb{S}^n$. Lévy's result has been generalized to Riemannian spaces by M. Gromov and V. D. Milman [GrMi]. They proved that for any closed Riemannian spaces (M, g),

$$\mathrm{ExDist}(M; \varepsilon) \leq \frac{C(\varepsilon)}{\sqrt{\lambda_1(M)}},$$

where $\lambda_1(M)$ denotes the first eigenvalue of X (see also [Gr4]. The same method can be carried over to Finsler spaces, due to Lemma 3.2.3.

Theorem 4.3.2 *Let $(M, F, d\mu)$ be a closed Finsler m space. For any* $0 < \varepsilon < 1/2$,

$$\mathrm{ExDist}(M; \varepsilon) \leq 2\sqrt{\frac{1}{2\varepsilon} - 1}\,\frac{1}{\sqrt{\lambda_1(M)}}.$$

Proof. Fix $\rho > 0$ and $0 < \varepsilon < 1/2$. Let $A \subset M$ be any closed subset and $U_\rho(A)$ the ρ-neighborhood of A. Define

$$u(x) := \begin{cases} \frac{\mathrm{dist}(A,x)}{\sqrt{V_\rho - V_0}} & x \in U_\rho(A) \\ \frac{\rho}{\sqrt{V_\rho - V_0}} & x \in M - U_\rho(A), \end{cases}$$

where $V_\rho := \mu(U_\rho(A))$ and $V_0 := \mu(A)$. Let $V = \mu(M)$.

By Lemma 3.2.3, we see that

$$F^*(du) = F(\nabla u) = \begin{cases} \frac{1}{\sqrt{V_\rho - V_0}} & x \in U_\rho(A) - A \\ 0 & x \in A \cup (M - U_\rho(A)), \end{cases}$$

This

$$\int_M [F^*(du)]^2 d\mu = \int_{U_\rho - A} \frac{1}{V_\rho - V_0} d\mu = 1.$$

Let $\lambda \in \mathrm{R}$ be an arbitrary number. Note that

$$\int_{U_\rho(A)} |u - \lambda|^2 d\mu \geq \int_A |u - \lambda|^2 d\mu = \lambda^2 \mu(A).$$

Thus

$$\begin{aligned}\int_M |u-\lambda|^2 d\mu &= \int_{M-U_\rho} |u-\lambda|^2 d\mu + \int_{U_\rho} |u-\lambda|^2 d\mu \\ &\geq \left(\lambda - \frac{\rho}{\sqrt{V_\rho - V_0}}\right)^2 (V-V_\rho) + \lambda^2 V_0 \\ &\geq \frac{(V-V_\rho)V_0}{(V_\rho - V_0)(V-V_\rho+V_0)}\rho^2.\end{aligned}$$

We obtain

$$\lambda_1(M) \leq \frac{\int_M [F^*(du)]^2 d\mu}{\inf_{\lambda\in \mathrm{R}} \int_M |u-\lambda|^2 d\mu} \leq \frac{(V_\rho - V_0)(V-V_\rho+V_0)}{(V-V_\rho)V_0}\rho^{-2}.$$

In other words

$$\rho \leq \sqrt{\frac{(V_\rho - V_0)(V-V_\rho+V_0)}{(V-V_\rho)V_0}} \frac{1}{\sqrt{\lambda_1(M)}}. \tag{4.22}$$

For any $\rho < \mathrm{ExDist}(M;\varepsilon)$, there is a subset A satisfying

$$V_0 \geq \varepsilon V, \qquad V_\rho < (1-\varepsilon)V.$$

Then by (4.22), we obtain

$$\rho \leq 2\sqrt{\frac{1}{2\varepsilon} - 1}\, \frac{1}{\sqrt{\lambda_1(M)}}.$$

This implies that

$$\mathrm{ExDist}(M;\varepsilon) \leq 2\sqrt{\frac{1}{2\varepsilon} - 1}\, \frac{1}{\sqrt{\lambda_1(M)}}.$$

Q.E.D.

The expansion distance is also controlled by the isoperimetric profile.

Theorem 4.3.3 *Let $(M, F, d\mu)$ be a closed Finsler m space. Suppose that $h_M \geq h$, where h is an isoperimetric function. Then for any $0 < \varepsilon < \frac{1}{2}$,*

$$\mathrm{ExDist}(M;\varepsilon) \leq \int_\varepsilon^{1-\varepsilon} \frac{1}{h(u)} du. \tag{4.23}$$

Proof. Let $A \subset M$ be a closed subset and

$$V_\rho := \mu(U_\rho(A)), \quad V_0 = \mu(A), \quad V = \mu(M).$$

Define $\varphi : [0, \rho_1) \to (-L, L)$ by

$$\varphi(\rho) := \int_{V_0/V}^{V_\rho/V} \frac{1}{h(u)} du.$$

where $\rho_1 := \sup_{x \in M} d(A, x)$.

Applying (3.25) to the special function $\varphi(x) = d(A, x)$, we obtain the following co-area formula

$$V_\rho = \int_0^\rho A_t dt, \tag{4.24}$$

where $A_t := \nu(\varphi^{-1}(t))$ denotes the induced measure of $\varphi^{-1}(t)$. Note that

$$\frac{A_\rho}{V} \geq h_M\left(\frac{V_\rho}{V}\right).$$

It follows from $h_M \geq h$ that

$$\varphi'(\rho) = \frac{A_\rho}{V} \frac{1}{h(V_\rho/V)} \geq \frac{h_M(V_\rho/V)}{h(V_\rho/V)} \geq 1, \quad \forall \rho < \rho_1.$$

This implies

$$\int_{V_0/V}^{V_\rho/V} \frac{1}{h(u)} du = \varphi(\rho) \geq \rho, \quad \forall \rho < \rho_1. \tag{4.25}$$

Let

$$\rho_o := \int_\varepsilon^{1-\varepsilon} \frac{1}{h(u)} du.$$

Assume that $\mu(A) \geq \varepsilon \mu(M)$, i.e.,

$$\frac{V_0}{V} \geq \varepsilon. \tag{4.26}$$

We assert that $\mu(U_{\rho_o}(A)) \geq (1 - \varepsilon)\mu(M)$, i.e.,

$$\frac{V_{\rho_o}}{V} \geq 1 - \varepsilon. \tag{4.27}$$

Note that if $\rho_o \geq \rho_1$, then

$$\frac{V_{\rho_o}}{V} = \frac{V_{\rho_1}}{V} = 1 > 1 - \varepsilon.$$

Now we assume that $\rho_o < \rho_1$. In this case, by (4.25) and (4.26), we have

$$\int_{\varepsilon}^{V_{\rho_o}/V} \frac{1}{h(u)} du \geq \int_{V_0/V}^{V_{\rho_o}/V} \frac{1}{h(u)} du = \varphi(\rho_o) \geq \rho_o = \int_{\varepsilon}^{1-\varepsilon} \frac{1}{h(u)} du.$$

This implies (4.27). Therefore, we can conclude that

$$\mathrm{ExDist}(M; \varepsilon) \leq \rho_o = \int_{\varepsilon}^{1-\varepsilon} \frac{1}{h(u)} du.$$

Q.E.D.

By (4.21) and (4.23), we see that for any isoperimetric function h,

$$\mathrm{ExDist}(M; \varepsilon) \leq \frac{1}{\mathrm{I}_h(M)} \int_{\varepsilon}^{1-\varepsilon} \frac{1}{h(u)} du.$$

In particular, for $h_\alpha(s) = \min(s, 1-s)^{1-\frac{1}{\alpha}}$, $1 < \alpha \leq \infty$,

$$\mathrm{ExDist}(M; \varepsilon) \leq \frac{C_\alpha}{\mathrm{I}_\infty(M)},$$

where

$$C_\alpha = \begin{cases} 2\alpha\left[\left(\frac{1}{2}\right)^{1/\alpha} - \varepsilon^{1/\alpha}\right] & \text{if } 1 < \alpha < \infty \\ -2\ln(2\varepsilon) & \text{if } \alpha = \infty \end{cases}$$

The expansion distance of Finsler m spaces can be estimated from below if they satisfy certain curvature bound. See Theorem 16.4.1 below.

4.4 Observable Diameter

To study the Lévy concentration of mm spaces, Gromov [Gr4] also introduce the notion of observable diameter.

Let (M, d, μ) be an mm space. Denote by $\mathrm{Lip}_1(M)$ the set of all Lipschitz functions $f : M \to \mathrm{R}$ satisfying

$$-d(x_1, x_2) \leq f(x_1) - f(x_2) \leq d(x_2, x_1), \quad \forall x_1, x_2 \in M.$$

By (3.21), we know that for any compact subset $A \subset M$, the distance function $\rho_+(x) := d(A,x)$ belongs to $\mathrm{Lip}_1(M)$, so does $\rho_-(x) := -d(x,A)$.

For $0 < \varepsilon < 1$, define

$$\mathrm{Diam}(f_*\mu,\varepsilon) := \inf\Big\{\mathrm{Diam}(I) : \ I \subset \mathrm{R}, \ \mu(f^{-1}(I)) \geq (1-\varepsilon)\mu(M)\Big\}$$

The *observable diameter* is defined by

$$\mathrm{ObsDiam}(M;\varepsilon) := \sup_{f\in \mathrm{Lip}_1(M)} \mathrm{Diam}(f_*\mu,\varepsilon).$$

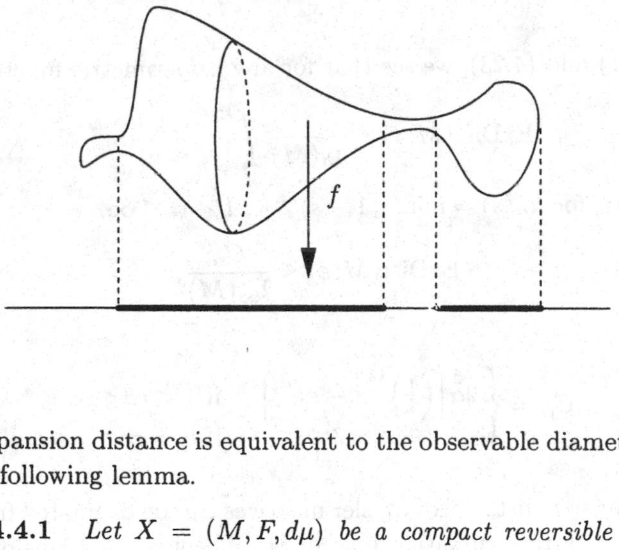

The expansion distance is equivalent to the observable diameter according to the following lemma.

Lemma 4.4.1 *Let $X = (M, F, d\mu)$ be a compact reversible Finsler m space with $\mu(M) < \infty$. Then for any $0 < \varepsilon < 1/2$,*

$$\mathrm{ExDist}(M;\varepsilon) \leq \mathrm{ObsDiam}(M;\varepsilon). \tag{4.28}$$

For any $0 < \varepsilon < 1$,

$$\mathrm{ObsDiam}(M;\varepsilon) \leq 2 \cdot \mathrm{ExDist}(M;\varepsilon/2). \tag{4.29}$$

Proof. First we prove (4.28). Let $\rho < \mathrm{ExDist}(M;\varepsilon)$. There exists a subset $A \subset M$ such that

$$\mu(A) \geq \varepsilon\mu(M), \qquad \mu\Big(U_\rho(A)\Big) < (1-\varepsilon)\mu(M).$$

Define

$$f(x) := \begin{cases} \operatorname{dist}(A, x) & x \in U_\rho(A) \\ \rho & x \in M - U_\rho(A). \end{cases}$$

Let $I = [a, b]$ be an arbitrary interval such that

$$\mu(f^{-1}(I)) \geq (1 - \varepsilon)\mu(M). \tag{4.30}$$

We assert that $a \leq 0$ and $b \geq \rho$. Hence $\operatorname{Diam}(I) = b - a \geq \rho$.

If $b < \rho$, then $f^{-1}(I) \subset U_\rho(A)$. This implies

$$\mu(f^{-1}(I)) \leq \mu\Big(U_\rho(A)\Big) < (1 - \varepsilon)\mu(M),$$

that contradicts (4.30).

If $a > 0$, then $f^{-1}(I) \subset M - U_a(A)$. Thus

$$\mu(f^{-1}(I)) \leq \mu(M) - \mu\Big(U_a(A)\Big) < \mu(M) - \mu(A) \leq (1 - \varepsilon)\mu(M),$$

that contradicts (4.30). Therefore $\operatorname{Diam}(I) = b - a \geq \rho$. This implies

$$\operatorname{Diam}(f_*\mu; \varepsilon) = \inf \operatorname{Diam}(I) \geq \rho.$$

Thus

$$\operatorname{ObsDiam}(M; \varepsilon) \geq \rho.$$

Letting $\rho \to \operatorname{ExDist}(M; \varepsilon)$ yields (4.28).

We now prove (4.29). Let $\rho < \operatorname{ObsDiam}(M; \varepsilon)$. There is a Lipschitz function $f \in \operatorname{Lip}_1(X)$ such that

$$\operatorname{Diam}(f_*\mu; \varepsilon) > \rho.$$

Let a be the supremum of r such that

$$\mu(f^{-1}(-\infty, r)) < \frac{\varepsilon}{2}\mu(M).$$

Let b be the infimum of R such that

$$\mu(f^{-1}(R, \infty)) < \frac{\varepsilon}{2}\mu(M).$$

Then for any $r < a$ and $R > b$,

$$\mu(f^{-1}[r, R]) = \mu(M) - \mu(f^{-1}(-\infty, r)) - \mu(f^{-1}(R, \infty)) > (1 - \varepsilon)\mu(M).$$

Thus

$$\mathrm{Diam}([r, R]) \geq \mathrm{Diam}(f_*\mu; \varepsilon) \geq \rho.$$

This implies

$$b - a \geq \rho.$$

Let $A = f^{-1}(-\infty, a)$ and $B = f^{-1}(b, \infty)$. Then

$$\mu(A) \geq \frac{\varepsilon}{2}\mu(M), \quad \mu(B) \geq \frac{\varepsilon}{2}\mu(M).$$

Let

$$\tilde{A} := f^{-1}\Big(-\infty, \frac{a+b}{2}\Big), \quad \tilde{B} := f^{-1}\Big(\frac{a+b}{2}, \infty\Big).$$

For any $x \in U_{\rho/2}(A)$ and $z \in A$ with $d(z, x) < \rho/2$,

$$f(x) \leq f(z) + d(z, x) < a + \frac{\rho}{2} \leq a + \frac{b-a}{2} = \frac{a+b}{2}.$$

Thus $x \in \tilde{A}$.

For $x \in U_{\rho/2}(B)$ and $z \in B$ with $d(z, x) < \rho/2$,

$$f(x) \geq f(z) - d(x, z) > b - \frac{\rho}{2} \geq b - \frac{b-a}{2} = \frac{a+b}{2}.$$

By the above argument, we conclude that

$$U_{\rho/2}(A) \subset \tilde{A}, \quad U_{\rho/2}(A) \subset \tilde{B}.$$

Since $\tilde{A} \cap \tilde{B} = \emptyset$,

$$\mu\Big(U_{\rho/2}(A)\Big) + \mu\Big(U_{\rho/2}(B)\Big) \leq \mu(\tilde{A}) + \mu(\tilde{B}) = \mu(\tilde{A} \cap \tilde{B}) \leq \mu(M).$$

We may assume that

$$\mu\Big(U_{\rho/2}(A)\Big) \leq \frac{1}{2}\mu(M) < (1 - \varepsilon)\mu(M).$$

This implies

$$\mathrm{ExDist}(M; \varepsilon/2) \geq \rho/2.$$

Letting $\rho \to \mathrm{ObsDiam}(M; \varepsilon)$, we obtain (4.29). Q.E.D.

It follows from (4.28) and (4.29) that for any $0 < \varepsilon < 1/2$,

$$\mathrm{ExDist}(M;\varepsilon) \le 2^m \mathrm{ExDist}(M;\varepsilon/2^m).$$

and for $0 < \varepsilon < 1$,

$$\mathrm{ObsDiam}(M;\varepsilon) \le 2^m \mathrm{ObsDiam}(M;\varepsilon/2^m).$$

Corollary 4.4.2 *Let $(M, F, d\mu)$ be a closed Finsler m space. For and* $0 < \varepsilon < 1$,

$$\mathrm{ObsDiam}(M;\varepsilon) \le 4\sqrt{\frac{1}{\varepsilon} - 1} \cdot \frac{1}{\lambda_1(M)}.$$

Chapter 5

Geodesics and Connection

In a metric space, minimizing curves are of special interest to us. Locally minimizing curves in a Finsler space are determined by a system of second ordinary differential equations (geodesic equations). With the geodesic coefficients, we introduce the Chern connection and the notion of covariant derivative.

5.1 Geodesics

In this section, we shall derive the Euler-Lagrange equations for locally minimizing curves in a Finsler space (M, F). Let $c : [a, b] \to M$ be a constant speed piecewise C^∞ curve $F(\dot{c}) = \lambda = constant$. By definition, there is a partition of $[a, b]$,

$$a = t_0 < \cdots < t_k = b,$$

such that c on each $[t_{i-1}, t_i]$ is C^∞. Fix the above partition and consider a piecewise C^∞ map $H : (-\varepsilon, \varepsilon) \times [a, b] \to M$ such that

(a) H is C^0 on $(-\varepsilon, \varepsilon) \times [a, b]$;
(b) H is C^∞ on each $(-\varepsilon, \varepsilon) \times [t_{i-1}, t_i]$, $i = 1, \cdots, k$;
(c) $c(t) = H(0, t)$, $a \le t \le b$.

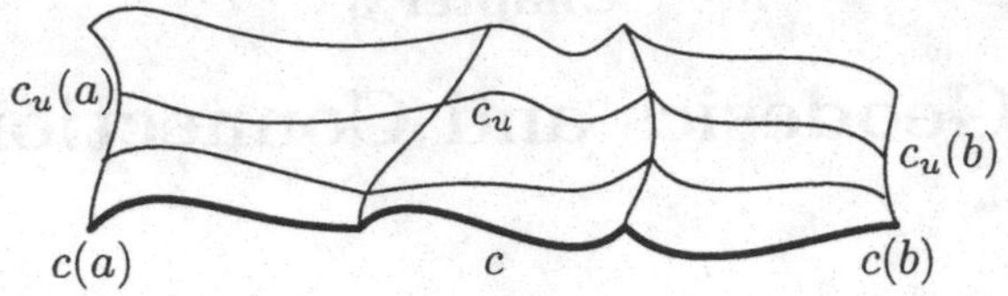

The vector field

$$V(t) = V^i(t)\frac{\partial}{\partial x^i}|_{c(t)} := \frac{\partial H}{\partial u}(0,t)$$

is called the *variation field* of H. The length of $c_u(t) := H(u,t)$ is given by

$$L(u) := \int_a^b F\Big(\dot{c}_u(t)\Big)dt = \sum_{i=1}^k \int_{t_{i-1}}^{t_i} F\Big(\frac{\partial H}{\partial t}(u,t)\Big)dt$$

Observe that

$$\begin{aligned}
L'(0) &= \int_a^b \frac{1}{2F}\Big\{[F^2]_{x^k}V^k + [F^2]_{y^k}\frac{dV^k}{dt}\Big\}dt \\
&= \int_a^b \Big\{\frac{1}{2F}[F^2]_{x^k} - \frac{d}{dt}\Big(\frac{1}{2F}[F^2]_{y^k}\Big)\Big\}V^k dt \\
&\quad + \sum_{i=1}^k \frac{1}{2F}[F^2]_{y^k}V^k\Big|_{t_{i-1}}^{t_i} \\
&= \int_a^b \frac{1}{2F}\Big\{[F^2]_{x^k} - [F^2]_{x^l y^k}\dot{c}^l - [F^2]_{y^k y^l}\ddot{c}^l\Big\}V^k dt \\
&\quad + \sum_{i=1}^k \frac{1}{F}g_{jk}\dot{c}^j V^k\Big|_{t_{i-1}}^{t_i}, \\
&= -\int_a^b \frac{1}{F}g_{jk}\Big\{\ddot{c}^j + 2G^j(\dot{c})\Big\}V^k dt \\
&\quad + \sum_{i=1}^k \frac{1}{F}g_{jk}\dot{c}^j V^k\Big|_{t_{i-1}}^{t_i}, \qquad (5.1)
\end{aligned}$$

where $g_{ij}(y) := \frac{1}{2}[F^2]_{y^i y^j}(y)$ and

$$G^i(y) := \frac{1}{4}g^{il}(y)\Big\{[F^2]_{x^k y^l}(y)y^k - [F^2]_{x^l}(y)\Big\}. \qquad (5.2)$$

Let

$$\kappa(t) := \frac{1}{F(\dot{c}(t))^2}\Big\{\ddot{c}^i + 2G^i(\dot{c})\Big\}\frac{\partial}{\partial x^i}|_{c(t)}. \tag{5.3}$$

$\kappa(t)$ is called the *geodesic curvature* of c at $c(t)$. We can express (5.1) in index-free form

$$\begin{aligned} L'(0) &= -\lambda\int_a^b \mathbf{g}_{\dot{c}}\Big(\kappa, V\Big)dt + \lambda^{-1}\mathbf{g}_{\dot{c}(b)}\Big(\dot{c}(b), V(b)\Big) - \mathbf{g}_{\dot{c}(a)}\Big(\dot{c}(a), V(a)\Big) \\ &\quad + \lambda^{-1}\sum_{i=1}^{k-1}\Big\{\mathbf{g}_{\dot{c}(t_i^-)}\Big(\dot{c}(t_i^-), V(t_i)\Big) - \mathbf{g}_{\dot{c}(t_i^+)}\Big(\dot{c}(t_i^+), V(t_i)\Big)\Big\}, \end{aligned} \tag{5.4}$$

where $\lambda = F(\dot{c}(t))$ is a constant by assumption.

Assume that c has minimal length. Then $L'(0) = 0$ for any piecewise C^∞ variation H of c fixing endpoints.

First, we take an arbitrary piecewise C^∞ variation H of c with $H(u, t_i) = c(t_i)$ (hence $V(t_i) = 0$), $i = 0, \cdots, k$. By (5.4), we obtain

$$L'(0) = -\lambda\int_a^b \mathbf{g}_{\dot{c}}\Big(\kappa, V\Big)dt = 0.$$

This implies

$$\kappa(t) = 0.$$

Now for any $1 \le i_o \le k-1$ and $v \in T_{c(t_{i_o})}M$, we take a piecewise C^∞ variation H of c, that fixes two endpoints of c with

$$V(t_{i_o}) = v, \qquad H(u, t_i) = c(t_i), \qquad i \ne i_o.$$

By (5.4), we obtain

$$L'(0) = \lambda^{-1}\Big\{\mathbf{g}_{\dot{c}(t_{i_o}^+)}\Big(\dot{c}(t_{i_o}^+), v\Big) - \mathbf{g}_{\dot{c}(t_{i_o}^-)}\Big(\dot{c}(t_{i_o}^-), v\Big)\Big\} = 0.$$

By Lemma 1.2.4, we conclude that

$$\dot{c}(t_{i_o}^-) = \dot{c}(t_{i_o}^+).$$

That is, c is C^1 at each t_i. In local coordinates, $\kappa = 0$ is equivalent to the following system

$$\ddot{c}^i + 2G^i(\dot{c}) = 0. \tag{5.5}$$

(5.5) is a system of second order ordinary differential equations. Thus c must be C^∞ at each t_i.

We have proved the following

Proposition 5.1.1 *Let c be a constant speed piecewise C^∞ curve in a Finsler space (M, F). If c has minimal length, then c is a C^∞ curve with vanishing geodesic curvature $\kappa = 0$.*

In virtue of Proposition 5.1.1, we make the following

Definition 5.1.2 A C^∞ curve in a Finsler space (M, F) is called a geodesic if it has constant speed and its geodesic curvature $\kappa = 0$.

First Variation Formula: Let $c : [a, b] \to M$ be a unit speed geodesic and $\sigma : (-\varepsilon, \varepsilon) \to M$ a C^1 curve with $\sigma(0) = c(b)$. Consider a C^∞ variation $H : (-\varepsilon, \varepsilon) \times [a, b] \to M$ with $H(u, a) = c(a)$ and $H(u, b) = \sigma(u)$. By (5.4), the length function $L(u)$ of the curve $c_u(t) := H(u, t)$, $a \le t \le b$, satisfies

$$L'(0) = \mathbf{g}_{\dot{c}(b)}\Big(\dot{c}(b), \dot{\sigma}(0)\Big). \tag{5.6}$$

Equation (5.6) is called the *first variation formula*.

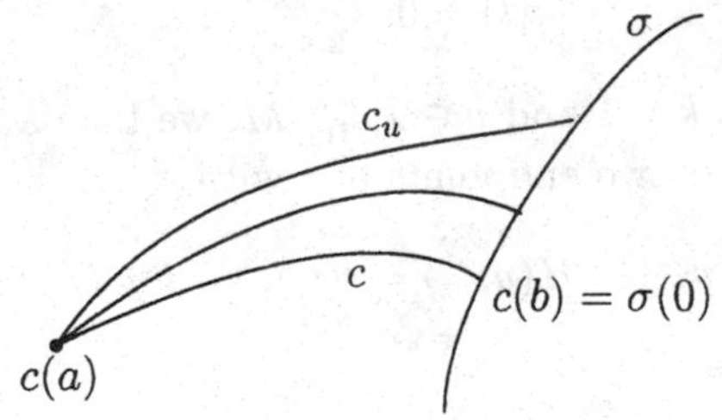

The local functions G^i in (5.2) can be expressed by

$$G^i(y) = \frac{1}{4} g^{il}(y) \Big\{ 2 \frac{\partial g_{jl}}{\partial x^k}(y) - \frac{\partial g_{jk}}{\partial x^l}(y) \Big\} y^j y^k. \tag{5.7}$$

We call G^i the *geodesic coefficients*. The geodesic coefficients G^i give rise to a globally defined vector field on $TM \setminus \{0\}$

$$\mathbf{G} := y^i \frac{\partial}{\partial x^i} - 2G^i(y) \frac{\partial}{\partial y^i}. \tag{5.8}$$

$\mathbf{G}$ is C^∞ on $TM \setminus \{0\}$ and C^1 at zero tangent vectors in TM. We call $\mathbf{G}$ the *spray* induced by F. A curve c is a geodesic if and only if it is the projection of an integral curve of $\mathbf{G}$. See [Sh9] for more detailed discussion on sprays.

From the definition, the geodesic coefficients $G^i(y)$ satisfy the following homogeneity condition

$$G^i(\lambda y) = \lambda^2 G^i(y), \qquad \lambda > 0. \tag{5.9}$$

But, $G^i(y)$ are not quadratic in $y \in T_xM$ in general.

Definition 5.1.3 A Finsler metric is called a *Berwald metric* if in standard local coordinate systems (x^i, y^i), the geodesic coefficients $G^i(y)$ are quadratic in $y \in T_xM$ for all $x \in M$, that is, there are local functions $\Gamma^i_{jk}(x)$ on M such that

$$G^i(y) = \frac{1}{2}\Gamma^i_{jk}(x) y^j y^k.$$

Berwald metrics are special Finsler metrics. We are going to show that Riemannian metrics are special Berwald metrics.

Example 5.1.1 Let $F(y) = \sqrt{g_{ij}(x)y^iy^j}$ be a Riemannian metric on a manifold M. By (5.7), we have

$$G^i(y) = \frac{1}{4} g^{il}(x)\Big\{2\frac{\partial g_{jl}}{\partial x^k}(x) - \frac{\partial g_{jk}}{\partial x^l}(x)\Big\} y^j y^k,$$

where $(g^{ij}(x)) := (g_{ij}(x))^{-1}$. Clearly, $G^i(y)$ are quadratic in $y \in T_xM$. Hence F is a Berwald metric. ♮

There are many non-Riemannian Berwald metrics [Sz]. Below is a simple example.

Example 5.1.2 Consider a Randers metric $F = \alpha + \beta$ on a manifold M, where $\alpha(y) = \sqrt{a_{ij}(x)y^iy^j}$ is a Riemannian metric and $\beta(y) = b_i(x)y^i$ is a 1-form with

$$\|\beta\|_x = \sup_{\alpha_x(y)=1} \beta_x(y) = \sqrt{a^{ij}(x)b_i(x)b_j(x)} < 1.$$

According to Example 5.1.1, the geodesic coefficients of α can be expressed in the following form

$$\bar{G}^i(y) = \frac{1}{2}\bar{\Gamma}^i_{jk}(x)y^j y^k,$$

where $\bar{\Gamma}^i_{jk}(x) = \bar{\Gamma}^i_{kj}(x)$ are local functions of $x \in M$. Define $b_{i|j}$ by

$$b_{i|j}dx^j := db_i - b_j\bar{\Gamma}^j_{ik}dx^k. \tag{5.10}$$

Let

$$r_{ij} := \frac{1}{2}\Big(b_{i|j} + b_{j|i}\Big), \qquad s_{ij} := \frac{1}{2}\Big(b_{i|j} - b_{j|i}\Big).$$

The geodesic coefficients G^i of F are given by

$$G^i = \bar{G}^i + Py^i + Q^i,$$

where

$$\begin{aligned} P(y): &= \frac{1}{2F(y)}\Big\{r_{ij}y^i y^j - 2\alpha(y)b_r a^{rp}s_{pl}y^l\Big\} \\ Q^i(y): &= \alpha a^{ir}s_{rl}y^l. \end{aligned}$$

Assume that β is parallel with respect to α, i.e., $b_{i|j} = 0$. Then

$$r_{ij} = 0 = s_{ij}.$$

Thus $P = 0$ and $Q^i = 0$. This implies that

$$G^i(y) = \bar{G}^i(y)$$

are quadratic in $y \in T_xM$ for all $x \in M$. By definition, $F = \alpha + \beta$ is a Berwald metric [HaIc]. In this case, the geodesics of F coincide with that of α as point sets.

By an elementary algebraic argument, one can show that if F is a Berwald metric, then $b_{i|j} = 0$. See [Ma5][HaIc][Ki2][SSAY] for details. ♮

5.2 Chern Connection

In 1943, S.S. Chern studied the equivalence problem for Finsler spaces using the Cartan's exterior differentiation method [Ch1][Ch2][BaCh]. He discovered a very simple connection. Later on, H. Rund independently introduced this connection in a different setting. Thus, Chern's connection was also called the Rund connection in literatures. In this book, we will introduce Chern's connection using the exterior differential method.

Let M be an n-manifold and $\pi : TM \to M$ the natural projection. Denote by π^*TM the pull-back tangent bundle over the manifold $TM \setminus \{0\}$. The vectors in π^*TM are denoted by (y, v), where $y, v \in T_xM$.

Take a local coordinate system (x^i) in M. The local natural frame $\{\frac{\partial}{\partial x^i}\}$ for TM determines a local natural frame $\{\partial_i\}$ for π^*TM,

$$\partial_i|_y := \Big(y, \frac{\partial}{\partial x^i}|_x\Big), \qquad y \in T_xM.$$

This gives rise to a linear isomorphism between $\pi^*TM|_y$ and T_xM for every $y \in T_xM$.

Given a Finsler metric F on M, $F(y) = F\Big(y^i \frac{\partial}{\partial x^i}|_x\Big)$ is a function of $(y^i) \in \mathrm{R}^n$ at each point $x \in M$. Let

$$g_{ij}(y) := \frac{1}{2}[F^2]_{y^iy^j}(y), \qquad C_{ijk}(y) := \frac{1}{4}[F^2]_{y^iy^jy^k}(y). \tag{5.11}$$

We obtain two tensors g and C on π^*TM defined respectively by

$$g\Big(U, V\Big) := g_{ij}(y)U^iV^j, \tag{5.12}$$

$$C\Big(U, V, W\Big) := C_{ijk}(y)U^iV^jW^k, \tag{5.13}$$

where $U = U^i\partial_i|_y, V = V^j\partial_j|_y$ and $W = W^k\partial_k|_y$. We call g the *fundamental tensor* and C the *Cartan tensor* of F. The Cartan torsion was actually appeared in P. Finsler's thesis [Fi]. But it was E. Cartan who first gave a geometric interpretation of this quantity [Ca].

Let

$$\delta y^i := dy^i + N^i_j(y)dx^j, \tag{5.14}$$

where

$$N^i_j(y) := \frac{\partial G^i}{\partial y^j}(y). \tag{5.15}$$

Then

$$\mathcal{H}^*TM := \mathrm{span}\Big\{dx^i\Big\}, \qquad \mathcal{V}^*TM := \mathrm{span}\Big\{\delta y^i\Big\}$$

is a well-defined subbundle of $T^*(TM \setminus \{0\})$. We obtain a decomposition for $T^*(TM \setminus \{0\})$,

$$T^*(TM \setminus \{0\}) = \mathcal{H}^*TM \oplus \mathcal{V}^*TM.$$

Let $\{\frac{\delta}{\delta x^i}, \frac{\partial}{\partial y^i}\}$ denote the local frame dual to $\{dx^i, \delta y^i\}$, that is

$$\frac{\delta}{\delta x^i} := \frac{\partial}{\partial x^i} - N_i^j(y)\frac{\partial}{\partial y^j}. \tag{5.16}$$

Then

$$\mathcal{H}TM := \mathrm{span}\Big\{\frac{\delta}{\delta x^i}\Big\}, \qquad \mathcal{V}TM := \mathrm{span}\Big\{\frac{\partial}{\partial y^i}\Big\}$$

are well-defined subbundles of $T(TM \setminus \{0\})$. We obtain a decomposition for $T(TM \setminus \{0\})$,

$$T(TM \setminus \{0\}) = \mathcal{H}TM \oplus \mathcal{V}TM.$$

Observe that

$$d[F^2] = 2g_{kl}(y)y^k\, \delta y^l \in \mathcal{V}^*TM.$$

Thus, $d[F^2] = 2FdF$ vanishes on horizontal tangent vectors. We obtain the following

Lemma 5.2.1 *F is horizontally constant, i.e., for any horizontal vector* $\mathbf{X} = X^i\frac{\delta}{\delta x^i} \in \mathcal{H}TM$,

$$\mathbf{X}(F) = dF(\mathbf{X}) = 0. \tag{5.17}$$

Let $\{e_i\}_{i=1}^n$ be an arbitrary local frame for π^*TM and $\{\omega^i, \omega^{n+i}\}_{i=1}^n$ denote the corresponding local coframe for $\mathcal{H}^*TM \oplus \mathcal{V}^*TM$. The correspondence is determined by

$$\partial_i \leftrightarrow dx^i \leftrightarrow \delta y^i. \tag{5.18}$$

If $e_i = \partial_i$, then $\omega^i = dx^i$ and $\omega^{n+i} = \delta y^i$. Set

$$y^i e_i = (y, y) \tag{5.19}$$

and

$$g_{ij} := g(e_i, e_j), \qquad C_{ijk} := C(e_i, e_j, e_k). \tag{5.20}$$

We obtain a set of local function y^i, g_{ij} and C_{ijk} on $TM \setminus \{0\}$.

Theorem 5.2.2 (Chern) *There is a unique set of local* 1*-forms* $\{\omega_j{}^i\}$ *on* $TM \setminus \{0\}$ *such that*

$$d\omega^i = \omega^j \wedge \omega_j{}^i \tag{5.21}$$

$$dg_{ij} = g_{kj}\,\omega_i{}^k + g_{kj}\,\omega_i{}^k + 2C_{ijk}\,\omega^{n+k}, \tag{5.22}$$

$$\omega^{n+i} = dy^i + y^j\omega_j{}^i. \tag{5.23}$$

Proof: Without loss of generality, we may take a natural local frame $\{e_i = \partial_i\}_{i=1}^n$ for π^*TM so that the corresponding local coframe for $T^*(TM \setminus \{0\})$ is $\{\omega^i = dx^i, \omega^{n+i} = \delta y^i\}_{i=1}^n$. In this case,

$$g_{ij}(y) = \frac{1}{2}[F^2]_{y^iy^j}(y), \qquad C_{ijk}(y) = \frac{1}{4}[F^2]_{y^iy^jy^k}(y).$$

Let

$$L_{ijk} := \frac{\partial C_{ijk}}{\partial x^l}y^l - 2C_{ijkl}G^l - C_{ljk}N_i^l - C_{ilk}N_j^l - C_{ijl}N_k^l, \tag{5.24}$$

where

$$C_{ijkl}(y) = \frac{\partial C_{ijk}}{\partial y^l}(y) = \frac{1}{4}[F^2]_{y^iy^jy^ky^l}(y).$$

We claim that the following set of 1-forms are the unique set of solutions to (5.21) and (5.22),

$$\omega_j{}^i := \Gamma^i_{jk}\,dx^k,$$

where

$$\Gamma^i_{jk} := \frac{\partial^2 G^i}{\partial y^j \partial y^k} - g^{il}L_{jkl}. \tag{5.25}$$

Observe that

$$d\omega^i - \omega^j \wedge \omega_j{}^i = -dx^j \wedge \Gamma^i_{jk}dx^k = -\frac{1}{2}\Big\{\Gamma^i_{jk} - \Gamma^i_{kj}\Big\}dx^j \wedge dx^k = 0.$$

Thus (5.21) holds.

Rewrite (5.7) as follows

$$g_{kl}G^k = \frac{1}{4}\Big\{2\frac{\partial g_{jl}}{\partial x^k} - \frac{\partial g_{jk}}{\partial x^l}\Big\}y^jy^k. \tag{5.26}$$

Differentiating (5.26) with respect to y^j yields

$$g_{kl}N^k_j = \frac{1}{2}\Big\{\frac{\partial g_{jl}}{\partial x^k} + \frac{\partial g_{kl}}{\partial x^j} - \frac{\partial g_{jk}}{\partial x^l}\Big\}y^k - 2C_{jkl}G^k. \tag{5.27}$$

Differentiating (5.27) with respect to y^m, we obtain

$$\begin{aligned} g_{kl}\Gamma^k_{jm} &= \frac{1}{2}\Big\{\frac{\partial g_{jl}}{\partial x^m} + \frac{\partial g_{ml}}{\partial x^j} - \frac{\partial g_{jm}}{\partial x^l}\Big\} + \frac{\partial C_{jlm}}{\partial x^k}y^k \\ &\quad -2C_{jklm}G^k - 2C_{jkl}N^k_m - 2C_{klm}N^k_j - L_{jlm}. \end{aligned} \tag{5.28}$$

From (5.28), we obtain

$$\frac{\partial g_{jl}}{\partial x^m} = g_{kl}\Gamma^k_{jm} + g_{kj}\Gamma^k_{lm} + 2C_{jkl}N^k_m. \tag{5.29}$$

This implies to (5.22). The proof of uniqueness is omitted.

By the homogeneity of F, we have

$$C_{ijk}y^i = C_{ijk}y^j = C_{ijk}y^k = 0 \tag{5.30}$$

and

$$L_{ijk}y^i = L_{ijk}y^j = L_{ijk}y^k = 0. \tag{5.31}$$

Thus

$$y^j\Gamma^i_{jk} = y^j\frac{\partial^2 G^i}{\partial y^j \partial y^k} - y^j g^{il}L_{jkl} = \frac{\partial G^i}{\partial y^j} = N^i_j.$$

Thus

$$\delta y^i = dy^i + y^j\Gamma^i_{jk}dx^k = dy^i + y^j\omega_j{}^i.$$

This gives (5.23). Q.E.D.

From the above argument, we obtain a new tensor $\mathcal{L}$ on π^*TM defined by

$$\mathcal{L}(U,V,W) := L_{ijk}(y)U^iV^jW^k, \tag{5.32}$$

where $U = U^i\partial_i|_y, V = V^j\partial_j|_y$ and $W = W^k\partial_k|_y$. We call $\mathcal{L}$ the *Landsberg tensor*. The Cartan tensor and Landsberg tensor play an important role in Finsler geometry.

Using the set of local 1-forms $\{\omega_j{}^i\}$ in Theorem 5.2.2, we define a map

$$\nabla : T(TM) \times C^\infty(\pi^*TM) \to \pi^*TM$$

by

$$\nabla_{\hat{X}} U := \Big\{ dU^i(\hat{X}) + U^j \omega_j{}^i(\hat{X}) \Big\} \otimes e_i,$$

or simply

$$\nabla U := \Big\{ dU^i + U^j \omega_j{}^i \Big\} \otimes e_i.$$

∇ is a linear connection on π^*TM. We call it the *Chern connection.*

For a vector field $X = (X^1, \cdots, X^n)$ on an open subset $\mathcal{U} \subset \mathrm{R}^n$, the directional derivative $\mathrm{D}_v X$ in a direction $v \in T_x\mathrm{R}^n = \mathrm{R}^n$ is defined by

$$\mathrm{D}_v X := \Big(dX^1(v), \cdots, dX^n(v) \Big) = v^i \frac{\partial X}{\partial x^i}.$$

We can extend the notion of directional derivative to vector fields on a Finsler space.

Let (M, F) be a Finsler space. At each point $x \in M$, define a map

$$\mathrm{D} : T_x M \times C^\infty(TM) \to T_x M$$

by

$$\mathrm{D}_y U := \Big\{ dU^i(y) + U^j(x) N^i_j(y) \Big\} \frac{\partial}{\partial x^i}\Big|_x \tag{5.33}$$

where $y \in T_x M$ and $U \in C^\infty(TM)$. Here $N^i_j(y)$ are local functions on TM defined in (5.15). We call $\mathrm{D}_y U(x)$ the *covariant derivative* of U at x in the direction y. Set $\mathrm{D}_y U(x) = 0$ for $y = 0$. D has the following properties:

(a) $\mathrm{D}_y(U + V) = \mathrm{D}_y U + \mathrm{D}_y V$;
(b) $\mathrm{D}_y(fU) = df_x(y)U + f(x)\mathrm{D}_y U$;
(c) $\mathrm{D}_{\lambda y} U = \lambda \mathrm{D}_y U$, $\lambda > 0$.

The family $\mathrm{D} := \{\mathrm{D}_y\}_{y \in TM}$ is called the *connection* of F. D is not a connection in a usual sense. If, in addition, D is linear, i.e.,

(d) $\mathrm{D}_{y+v} U = \mathrm{D}_y U + \mathrm{D}_v U$,

then D is called an *affine connection* on TM (or M).

The following proposition is obvious.

Proposition 5.2.3 *The connection* D *of a Finsler metric* F *is affine if and only if* F *is a Berwald metric.*

Proof. In a standard local coordinate system (x^i, y^i) in TM,

$$N^i_j(y) = \frac{\partial G^i}{\partial y^j}(y), \qquad N^i_j(y)y^j = 2G^i(y).$$

Thus $G^i(y)$ are quadratic in $y \in T_xM$ if and only if $N^i_j(y)$ are linear in $y \in T_xM$. This proves the proposition. Q.E.D.

For Berwald metrics, the connection D is an affine connection on the tangent bundle TM. We call it the *Levi-Civita connection.*

Example 5.2.1 Consider a Riemannian metric $F(y) = \sqrt{g_{ij}(x)y^iy^j}$ on a manifold M. The geodesic coefficients $G^i(y)$ are given by

$$G^i(y) = g^{il}(x)\Big\{\frac{\partial g_{kl}}{\partial x^j}(x) - \frac{1}{2}\frac{\partial g_{jk}}{\partial x^l}(x)\Big\}y^jy^k. \tag{5.34}$$

Thus $G^i(y)$ are quadratic in $y \in T_xM$ for all $x \in M$ and D is an affine connection.

The Levi-Civita connection D satisfies the torsion-free condition

$$\mathrm{D}_UV - \mathrm{D}_VU = [U, V] \tag{5.35}$$

and

$$d\Big[g(U,V)\Big](y) = g\Big(\mathrm{D}_yU, V\Big) + g\Big(U, \mathrm{D}_yV\Big), \tag{5.36}$$

where $y \in T_xM$ and $U, V \in C^\infty(TM)$. It can be easily proved that the Levi-Civita connection is the unique linear connection satisfying (5.35) and (5.36). ♯

Now we discuss the relationship between the Levi-Civita connection D on TM and the Chern connection ∇ on π^*TM for Berwald metrics.

For a Berwald metric, in a local coordinate system (x^i) in M, the Christoffel symbols $\Gamma^i_{jk}(x) := \frac{\partial^2 G^i}{\partial y^j \partial y^k}(y)$ are functions of x only, so that

$$N^i_j(y) = \Gamma^i_{jk}(x)y^k.$$

The Levi-Civita connection is given by

$$\mathrm{D}U = \left\{ dU^i + U^j \Gamma^i_{jk} dx^k \right\} \otimes \frac{\partial}{\partial x^i},$$

where $U = U^i \frac{\partial}{\partial x^i} \in C^\infty(TM)$.

Take an arbitrary local frame $\{\mathbf{b}_i\}_{i=1}^n$ for TM and let $\{\theta^i\}_{i=1}^n$ denote the local dual frame for T^*M. Write

$$\mathrm{D}U = \{dU^i + \theta_j{}^i\} \otimes \mathbf{b}_i,$$

where $U = U^i \mathbf{b}_i$. $\{\theta_j{}^i\}$ are called the *Levi-Civita connection forms* with respect to $\{\mathbf{b}_i\}_{i=1}^n$.

Let $\pi : TM \to M$ denote the natural projection. We can lift $\{\mathbf{b}_i\}_{i=1}^n$ to a local frame $\{e_i\}_{i=1}^n$ for π^*TM, where

$$e_i|_y := (y, \mathbf{b}_i|_x), \qquad y \in T_xM.$$

The Chern connection forms $\{\omega_j{}^i\}$ with respect to $\{e_i\}_{i=1}^n$ are given by

$$\omega_j{}^i = \pi^*\theta_j{}^i.$$

Let y^i denote the local functions on TM determined by

$$(y, y) = y^i e_i = (y, y^i \mathbf{b}_i).$$

The local coframe $\{\omega^i, \omega^{n+i}\}_{i=1}^n$ for $T^*(TM \setminus \{0\})$ are given by

$$\omega^i = \pi^*\theta^i, \qquad \omega^{n+i} = dy^i + y^j \pi^*\theta_j{}^i.$$

Consider a tensor $S = S^i_j \, \mathbf{b}_i \otimes \theta^j$ on M. Define $Q = Q^i e_i$ by

$$Q^i(y) := y^j S^i_j(x), \qquad y = y^i \mathbf{b}_i.$$

Then Q is a well-defined tensor on π^*TM. On M we define $S^i_{j|k}$ by

$$dS^i_j + S^k_j \theta_k{}^i - S^i_k \theta_j{}^k =: S^i_{j|k} \theta^k. \tag{5.37}$$

While on $TM \setminus \{0\}$, we define $Q^i_{|k}$ and $Q^i_{\cdot k}$ by

$$dQ^i + Q^k \omega_k{}^i =: Q^i_{|k} \omega^k + Q^i_{\cdot k} \omega^{n+k}. \tag{5.38}$$

Observe that

$$\begin{aligned} dQ^i + Q^k\omega_k^{\ i} &= y^j\pi^*(dS^i_j) + S^i_k dy^k + y^j\pi^*(S^k_j\theta_k^{\ i}) \\ &= y^j\pi^*\Big[dS^i_j + S^k_j\theta_k^{\ i}\Big] + S^i_k\Big[\omega^{n+k} - y^j\pi^*\theta_j^{\ k}\Big] \\ &= y^j\pi^*\Big[dS^i_j + S^k_j\theta_k^{\ i} - S^i_k\theta_j^{\ k}\Big] + S^i_k\omega^{n+k} \\ &= y^j\pi^*(S^i_{j|k}\theta^k) + S^i_k\omega^{n+k}. \end{aligned}$$

We obtain

$$Q^i_{|k} = y^j S^i_{j|k} \tag{5.39}$$

$$Q^i_{\cdot k} = S^i_k. \tag{5.40}$$

5.3 Covariant Derivatives

Let (M, F) be a Finsler space and $c(t)$, $a \le t \le b$, a geodesic. For a vector field $V = V^i(t)\frac{\partial}{\partial x^i}|_{c(t)}$ along c, define

$$\mathrm{D}_{\dot c}V(t) := \Big\{\frac{dV^i}{dt}(t) + V^j N^i_j(\dot c(t))\Big\}\frac{\partial}{\partial x^i}|_{c(t)}.$$

$\mathrm{D}_{\dot c}V(t)$ is called the *covariant derivative* of $V(t)$ along c. A vector $V(t)$ is said to be *parallel* along c, if

$$\mathrm{D}_{\dot c}V(t) = 0.$$

Let $U = U^i(t)\frac{\partial}{\partial x^i}|_{c(t)}$ and $V = V^i(t)\frac{\partial}{\partial x^i}|_{c(t)}$ be parallel vector fields along c. They satisfy the following equations

$$\frac{dU^m}{dt} = -U^j N^m_j(\dot c), \qquad \frac{dV^m}{dt} = -V^l N^m_l(\dot c). \tag{5.41}$$

Consider the function

$$\mathbf{g}_{\dot c}(U, V) = g_{jl}U^jV^l.$$

By (5.41), we obtain

$$\frac{d}{dt}\Big[g_{jl}U^jV^l\Big] = \Big\{\frac{\partial g_{jl}}{\partial x^m}\dot c^m - g_{jm}N^m_l - g_{jm}N^m_l - 4C_{jlm}G^m\Big\}U^jV^l.$$

Contracting (5.29) with y^m yields

$$y^m \frac{\partial g_{jl}}{\partial x^m} = g_{ml} N_j^m + g_{jm} N_l^m + 4 C_{jlm} G^m. \tag{5.42}$$

It follows from (5.42) that

$$\frac{d}{dt}\Big[\mathbf{g}_{\dot c}(U,V)\Big] = \frac{d}{dt}\Big[g_{jl} U^j V^l\Big] = 0.$$

Thus

$$\mathbf{g}_{\dot c}(U,V) = \mathit{constant}.$$

Let $c(t)$, $a \le t \le b$, be a geodesic. Define a map $P_c : T_{c(a)}M \to T_{c(b)}M$ by

$$P_c(v) := V(b), \qquad v \in T_{c(a)}M,$$

where $V(t)$ is the parallel vector field along c with $V(a) = v$. P_c is called the *parallel translation* along c.

The above argument proves the following

Lemma 5.3.1 *For any geodesic $c(t)$, $a \le t \le b$, in a Finsler space (M, F), the parallel translation P_c preserves the inner products $\mathbf{g}_{\dot c}$ along c,*

$$\mathbf{g}_{c(b)}\Big(P_c(u), P_c(v)\Big) = \mathbf{g}_{c(a)}\Big(u, v\Big), \qquad u, v \in T_{c(a)}M.$$

In general, the parallel translation does not preserve the Minkowski norms. Y. Ichijō [Ic] proved that on a Berwald space, the parallel translation along any geodesic preserves the Minkowski norms. Thus Berwald spaces can be viewed as Finsler spaces modeled on a single Minkowski space.

Lemma 5.3.2 ([Ic]) *Let (M, F) be a Berwald space and $c(t)$, $a \le t \le b$, a geodesic. Then the parallel translation P_c preserves the Minkowski norm, i.e.,*

$$F_{c(b)}\Big(P_c(v)\Big) = F_{c(a)}(v), \qquad v \in T_{c(a)}M.$$

Proof. Let $V = V^i(t)\frac{\partial}{\partial x^i}|_{c(t)}$ be parallel along c. V satisfies

$$\frac{dV^i}{dt} = -V^j N^i_j(\dot{c}). \tag{5.43}$$

Consider the function

$$F(t) := F(V(t)).$$

By assumption,

$$N^i_j(y)v^j = \Gamma^i_{jk}(x)v^j y^k = N^i_j(v)y^j.$$

By Lemma 5.2.1 and (5.43), we obtain

$$\begin{aligned} F'(t) &= dF\Big[\dot{c}^i\frac{\partial}{\partial x^i} + \frac{dV^i}{dt}\frac{\partial}{\partial y^i}\Big] \\ &= dF\Big[\dot{c}^i\frac{\delta}{\delta x^i} + \Big(\dot{c}^j N^i_j(V) - V^j N^i_j(\dot{c})\Big)\frac{\partial}{\partial y^i}\Big] \\ &= dF\Big[\Big(\dot{c}^j\Gamma^i_{jk}(c)V^k - V^j\Gamma^i_{jk}(c)\dot{c}^k\Big)\frac{\partial}{\partial y^i}\Big] = 0. \end{aligned}$$

Thus $F(t) = constant$ Q.E.D.

On a Berwald space, the canonical connection D is an affine connection. According to Szabó [Sz], if a Finsler metric F is Berwaldian, then there is a Riemannian metric g whose Levi-Civita connection coincides with the canonical connection. There are many non-Riemannian Berwald metrics. See Example 5.1.2 above.

5.4 Geodesic Flow

Let (M, F) be a Finsler space and

$$\mathbf{G} = y^i\frac{\partial}{\partial x^i} - 2G^i(y)\frac{\partial}{\partial y^i}$$

the spray of F. For a vector $y \in TM$, denote by $\phi_t(y)$ the integral curve of $\mathbf{G}$ with $\phi_0(y) = y$. From the definition, we see that the curve $c(t) := \phi_t(y)$ is a geodesic with $\dot{c}(0) = y$. We call ϕ_t the *geodesic flow* of F. It follows from Lemma 5.2.1 that

$$\frac{d}{dt}\Big[F(\phi_t(y))\Big] = dF\Big(\mathbf{G}_{\phi_t(y)}\Big) = 0.$$

Thus

$$F(\phi_t(y)) = constant. \tag{5.44}$$

This means that the geodesic flow ϕ_t preserves the Finsler metric.

Recall the Hilbert form on $TM \setminus \{0\}$ that is given by

$$\omega := \frac{1}{2}[F^2]_{y^i}(y)dx^i = g_{ij}(y)y^j dx^i, \tag{5.45}$$

where $g_{ij}(y) := \frac{1}{2}[F^2]_{y^i y^j}(y)$. The Hilbert form has the following property.

Lemma 5.4.1 *For any t,*

$$\frac{d}{dt}\Big[(\phi_t)^*\omega\Big] = \frac{1}{2}d\Big[(\phi_t)^*[F^2]\Big]. \tag{5.46}$$

Hence

$$(\phi_t)^* d\omega = d\omega.$$

Proof. Let $(\varphi^i(t), \psi^i(t))$ be the standard local coordinates of $y(t) := \phi_t(y)$ in TM. $y(t)$ is an integral curve of $\mathbf{G}$, hence

$$\begin{cases} \frac{d\varphi^i_t}{dt}(t) = \psi^i_t \\ \frac{d\psi^i_t}{dt}(t) = -2G^i(y(t)). \end{cases} \tag{5.47}$$

By (5.47),

$$(\phi_t)^*\omega = \frac{1}{2}[F^2]_{y^i}(y(t))\Big[\frac{\partial\varphi^i_t}{\partial x^j}dx^j + \frac{\partial\varphi^i_t}{\partial y^j}dy^j\Big].$$

Note that

$$\Big[\frac{\partial\varphi^i_t}{\partial x^j}dx^j + \frac{\partial\varphi^i_t}{\partial y^j}dy^j\Big]\Big|_{t=0} = dx^i, \qquad \frac{d}{dt}\Big[\frac{\partial\varphi^i_t}{\partial x^j}dx^j + \frac{\partial\varphi^i_t}{\partial y^j}dy^j\Big]\Big|_{t=0} = dy^i.$$

We immediately obtain

$$\begin{aligned} \frac{d}{dt}\Big[(\phi_t)^*\omega\Big]\Big|_{t=0} &= \frac{1}{2}\Big\{[F^2]_{x^k y^i}y^k - 2[F^2]_{y^i y^k}G^k(y)\Big\}dx^i \\ &\quad + \frac{1}{2}[F^2]_{y^i}dy^i. \end{aligned} \tag{5.48}$$

It follows from (5.2) that

$$2[F^2]_{y^i y^k}G^k = [F^2]_{x^k y^i}y^k - [F^2]_{x^i}. \tag{5.49}$$

Substituting (5.49) into (5.48) yields

$$\frac{d}{dt}\Big[(\phi_t)^*\omega\Big]\Big|_{t=0} = \frac{1}{2}\Big\{[F^2]_{x^i}dx^i + [F^2]_{y^i}dy^i\Big\} = \frac{1}{2}d[F^2](y).$$

Since $\phi_{t+s} = \phi_t \circ \phi_s$, one obtains (5.46). Using (5.46), we obtain

$$\frac{d}{dt}\Big[(\phi_t)^*d\omega\Big] = d\Big[\frac{d}{dt}\Big((\phi_t)^*\omega\Big)\Big] = \frac{1}{2}d^2\Big[(\phi_t)^*[F^2]\Big] = 0.$$

Thus

$$(\phi_t)^*d\omega = (\phi_0)^*d\omega = \omega.$$

Q.E.D.

The Finsler metric F induces a Riemannian metric of Sasaki type on $TM \setminus \{0\}$,

$$\hat{g} := g_{ij}(y)dx^i \otimes dx^j + g_{ij}(y)\delta y^i \otimes \delta y^j,$$

where $\delta y^i = dy^i + N^i_j(y)dy^j$. The Riemannian metric $\hat{g}$ determines a non-degenerate $2n$-form $dV_{\hat{g}}$ on $TM \setminus \{0\}$ by

$$dV_{\hat{g}} = \det\big(g_{ij}\big)dx^1 \cdots dx^n dy^1 \cdots dy^n. \tag{5.50}$$

Observe that

$$d\omega = \frac{\partial g_{ik}}{\partial x^l}y^k dx^l \wedge dx^i - g_{ij}dx^i \wedge dy^j = -g_{ij}dx^i \wedge \delta y^j.$$

Thus, we can express $dV_{\hat{g}}$ as follows

$$dV_{\hat{g}} = (-1)^{\frac{n(n+1)}{2}}\frac{1}{n!}d\omega \wedge \cdots \wedge d\omega.$$

$dV_{\hat{g}}$ is just the volume form we have mentioned in (2.16).

By Lemma 5.4.1, we obtain the following

Proposition 5.4.2 ([Da1]) *The geodesic flow ϕ_t preserves the Riemannian volume on $TM \setminus \{0\}$.*

Let $i : SM \to TM \setminus \{0\}$ be the natural embedding. By (5.44), $\phi_t : SM \to SM$. Let $\dot{\omega} := i^*\omega$. It follows from (5.44) and (5.46) that

$$\frac{d}{dt}\Big[(\phi_t)^*\dot{\omega}\Big] = 0.$$

In other words, ϕ_t preserves $\dot{\omega}$.

Let $\dot{g} := i^*\hat{g}$ denote the induced Riemannian metric on SM. Then the induced Riemannian volume form $dV_{\dot{g}}$ on SM is given by

$$dV_{\dot{g}} = (-1)^{\frac{n(n+1)}{2}-1} \frac{1}{(n-1)!} \dot{\omega} \wedge d\dot{\omega} \wedge \cdots \wedge d\dot{\omega}.$$

Proposition 5.4.3 ([Da1]) *The geodesic flow ϕ_t preserves $dV_{\dot{g}}$ on* SM.

It is worth making further investigation on the geodesic flow on a Finsler space. The geodesic flow ϕ_t on a reversible Finsler space (M, F) is of *Anosov type* if there exists a decomposition of $T(\mathrm{S}M)$ into

$$T(\mathrm{S}M) = E^s \oplus E^u \oplus E^0,$$

where $E^0 = \mathrm{R} \cdot \mathbf{G}$ and $E^s \neq 0, E^u \neq 0$ are invariant under the flow in the sense that

$$(\phi_t)_*(E^s_y) = E^s_{\phi_t(y)}, \qquad (\phi_t)_*(E^u_y) = E^u_{\phi_t(y)}.$$

Further, there are constants $A, B > 0$ such that

$$\|(\phi_t)_*(\hat{X})\| \leq Ae^{-Bt}\|\hat{X}\|, \qquad \forall \hat{X} \in E^s, \ \forall t > 0,$$
$$\|(\phi_{-t})_*(\hat{X})\| \leq Ae^{-Bt}\|\hat{X}\|, \qquad \forall \hat{X} \in E^u, \ \forall t > 0.$$

P. Foulon [Fo1] proved that the geodesic flow on a closed reversible Finsler space of negative curvature must be of Anosov type. Recently, Foulon [Fo2] has done some work on contact Anosov flows on closed manifolds. In particular, he shows that for a smooth contact Anosov flow on a closed three manifold, the measure of maximal entropy is in the Lebesgue class if and only if the flow is up to finite covers conjugate to the geodesic flow of a metric of constant negative curvature on a closed surface. This shows that the ratio between the measure theoretic entropy and the topological entropy of a contact Anosov flow is strictly smaller than one on any closed three manifold which is not a Seifert bundle.

Chapter 6
Riemann Curvature

Riemann curvature is the central concept in Riemannian geometry. This quantity was first introduced by B. Riemann in 1854 for Riemannian metrics. Since then, Riemannian geometry has become one of the important branches in modern mathematics. In 1926, L. Berwald extended the Riemann curvature to Finsler metrics [Bw1] [Bw2]. In this chapter, we will introduce and discuss this very important quantity in Finsler geometry from various points of view.

6.1 Birth of the Riemann Curvature

The Riemann curvature of a Finsler space is a family of linear transformations on tangent spaces. It can be defined via the variations of geodesics. This approach is different from Riemann's and Berwald's approaches.

Let (M, F) be a Finsler space. Consider a geodesic $c(t)$, $a \le t \le b$. A C^∞ map $H : (-\varepsilon, \varepsilon) \times [a, b] \to M$ is called a *geodesic variation* of c if

$$H(0, t) = c(t)$$

and for each $u \in (-\varepsilon, \varepsilon)$, the curve

$$c_u(t) := H(u, t)$$

is a geodesic. We will show that for any geodesic variation, its variation field $J(t) := \frac{\partial H}{\partial u}(0, t)$ satisfies a special system of second order ordinary differential equations.

Lemma 6.1.1 *Let (M, F) be a Finsler space. There is a family of transformations $\mathbf{R}_y : T_xM \to T_xM$, $y \in T_xM \setminus \{0\}$, such that for any geodesic variation H of a geodesic c, the variation vector field $J(t) := \frac{\partial H}{\partial u}(0, t)$ along c satisfies the following equation*

$$\mathrm{D}_{\dot c}\mathrm{D}_{\dot c}J + \mathbf{R}_{\dot c}(J) = 0. \tag{6.1}$$

Proof. By assumption, each $c_u(t) = H(u, t)$ is a geodesic. Thus

$$\frac{\partial^2 H^i}{\partial t^2} + 2G^i\Big(\frac{\partial H}{\partial t}\Big) = 0. \tag{6.2}$$

For simplicity, let

$$T = T^i\frac{\partial}{\partial x^i} := \frac{\partial H}{\partial t}, \qquad U = U^i\frac{\partial}{\partial x^i} := \frac{\partial H}{\partial u}.$$

Equation (6.2) becomes

$$\frac{\partial T^i}{\partial t} + 2G^i(T) = 0. \tag{6.3}$$

Note that

$$\frac{\partial T^i}{\partial u} = \frac{\partial}{\partial u}\Big(\frac{\partial H^i}{\partial t}\Big) = \frac{\partial}{\partial t}\Big(\frac{\partial H^i}{\partial u}\Big) = \frac{\partial U^i}{\partial t}.$$

Differentiating (6.3) with respect to u yields

$$\frac{\partial^2 U^i}{\partial t^2} = -2U^k\frac{\partial G^i}{\partial x^k}(T) - 2\frac{\partial U^j}{\partial t}\frac{\partial G^i}{\partial y^j}(T).$$

Observe that

$$\begin{aligned}
\frac{\partial}{\partial u}\Big[G^i(T)\Big] &= U^k\frac{\partial G^i}{\partial x^k}(T) + \frac{\partial U^j}{\partial t}\frac{\partial G^i}{\partial y^j}(T),\\
\frac{\partial}{\partial t}\Big[\frac{\partial G^i}{\partial y^j}(T)\Big] &= T^k\frac{\partial^2 G^i}{\partial x^k \partial y^j}(T) + \frac{\partial T^k}{\partial t}\frac{\partial^2 G^i}{\partial y^j \partial y^k}(T)\\
&= T^k\frac{\partial^2 G^i}{\partial x^k \partial y^j}(T) - 2G^k(T)\frac{\partial^2 G^i}{\partial y^j \partial y^k}(T).
\end{aligned}$$

By the above identities, one obtains

$$\mathrm{D}_T\mathrm{D}_T U = \mathrm{D}_T\Big[\Big(\frac{\partial U^i}{\partial t} + U^j\frac{\partial G^i}{\partial y^j}(T)\Big)\frac{\partial}{\partial x^i}\Big]$$

$$= -U^k\Big\{2\frac{\partial G^i}{\partial x^k} - T^j\frac{\partial^2 G^i}{\partial x^j \partial y^k} + 2G^j\frac{\partial^2 G^i}{\partial y^j \partial y^k} - \frac{\partial G^i}{\partial y^j}\frac{\partial G^j}{\partial y^k}\Big\}\frac{\partial}{\partial x^i}$$
$$= -U^k R^i{}_k(T)\frac{\partial}{\partial x^i},$$

where

$$R^i{}_k(y) := 2\frac{\partial G^i}{\partial x^k} - y^j\frac{\partial^2 G^i}{\partial x^j \partial y^k} + 2G^j\frac{\partial^2 G^i}{\partial y^j \partial y^k} - \frac{\partial G^i}{\partial y^j}\frac{\partial G^j}{\partial y^k}. \tag{6.4}$$

For every vector $y \in T_xM \setminus \{0\}$, define a linear transformation

$$\mathbf{R}_y = R^i{}_k(y)\frac{\partial}{\partial x^i} \otimes dx^k\Big|_x : T_xM \to T_xM.$$

We obtain

$$\mathrm{D}_T\mathrm{D}_T U + \mathbf{R}_T(U) = 0.$$

Restricting the above equation to c, we obtain an equation for $J(t) := U(0,t)$,

$$\mathrm{D}_{\dot c}\mathrm{D}_{\dot c} J + \mathbf{R}_{\dot c}(J) = 0. \tag{6.5}$$

This completes the proof. Q.E.D.

The geodesic variations give rise to a family of transformations

$$\mathbf{R} = \Big\{\mathbf{R}_y : T_xM \to T_xM \;\Big|\; y \in T_xM \setminus \{0\},\ x \in M\Big\}.$$

We call $\mathbf{R}$ the *Riemann curvature.*

It follows from (6.4) that

$$\mathbf{R}_y(y) = 0. \tag{6.6}$$

Moreover, the $\mathbf{R}_y$ is self-adjoint with respect to $\mathbf{g}_y$,

$$\mathbf{g}_y\Big(\mathbf{R}_y(u), v\Big) = \mathbf{g}_y\Big(u, \mathbf{R}_y(v)\Big), \qquad u, v \in T_xM. \tag{6.7}$$

This fact will be proved in Section 8.1 below. By granting (6.7), we obtain

$$\mathbf{g}_y\Big(\mathbf{R}_y(u), y\Big) = \mathbf{g}_y\Big(u, \mathbf{R}_y(y)\Big) = 0. \tag{6.8}$$

Let $P \subset T_xM$ be a tangent plane. For a vector $y \in P \setminus \{0\}$, define

$$\mathbf{K}(P, y) := \frac{\mathbf{g}_y\Big(\mathbf{R}_y(u), u\Big)}{\mathbf{g}_y(y, y)\mathbf{g}_y(u, u) - \mathbf{g}_y(y, u)\mathbf{g}_y(y, u)}, \tag{6.9}$$

where $u \in P$ such that $P = \mathrm{span}\{y, u\}$. By (6.6) and (6.8), we can easily show that $\mathbf{K}(P, y)$ is independent of $u \in P$ with $P = \mathrm{span}\{y, u\}$. The number $\mathbf{K}(P, y)$ is called the *flag curvature* of the flag (P, y) in T_xM.

When $n = \dim M = 2$,

$$\mathbf{K}(y) := \mathbf{K}(T_xM, y), \qquad y \in T_xM \setminus \{0\},$$

is a scalar function $TM \setminus \{0\}$. We call $\mathbf{K}(y)$ the *Gauss curvature.*

For a vector $y \in T_xM \setminus \{0\}$, there are infinitely many tangent planes $P \subset T_xM$ containing y. The flag curvature $\mathbf{K}(p, y)$ depends on the tangent planes P containing y. A Finsler metric F on a manifold M is said to be of *scalar curvature* $\mathbf{K}(y)$ if $\mathbf{K}(P, y) = \mathbf{K}(y)$ is independent of the tangent plane P containing y for all $y \in T_xM$. This is equivalent to the following

$$\mathbf{R}_y(u) = \mathbf{K}(y)\Big\{\mathbf{g}_y(y, y)\, u - \mathbf{g}_y(y, u)\, y\Big\}, \qquad y, u \in T_xM \setminus \{0\}.$$

By definition, every two-dimensional Finsler metric is of scalar curvature.

Define

$$\mathbf{Ric}(y) := \sum_{i=1}^{n} R^i{}_i(y). \tag{6.10}$$

Ric is a scalar function on $TM \setminus \{0\}$ with the following homogeneity:

$$\mathbf{Ric}(\lambda y) = \lambda^2 \mathbf{Ric}(y), \qquad \lambda > 0.$$

We call **Ric** the *Ricci curvature.* For the sake of convenience, let

$$R(y) := \frac{1}{n-1}\mathbf{Ric}(y).$$

We call $R(y)$ the *Ricci scalar.* By the homogeneity of R, we have

$$R(y) = \frac{1}{2} R_{y^i y^j}(y) y^i y^j.$$

6.2 Geodesic Fields

In this section, we will give an alternative formula for the Riemann curvature. This formula is not good for computing the Riemann curvature, but it is very useful in comparison geometry.

Lemma 6.2.1 *Let Y be a geodesic field on an open subset $\mathcal{U}$ in a Finsler space (M, F). Let $\hat{g} := \mathbf{g}_Y$ denote the Riemannian metric on $\mathcal{U}$. Then $\hat{F} := \sqrt{\hat{g}}$ is Y-related to F in the following sense*

$$N^i_j(Y) = \hat{N}^i_j(Y). \tag{6.11}$$

In particular, Y is a geodesic field of $\hat{F}$.

Proof: Equation (5.27) gives

$$N^i_j(y) = \frac{1}{2} g^{il} \Big\{ \frac{\partial g_{jl}}{\partial x^k} + \frac{\partial g_{kl}}{\partial x^j} - \frac{\partial g_{jk}}{\partial x^l} \Big\} y^k - 2 g^{il} C_{jkl} G^k. \tag{6.12}$$

Note that

$$\frac{\partial \hat{g}_{ij}}{\partial x^k} = \frac{\partial g_{ij}}{\partial x^k}(Y) + 2 C_{ijl}(Y) \frac{\partial Y^l}{\partial x^k}. \tag{6.13}$$

It follows from (6.12) and (6.13) that

$$\begin{aligned}
\hat{N}^i_j(Y) &= \frac{1}{2} \hat{g}^{il} \Big\{ \frac{\partial \hat{g}_{jl}}{\partial x^k} + \frac{\partial \hat{g}_{kl}}{\partial x^j} - \frac{\partial \hat{g}_{jk}}{\partial x^l} \Big\} Y^k \\
&= \frac{1}{2} g^{il}(Y) \Big\{ \frac{\partial g_{jl}}{\partial x^k}(Y) + \frac{\partial g_{kl}}{\partial x^j}(Y) - \frac{\partial g_{jk}}{\partial x^l}(Y) \Big\} Y^k \\
&\quad - 2 g^{il}(Y) C_{jkl}(Y) G^k(Y) \\
&= N^i_j(Y).
\end{aligned}$$

This gives (6.11). By (6.11), we obtain

$$2G^i(Y) = N^i_j(Y) Y^j = \hat{N}^i_j(Y) Y^j = 2\hat{G}^i(Y).$$

Thus Y also satisfies

$$Y^j \frac{\partial Y^i}{\partial x^j} + 2\hat{G}^i(Y) = 0.$$

Thus Y is also a geodesic field of $\hat{F}$. Q.E.D.

Let Y be a geodesic field on an open subset $\mathcal{U} \subset M$ and $\hat{g} := \mathbf{g}_Y$ denote the induced Riemannian metric on $\mathcal{U}$. It follows from Lemma 6.2.1 that the connection D of F and the Levi-Civita connection of $\hat{F} := \sqrt{\hat{g}}$ are related by

$$\mathrm{D}_Y U = \hat{\mathrm{D}}_Y U. \tag{6.14}$$

Further, we have the following

Proposition 6.2.2 *Let Y be a geodesic field on an open subset $\mathcal{U}$ in a Finsler space (M, F) and $\hat{g} := \mathbf{g}_Y$. Let $\mathbf{R}$ and $\hat{\mathbf{R}}$ denote the Riemann curvature of F and $\hat{F} = \sqrt{\hat{g}}$ respectively. Then for $y = Y_x \in T_xM$*

$$\mathbf{R}_y = \hat{\mathbf{R}}_y. \tag{6.15}$$

Proof. Observe that

$$\begin{aligned}
\frac{\partial \hat{G}^i}{\partial x^k}(Y) &= \frac{\partial}{\partial x^k}\Big[\hat{G}^i(Y)\Big] - \hat{N}^i_j(Y)\frac{\partial Y^j}{\partial x^k} \\
&= \frac{\partial}{\partial x^k}\Big[G^i(Y)\Big] - N^i_j(Y)\frac{\partial Y^j}{\partial x^k} \\
&= \frac{\partial G^i}{\partial x^k}(Y)
\end{aligned} \tag{6.16}$$

and

$$\begin{aligned}
Y^j\frac{\partial \hat{N}^i_k}{\partial x^j}(Y) &= Y^j\Big\{\frac{\partial}{\partial x^j}\Big[\hat{N}^i_k(Y)\Big] - \hat{\Gamma}^i_{kl}(Y)\frac{\partial Y^l}{\partial x^j}\Big\} \\
&= Y^j\frac{\partial}{\partial x^j}\Big[N^i_k(Y)\Big] + 2\hat{G}^l(Y)\hat{\Gamma}^i_{kl}(Y) \\
&= Y^j\frac{\partial N^i_k}{\partial x^j}(Y) + \Gamma^i_{kl}(Y)Y^j\frac{\partial Y^l}{\partial x^j} + 2\hat{G}^l(Y)\hat{\Gamma}^i_{kl}(Y) \\
&= Y^j\frac{\partial N^i_k}{\partial x^j}(Y) - 2G^l(Y)\Gamma^i_{kl}(Y) + 2\hat{G}^l(Y)\hat{\Gamma}^i_{kl}(Y).
\end{aligned}$$

This implies

$$Y^j\frac{\partial \hat{N}^i_k}{\partial x^j}(Y) - 2\hat{G}^l(Y)\hat{\Gamma}^i_{kl}(Y) = Y^j\frac{\partial N^i_k}{\partial x^j}(Y) - 2G^l(Y)\Gamma^i_{kl}(Y). \tag{6.17}$$

Plugging (6.16) and (6.17) into (6.4) yields

$$\hat{R}^i{}_k(Y) = R^i{}_k(Y).$$

Q.E.D.

For a Riemannian metric $F(y) = \sqrt{g(y,y)}$ on a manifold M, the Riemann curvature $\mathbf{R}_y : T_xM \to T_xM$ is quadratic in $y \in T_xM$. Moreover, $\mathbf{R}_y$ is self-adjoint with respect to g,

$$g\Big(\mathbf{R}_y(u),\ v\Big) = g\Big(u,\ \mathbf{R}_y(v)\Big), \qquad u, v \in T_xM. \tag{6.18}$$

The proof of (6.18) is given in standard text books on Riemannian geometry. We will prove it later in Section 8.1.

Granting that (6.18) holds for Riemannian metrics, we consider a Finsler space (M, F). For a vector $y \in T_xM \setminus \{0\}$, extend y to a geodesic field Y in a neighborhood $\mathcal{U}$ of x. Let $\hat{\mathbf{R}}_w : T_zM \to T_zM$, $w \in T_zM \setminus \{0\}$, denote the Riemann curvature of $\hat{g} := \mathbf{g}_Y$. By (6.18),

$$\hat{g}\Big(\mathbf{R}_Y(u),\ v\Big) = \hat{g}\Big(u,\ \hat{\mathbf{R}}_Y(v)\Big), \qquad u, v \in T_zM.$$

Restricting the above equation to x yields

$$\mathbf{g}_y\Big(\mathbf{R}_y(u),\ v\Big) = \mathbf{g}_y\Big(u,\ \mathbf{R}_y(v)\Big), \qquad u, v \in T_xM \tag{6.19}$$

Thus if (6.19) holds for Riemannian metrics, then it holds for Finsler metrics.

6.3 Projectively Related Finsler Metrics

Two Finsler metrics on a manifold are said to be *pointwise projectively related* if they have the same geodesics as point sets. The Riemann curvature is determined by geodesics. A natural question arises: What happens to the Riemann curvature if two Finsler metrics on a manifold have the same geodesics as point sets? In this section, we will discuss this problem. See [Dg][BuKe][Ma4] for related discussion.

Given a Finsler space (M, F) and a Finsler metric $\tilde{F}$ on M. We view $\tilde{F}$ as a scalar function on TM. Let $\{\omega^i, \omega^{n+i}\}$ be a local coframe for $T^*(TM \setminus \{0\})$ corresponding to a local frame $\{e_i\}_{i=1}^n$ for π^*TM. See (5.18) for the correspondence. Define

$$d\tilde{F} = \tilde{F}_{|i}\omega^i + \tilde{F}_{\cdot i}\omega^{n+i},$$

$$d\tilde{F}_{|i} - \tilde{F}_{|j}\omega_i^{\ j} = \tilde{F}_{|i|j}\omega^j + \tilde{F}_{|i\cdot j}\omega^{n+j}.$$

In a standard local coordinate system (x^i, y^i) in TM, $\omega^i = dx^i$ and $\omega^{n+i} = \delta y^i$. $\tilde{F}_{|k}$ and $\tilde{F}_{|k\cdot l}$ are given by

$$\tilde{F}_{|k} = \tilde{F}_{x^k} - N^l_k \tilde{F}_{y^k}, \qquad \tilde{F}_{|k\cdot l} = (\tilde{F}_{|k})_{y^l}.$$

Let $G^i(y)$ and $\tilde{G}^i(y)$ denote the geodesic coefficients of F and $\tilde{F}$ in a *common* standard coordinate system (x^i, y^i) in TM. They are defined by (5.2). From the geodesic equations, one can easily see that F and $\tilde{F}$ have the same geodesics as point sets if and only if

$$\tilde{G}^i(y) = G^i(y) + P(y)y^i,$$

where $P(y)$ is a scalar function satisfying

$$P(\lambda y) = \lambda P(y), \qquad \lambda > 0.$$

The following important result is due to A. Rapcsák [Ra].

Lemma 6.3.1 (Rapcsák) *For two Finsler metrics F and $\tilde{F}$ on a manifold M, the following conditions are equivalent*

(a) $\tilde{F}$ and F have the same geodesics as point sets;
(b) There is a scalar function $P(y)$ on TM such that

$$\tilde{G}^i(y) = G^i(y) + P(y)y^i.$$

In this case, P is given by

$$P(y) = \frac{\tilde{F}_{|k}y^k}{2\tilde{F}}.$$

(c) $\tilde{F}$ satisfies

$$\tilde{F}_{|k\cdot l}y^k - \tilde{F}_{|l} = 0. \tag{6.20}$$

Proof: By a direct computation, one can verify that

$$\tilde{G}^i = G^i + Py^i + Q^i,$$

where

$$P = \frac{\tilde{F}_{|k}y^k}{2\tilde{F}},$$
$$Q^i = \frac{1}{2}\tilde{F}\tilde{g}^{il}\Big\{\tilde{F}_{|k\cdot l}y^k - \tilde{F}_{|l}\Big\}.$$

Thus F and $\tilde{F}$ are pointwise projectively related if and only if $Q^i = 0$, i.e., (6.20) holds. Q.E.D.

Example 6.3.1 ([HaIc]) Let (M, F) be a Finsler space. Consider a 1-form $\beta(y) = b_i(x)y^i$ on M. Assume that

$$\|\beta\| := \sup_{F(y)=1} \beta(y) < 1.$$

Then $\tilde{F} := F + \beta$ is again a Finsler metric on M. Observe that

$$\beta_{|l} = \frac{\partial b_i}{\partial x^l} y^l - b_j N^j_l,$$

$$\beta_{|k \cdot l} y^k = \frac{\partial b_l}{\partial x^k} y^k - b_j \frac{\partial N^j_k}{\partial y^l} y^k$$

Note that

$$\frac{\partial N^j_k}{\partial y^l} y^k = \frac{\partial^2 G^j}{\partial y^k \partial y^l} y^k = \frac{\partial G^j}{\partial y^l} = N^j_l.$$

Thus

$$\beta_{|k \cdot l} y^k - \beta_{|l} = \Big(\frac{\partial b_l}{\partial x^k} - \frac{\partial b_k}{\partial x^l} \Big) y^k.$$

By Lemma 5.2.1,

$$F_{|k} = F_{x^k} - N^l_k F_{y^l} = 0.$$

We obtain

$$\tilde{F}_{|k \cdot l} y^k - \tilde{F}_{|l} = \beta_{|k \cdot l} y^k - \beta_{|l} = \Big(\frac{\partial b_l}{\partial x^k} - \frac{\partial b_k}{\partial x^l} \Big) y^k.$$

Thus $\tilde{F} = F + \beta$ is pointwise projective to F if and only if β is close. ♯

Given two Finsler metrics F and $\tilde{F}$ on an n-dimensional manifold M. Assume that F and $\tilde{F}$ are pointwise projectively related, hence by Lemma 6.3.1

$$\tilde{G}^i(y) = G^i(y) + P(y) y^i,$$

where

$$P(y) = \frac{\tilde{F}_{|k} y^k}{2\tilde{F}}. \tag{6.21}$$

Let $R^i{}_k(y)$ and $\tilde{R}^i{}_k(y)$ denote the coefficients of the Riemann curvature of F and $\tilde{F}$ respectively. By (6.4), we immediately obtain

$$\tilde{R}^i{}_k(y) = R^i{}_k(y) + \Xi(y)\delta^i_k + \tau_k(y)y^i, \tag{6.22}$$

where

$$\Xi = P^2 - P_{|i}y^i \qquad \tau_k = 3(P_{|k} - PP_{.k}) + \Xi_{.k}.$$

Equation (6.22) is proved in [MaWe]. In index-free form we can rewrite (6.22) as follows

$$\tilde{\mathbf{R}}_y(u) = \mathbf{R}_y(u) + \Xi(y)I + \tau_y(u)\, y, \qquad u \in T_xM,$$

where I denotes the identity map and $\tau_y(u) := \tau_k(y)u^k$, $u = u^k \frac{\partial}{\partial x^i}|_x$.

Consider a Finsler metric F on an open domain $\Omega \subset \mathrm{R}^n$. According to Lemma 6.3.1, the following conditions are equivalent

(a) The geodesics of F are straight lines in Ω;
(b) The geodesic coefficients G^i of F is in the following form

$$G^i(y) = P(y)y^i.$$

In this case, P is given by

$$P = \frac{F_{x^k}y^k}{2F}.$$

(c) F satisfies

$$F_{x^k y^i}y^k = F_{x^i}.$$

Assume that F is pointwise projectively flat. Then the Riemann curvature is in the following form

$$\mathbf{R}_y(u) = \Xi(y)u + \tau_y(u)\, y, \qquad u \in T_xM.$$

By (6.8), we obtain

$$0 = \mathbf{g}_y\Big(\mathbf{R}_y(u), y\Big) = \Xi(y)\mathbf{g}_y(y,u) + \tau_y(u)F^2(y).$$

This gives

$$\tau_y(u) = -\frac{\Xi(y)}{F^2(y)}\mathbf{g}_y(y,u).$$

Let

$$\mathbf{K}(y) := \frac{\Xi(y)}{F^2(y)}.$$

By the above formulas, we obtain

$$\mathbf{R}_y(u) = \mathbf{K}(y)\Big\{\mathbf{g}_y(y,y)\, u - \mathbf{g}_y(y,u)\, y\Big\}, \qquad y, u \in T_xM \setminus \{0\}.$$

Thus F is of scalar curvature.

Chapter 7

Non-Riemannian Curvatures

When one looks at a Finsler space, he not only sees the shape of the space, but also the "color" of the space. In this chapter, we will introduce and discuss the Cartan torsion and the Chern curvature, etc. Roughly speaking, the Cartan torsion describes the "color" of the space at a point and the Chern curvature tensor and its portion (the Landsberg curvature) describe the rate of changes of the color over the space. These quantities all vanish for Riemannian spaces. Thus Finsler spaces are much more "colorful" than Riemannian spaces.

7.1 Cartan Torsion

The Cartan torsion is a non-Euclidean quantity of Minkowski spaces. Since every tangent space on a Finsler space is a Minkowski space, the Cartan torsion is defined for a Finsler space.

Let (V, F) be an n-dimensional Minkowski space. Recall that the induced inner product $\mathbf{g}_y$ in V is independent of $y \in \mathrm{V} \setminus \{0\}$ if and only if F is Euclidean. For a vector $y \in \mathrm{V} \setminus \{0\}$, define

$$\mathbf{C}_y(u, v, w) := \frac{1}{2}\frac{d}{dt}\Big[\mathbf{g}_{y+tw}(u, v)\Big]\Big|_{t=0}, \qquad u, v, w \in \mathrm{V}.$$

We have

$$\mathbf{C}_y(u, v, w) = \frac{1}{4}\frac{\partial^3}{\partial r \partial s \partial t}\Big[F^2(y + ru + sv + tw)\Big]\Big|_{r=s=t=0}.$$

Thus for each $y \in \mathrm{V} \setminus \{0\}$, $\mathbf{C}_y$ is a symmetric multi-linear form on V. The homogeneity of F implies

$$\mathbf{C}_y(y, v, w) = 0, \qquad v, w \in \mathrm{V}.$$

The family $\mathbf{C} = \{\mathbf{C}_y\}_{y \in \mathrm{V}\setminus\{0\}}$ is called the *Cartan torsion.*

Let $\{\mathbf{b}_i\}_{i=1}^n$ be a basis for V. $F(y) = F(y^i\mathbf{b}_i)$ is a function of $(y^i) \in \mathrm{R}^n$. Let

$$g_{ij}(y) = \frac{1}{2}[F^2]_{y^iy^j}(y), \qquad C_{ijk}(y) = \frac{1}{4}[F^2]_{y^iy^jy^k}(y).$$

We have

$$\mathbf{g}_y(u, v) = g_{ij}(y)u^iv^j, \qquad \mathbf{C}_y(u, v, w) := C_{ijk}(y)u^iv^jw^k, \tag{7.1}$$

where $u = u^i\mathbf{b}_i$, $v = v^j\mathbf{b}_j$ and $w = w^k\mathbf{b}_k$.

The Cartan torsion is related to the Cartan tensor in (5.13) by

$$\mathbf{C}_y(u, v, w) = \mathcal{C}(U, V, W), \qquad u, v, w \in T_xM,$$

where $U = (y, u), V = (y, v), W = (y, w) \in \pi^*TM$.

Define

$$\mathbf{I}_y(u) := \sum_{ij=1}^{n} g^{ij}(y)\mathbf{C}_y(\mathbf{b}_i, \mathbf{b}_j, u) = \sum_{ij=1}^{n} g^{ij}(y)C_{ijk}(y)u^k, \qquad u = u^i\mathbf{b}_i, \tag{7.2}$$

where $(g^{ij}(y)) := (g_{ij}(y))^{-1}$. The family $\mathbf{I} = \{\mathbf{I}_y\}_{y \in V\setminus\{0\}}$ is called the *mean Cartan torsion.* Note that in dimension two, the mean Cartan torsion completely determines the Cartan torsion. $\mathbf{I}_y$ can be also expressed as

$$\mathbf{I}_y(u) = u^i \frac{\partial}{\partial y^i} \ln \sqrt{\det(g_{jk}(y))}, \qquad u = u^i\mathbf{b}_i. \tag{7.3}$$

Thus $\mathbf{I} = 0$ if and only if $\det(g_{ij}(y)) = constant$. Using the maximal principles of the Laplacian on $\mathrm{S} = F^{-1}(1)$, Deicke [De] proved the following important result

Theorem 7.1.1 (Deike) *A Minkowski norm F is Euclidean if and only if* $\mathbf{I} = 0$.

See also [Bk] and Chapter 14 in [BCS1] for a proof. We should point out that Theorem 7.1.1 does not hold for singular Minkowski spaces [JiSh]. There are infinitely many singular Minkowski norms with vanishing mean Cartan torsion.

Define

$$\|\mathbf{I}\| := \sup_{F(y)=1, v\neq 0} \frac{|\mathbf{I}_y(v)|}{[\mathbf{g}_y(v,v)]^{1/2}} \tag{7.4}$$

$$\|\mathbf{C}\| := \sup_{F(y)=1, v\neq 0} \frac{|\mathbf{C}_y(v,v,v)|}{[\mathbf{g}_y(v,v)]^{3/2}}. \tag{7.5}$$

Proposition 7.1.2 *Let $F = \alpha + \beta$ be a Randers norm on* V*, where α is an Euclidean norm and β is a linear functional with*

$$\|\beta\| := \sup_{\alpha(y)=1} |\beta(y)| < 1.$$

The Cartan torsion is uniformly bounded. More precisely,

$$\|\mathbf{I}\| \leq \frac{n+1}{\sqrt{2}}\sqrt{1-\sqrt{1-\|\beta\|^2}} \tag{7.6}$$

$$\|\mathbf{C}\| \leq \frac{3}{\sqrt{2}}\sqrt{1-\sqrt{1-\|\beta\|^2}}. \tag{7.7}$$

Proof. Fix a basis $\{\mathbf{b}_i\}_{i=1}^n$ for V. Let

$$\beta(y) = b_i y^i, \qquad \mathbf{I}_y(u) = I_i(y)u^i, \qquad y = y^i\mathbf{b}_i, \; u = u^i\mathbf{b}_i.$$

By (1.7) and (7.3), we obtain

$$\begin{aligned} I_i(y) &= \frac{\partial}{\partial y^i}\ln\sqrt{\det(g_{jk}(y))} \\ &= \frac{n+1}{2}\Big\{\frac{F_{y^i}}{F} - \frac{\alpha_{y^i}}{\alpha}\Big\} \\ &= \frac{n+1}{2F}\Big\{b_i - \frac{\beta}{\alpha}\alpha_{y^i}\Big\}. \end{aligned} \tag{7.8}$$

By (7.4), the norm of $\mathbf{I}_y$ is given by

$$\|\mathbf{I}_y\| = \sqrt{I_i(y)I_j(y)g^{ij}(y)}. \tag{7.9}$$

It follows from (1.8), (7.8) and (7.9) that

$$\|\mathbf{I}_y\| = \frac{n+1}{2F}\sqrt{\Big(\frac{\alpha}{F}\Big)\Big[\|\beta\|^2 - \Big(\frac{\beta}{\alpha}\Big)^2\Big]}. \tag{7.10}$$

Since $\beta(y) \le \|\beta\|\alpha(y)$, we can write $\beta(y) = \|\beta\|\cos\theta$ for some $0 \le \theta \le 2\pi$. Assume that $F(y) = 1$. Then

$$\alpha(y) = 1 - \beta(y) = 1 - \|\beta\|\alpha(y)\cos\theta.$$

This gives

$$\alpha(y) = \frac{1}{1 + b\cos\theta}.$$

Plugging it into (7.10) yields

$$\|\mathbf{I}_y\| = \frac{n+1}{2}\sqrt{\frac{\|\beta\|^2\sin^2\theta}{1+\|\beta\|\cos\theta}} \le \frac{n+1}{\sqrt{2}}\sqrt{1-\sqrt{1-\|\beta\|^2}}.$$

This gives (7.6).

To prove (7.7), we first reduce the problem to the two-dimensional case. Let y_o, v_o with $F(y_o) = 1$ and $g_{y_o}(v_o, v_o) = 1$ such that

$$\|\mathbf{C}\| = \mathbf{C}_{y_o}(v_o, v_o, v_o).$$

Let $\bar{\mathrm{V}} = \mathrm{span}\Big\{y_o, v_o\Big\}$ and $\bar{F} := F|_{\bar{\mathrm{V}}}$. We have

$$\mathbf{C}_{y_o}(v_o, v_o, v_o) = \frac{1}{4}\frac{\partial^3}{\partial s^3}\Big[F^2(y_o + sv_o)\Big]|_{s=0} = \bar{\mathbf{C}}_{y_o}(v_o, v_o, v_o).$$

Let $\bar{\alpha} := \alpha|_{\bar{\mathrm{V}}}$ and $\bar{\beta} = \beta_{|\bar{\mathrm{V}}}$. We have

$$\|\bar{\beta}\| = \sup_{\bar{\alpha}(y)=1} \bar{\beta}(y) \le \sup_{\alpha(y)=1} \beta(y) = \|\beta\|.$$

Let $\bar{\mathbf{I}}(y_o)$ denote the main scalar of $\bar{F}$ at y_o. It is easy to see that

$$\|\bar{\mathbf{C}}\| = \max|\bar{\mathbf{I}}|.$$

Taking a special orthonormal basis $\{\mathbf{b}_1, \mathbf{b}_2\}$ for $(\bar{\mathrm{V}}, \bar{\alpha})$ such that $\bar{\beta}(y) = \|\bar{\beta}\|u$ for $y = u\mathbf{b}_1 + v\mathbf{b}_2$. With respect to this basis,

$$\bar{F}(y) = \sqrt{u^2+v^2} + \|\bar{\beta}\|\,u, \qquad y = u\mathbf{b}_1 + v\mathbf{b}_2.$$

It follows from (7.6) that

$$|\bar{\mathbf{I}}| \leq \frac{3}{\sqrt{2}}\sqrt{1-\sqrt{1-\|\bar{\beta}\|^2}}. \tag{7.11}$$

Thus

$$\|\mathbf{C}\| \leq \|\bar{\mathbf{C}}\| = \max|\bar{\mathbf{I}}| \leq \frac{3}{\sqrt{2}}\sqrt{1-\sqrt{1-\|\bar{\beta}\|^2}} \leq \frac{3}{\sqrt{2}}\sqrt{1-\sqrt{1-\|\beta\|^2}}.$$

Q.E.D.

In dimension two, the bound (7.7) is just the bound (7.6). The bound for the Cartan torsion in dimension two is suggested by B. Lackey. See Exercise 11.2.6 in [BCS1].

The norm of the (mean) Cartan torsion can be as large as one wishes. For example, the norm of the Cartan torsion of the following Minkowski norms F_λ on R^n approaches ∞ as $\lambda \to \infty$.

$$F_\lambda(y) := \sqrt{\sum_{i=1}^n (y^i)^2 + \lambda\sqrt{\sum_{i=1}^n (y^i)^4}}.$$

Example 7.1.1 Let

$$F(u,v) := \Big\{u^4 + 3c\,u^2v^2 + v^4\Big\}^{1/4},$$

where $0 < c < 2$. F is a Minkowski norm on R^2 (Example 1.2.3). For any $\mathbf{y} = (u,v)$ with $u > 0$,

$$\mathbf{I}_\mathbf{y} = \frac{2\sqrt{3}(9c^2-4)(v^4-u^4)uv}{3\Big(2c\,u^4 + (4-3c^2)\,u^2v^2 + 2c\,v^4\Big)^{3/2}}.$$

Take $\mathbf{y} = (u,v)$ with $v = \sqrt{\frac{2}{3c}}u > 0$, we obtain

$$|\mathbf{I}_\mathbf{y}| = \frac{(9c^2-4)^2}{64c} \to \infty, \qquad \text{as } c \to 0^+.$$

♯

7.2 Chern Curvature

Besides the Cartan curvature, there are other quantities which always vanish on Riemannian spaces. In this section, we are going to discuss two non-Riemannian quantities — the Chern curvature and the Landsberg curvature.

Let (M, F) be a Finsler space. In a standard local coordinate system (x^i, y^i) in TM, the local functions $g_{ij}(y) = \frac{1}{2}[F^2]_{y^iy^j}(y)$ and $C_{ijk}(y) := \frac{1}{4}[F^2]_{y^iy^jy^k}(y)$ on TM define the inner product $\mathbf{g}_y$ and the Cartan torsion $\mathbf{C}_y$ on T_xM respectively. See (7.1). Let L_{ijk} be the set of local functions defined in (5.24).

$$L_{ijk}(y) = y^l \frac{\partial C_{ijk}}{\partial x^l} - 2C_{ijkl}G^l - C_{ljk}N^l_i - C_{ilk}N^l_j - C_{ijl}N^l_k, \tag{7.12}$$

where $C_{ijkl} := \frac{1}{4}[F^2]_{y^iy^jy^ky^l}(y)$. The Christoffel symbols of the Chern connection are given by

$$\Gamma^i_{jk} := \frac{\partial^2 G^i}{\partial y^j \partial y^k} - L^i{}_{jk},$$

where $L^i{}_{kl} := g^{ij}L_{jkl}$. Differentiating Γ^i_{jk} with respect to y^l yields a new quantity:

$$P_j{}^i{}_{kl}(y) := -\frac{\partial \Gamma^i_{jk}}{\partial y^l} = -\frac{\partial^3 G^i}{\partial y^j \partial y^k \partial y^l} + \frac{\partial L^i{}_{jk}}{\partial y^l}. \tag{7.13}$$

For a vector $y \in T_xM \setminus \{0\}$, define

$$\mathbf{L}_y(u, v, w) : = L_{ijk}(y)u^iv^jw^k, \tag{7.14}$$

$$\mathbf{P}_y(u, v, w) : = P_j{}^i{}_{kl}(y)u^jv^kw^l \frac{\partial}{\partial x^i}|_x, \tag{7.15}$$

where $u = u^i \frac{\partial}{\partial x^i}|_x, v = v^j \frac{\partial}{\partial x^j}|_x, w = w^k \frac{\partial}{\partial x^k}|_x \in T_xM$. $\mathbf{L} := \{\mathbf{L}_y\}_{y\in TM\setminus\{0\}}$ is called the *Landsberg curvature* [La1][La2] and $\mathbf{P} := \{\mathbf{P}_y\}_{y\in TM\setminus\{0\}}$ is called the *Chern curvature* [Ch1][Ch2].

The Landsberg curvature is related to the Landsberg tensor in (5.32) by

$$\mathbf{L}_y(u, v, w) = \mathcal{L}(U, V, W), \qquad u, v, w \in T_xM,$$

where $U = (y, u), V = (y, v), W = (y, w) \in \pi^*TM$.

Let $c(t)$ be an arbitrary geodesic in (M, F). Take arbitrary parallel vector fields $U(t), V(t), W(t)$ along c. According to (7.12),

$$\mathbf{L}_{\dot{c}(t)}\Big(U(t), V(t), W(t)\Big) = \frac{d}{dt}\Big[\mathbf{C}_{\dot{c}(t)}\Big(U(t), V(t), W(t)\Big)\Big]. \tag{7.16}$$

Thus the Landsberg curvature measures the rate of changes of the Cartan torsion along geodesics. A Finsler metric is called a *Landsberg metric* if $\mathbf{L} = 0$. By (7.16), we see that on a Landsberg space, the Cartan torsion is constant along geodesics.

Proposition 7.2.1 *The Landsberg curvature is related to the Chern curvature by*

$$\mathbf{L}_y(u, v, w) = -\mathbf{g}_y\Big(u,\ \mathbf{P}_y(y, v, w)\Big) \tag{7.17}$$

$$= \mathbf{g}_y\Big(\mathbf{P}_y(u, v, w),\ y\Big). \tag{7.18}$$

Proof: By the homogeneity of G^i, we obtain

$$y^s P_s{}^i{}_{kl} = y^s \frac{\partial L^i{}_{sk}}{\partial y^l} = \frac{\partial}{\partial y^l}\Big(y^s L^i{}_{sk}\Big) - L^i{}_{kl} = -L^i{}_{kl}.$$

Thus

$$g_{ij} y^s P_s{}^i{}_{kl} = -g_{ij} L^i{}_{kl} = -L_{jkl}.$$

This gives (7.17).

Differentiating (5.28) with y^i, then contacting the resulting identity with y^l, we obtain

$$\begin{aligned} y^l g_{kl} \frac{\partial \Gamma^k_{jm}}{\partial y^i} &= -2y^l \frac{\partial C_{ijm}}{\partial x^l} + 4C_{ijkm} G^k \\ &\quad + 2C_{ijk} N^k_m + 2C_{ikm} N^k_j + 2C_{ikm} N^k_j - y^l \frac{\partial L_{jlm}}{\partial y^i} \\ &= -2L_{ijm} - y^l \frac{\partial L_{jlm}}{\partial y^i} \\ &= -2L_{ijm} - \Big[\frac{\partial}{\partial y^i}\Big(y^l L_{jlm}\Big) - L_{ijm}\Big] = -L_{ijm}. \end{aligned}$$

Thus

$$y^l g_{kl} P_j{}^k{}_{mi} = -y^l g_{kl} \frac{\partial \Gamma^k_{jm}}{\partial y^i} = L_{ijm}.$$

This gives (7.18). Q.E.D.

Berwald metrics can be characterized by the Chern curvature.

Proposition 7.2.2 *A Finsler metric F is a Berwald metric if and only if* $\mathbf{P} = 0$.

Proof. Assume that F is Berwaldian. By Definition 5.1.3, $G^i(y)$ are quadratic in $y \in T_xM$. Then

$$\frac{\partial N^k_j}{\partial y^m \partial y^i} = \frac{\partial^3 G^i}{\partial y^j \partial y^k \partial y^l} = 0.$$

Differentiating (5.27) with respect to y^m and y^i, then contracting the resulting equation with y^l, we obtain

$$0 = -2y^l \frac{\partial C_{ijm}}{\partial x^l} + 4C_{ijkm} G^k + 2C_{jkm} N^k_i + 2C_{ijk} N^k_m + 2C_{ikm} N^k_j.$$

By (7.12), we see that $L_{ijm} = 0$. Thus

$$\Gamma^i_{jk} = \frac{\partial^2 G^i}{\partial y^j \partial y^k}$$

and

$$P_j{}^i{}_{kl} = -\frac{\partial \Gamma^i_{jk}}{\partial y^l} = -\frac{\partial^3 G^i}{\partial y^j \partial k \partial y^l} = 0.$$

Thus $G^i(y)$ are quadratic in $y \in T_xM$ and F is a Berwald metric.

Assume that $\mathbf{P} = 0$. Thus $\mathbf{L} = 0$ by (7.17). Equation (7.13) implies

$$P_j{}^i{}_{kl} = -\frac{\partial \Gamma^i_{jk}}{\partial y^l} = 0.$$

Thus $\Gamma^i_{jk}(y)$ are independent of $y \in T_xM$ and

$$G^i(y) = \frac{1}{2} N^i_j(y) y^j = \frac{1}{2} \Gamma^i_{jk}(x) y^j y^k$$

are quadratic in $y \in T_xM$. We conclude that F is a Berwald metric Q.E.D.

From (7.17) and Proposition 7.2.2, we immediately conclude that every Berwald space is a Landsberg space. It is an open problem in Finsler geometry whether or not there is a Landsberg metric which is not a Berwald metric. So far no example has been found.

Example 7.2.1 Consider the Funk metric F on a strongly convex domain $\Omega \subset \mathrm{R}^n$. F is a nonnegative function on $T\Omega = \Omega \times \mathrm{R}^n$. According to (1.21), F satisfies

$$F_{x^i} = FF_{y^i}. \tag{7.19}$$

By (7.19), one can easily show that

$$F_{x^k y^i} y^k = F_{x^i}$$

and the geodesic coefficients G^i of F are given by

$$G^i(y) = \frac{1}{2} F(y) y^i. \tag{7.20}$$

Hence

$$N^i_j = \frac{1}{2} F_{y^j} y^i + \frac{1}{2} F \delta^i_j. \tag{7.21}$$

We are going to compute the Landsberg curvature using (7.12).

By (7.19), we obtain

$$y^l \frac{\partial C_{ijk}}{\partial y^l} = \frac{1}{4} y^l [F^2]_{x^l y^i y^j y^k} = \frac{1}{2} y^l [F^2 F_{y^l}]_{y^i y^j y^k} == \frac{1}{6} y^l [F^3]_{y^l y^i y^j y^k} = 0.$$

Using (7.20) and (7.21), we obtain

$$C_{ijkl} G^l = \frac{1}{2} F y^l \frac{\partial C_{ijk}}{\partial y^l} = -\frac{1}{2} F C_{ijk}$$

and

$$C_{ljk} N^l_i = \frac{1}{2} F_{y^i} y^l C_{ljk} + \frac{1}{2} F \delta^l_i C_{ljk} = \frac{1}{2} F C_{ijk}.$$

Plugging them into (7.12) yields

$$L_{ijk} = -\frac{1}{2} F C_{ijk}.$$

In index-free form,

$$\mathbf{L}_y(u, v, w) = -\frac{F(y)}{2} \mathbf{C}_y(u, v, w). \tag{7.22}$$

Let $c(t)$ be a unit speed geodesic in Ω with $\dot{c}(0) = y \in T_x\Omega$ and $U(t)$ a parallel vector field along c. From (7.22), we see that the function

$$\mathbf{C}(t) := \mathbf{C}_{\dot{c}(t)}(U(t), U(t), U(t))$$

satisfies

$$\mathbf{C}'(t) + \frac{1}{2}\mathbf{C}(t) = 0. \tag{7.23}$$

The general solution of (7.23) is

$$\mathbf{C}(t) = \mathbf{C}(0)\exp(-\frac{1}{2}t).$$

The maximal interval of c is $(-\delta, \infty)$, where

$$\delta = F(y)\ln\left[1 + \frac{F(y)}{F(-y)}\right].$$

Thus $\mathbf{C}(t)$ is a bounded function on $(-\delta, \infty)$. It is not clear that if $\mathbf{C}$ is bounded on Ω, when Ω is not the unit ball in R^n. ♯

Let $\{\mathbf{b}_i\}_{i=1}^n$ be an arbitrary basis for T_xM. We define the mean of $\mathbf{L}_y$ by

$$\mathbf{J}_y(u) := \sum_{ij=1}^{n} g^{ij}(y)\mathbf{L}_y(u, \mathbf{b}_i, \mathbf{b}_j),$$

where $g_{ij}(y) = \mathbf{g}_y(\mathbf{b}_i, \mathbf{b}_j)$. The family $\mathbf{J} = \{\mathbf{J}_y\}_{y \in TM\setminus\{0\}}$ is called the *mean Landsberg curvature*. A Finsler metric is called a *weak Landsberg metric* if $\mathbf{J} = 0$. In dimension two, $\mathbf{J}$ completely determines $\mathbf{L}$.

Let (M, F) be a Finsler space and c an arbitrary geodesic. Take an arbitrary parallel vector field $V(t)$ along c. It follows from (7.16) that

$$\mathbf{J}_{\dot{c}(t)}\Big(V(t)\Big) = \frac{d}{dt}\Big[\mathbf{I}_{\dot{c}(t)}\Big(V(t)\Big)\Big]. \tag{7.24}$$

Thus on a weak Landsberg space, the mean Cartan torsion $\mathbf{I}$ is constant along any geodesic.

There is an induced Riemannian metric $\hat{g}$ of Sasaki type on $TM \setminus \{0\}$,

$$\hat{g} = g_{ij}(y)dx^i \otimes dx^j + g_{ij}\delta y^i \otimes \delta y^j.$$

T. Aikou proved that if $\mathbf{L} = 0$, then all the slit tangent spaces $T_xM \setminus \{0\}$ are totally geodesic in $TM \setminus \{0\}$ [Ai]. We can show that if $\mathbf{J} = 0$, then all the slit tangent spaces $T_xM \setminus \{0\}$ are minimal in $TM \setminus \{0\}$.

7.3 S-Curvature

In this section, we will introduce and discuss an important non-Riemannian curvature for Finsler m spaces. It measures the rate of changes of Minkowski tangent spaces over a Finsler m space.

Let $(M, F, d\mu)$ be a Finsler m space. Take an arbitrary basis $\{\mathbf{b}_i\}_{i=1}^n$ for T_xM and its dual basis $\{\theta^i\}_{i=1}^n$ for T_x^*M. Express $d\mu = \sigma(x)\,\theta^1 \wedge \cdots \wedge \theta^n$, For a vector $y \in T_xM \setminus \{0\}$, we define

$$\tau(y) := \ln \frac{\sqrt{\det\left(g_{ij}(y)\right)}}{\sigma}, \tag{7.25}$$

where $g_{ij}(y) := \mathbf{g}_y(\mathbf{b}_i, \mathbf{b}_j)$. τ is called the *distortion* of $(M, F, d\mu)$.

The distortion τ has the following homogeneity property

$$\tau(\lambda y) = \tau(y), \qquad \lambda > 0,\ y \in T_xM \setminus \{0\}.$$

The vertical derivative of τ is nothing but the mean Cartan torsion, namely,

$$\frac{d}{dt}\Big[\tau(y + tv)\Big]\Big|_{t=0} = \mathbf{I}_y(v), \qquad v \in T_xM. \tag{7.26}$$

First, let us give a proof for (7.26). In local coordinates

$$\tau(y) = \ln\left[\frac{\sqrt{\det(g_{ij}(y))}}{\sigma}\right], \tag{7.27}$$

where $g_{ij}(y) = \mathbf{g}_y(\frac{\partial}{\partial x^i}|_x, \frac{\partial}{\partial x^j}|_x)$ and $d\mu = \sigma(x)dx^1 \cdots dx^n$. Observe that

$$\frac{\partial}{\partial y^k}\left[\ln\sqrt{\det(g_{ij}(y))}\right] = \frac{1}{2}g^{ij}(y)\frac{\partial g_{ij}}{\partial y^k}(y) = g^{ij}(y)C_{ijk}(y) = I_k(y).$$

This gives (7.26).

According to Deike's theorem (Theorem 7.1.1), $\mathbf{C} = 0$ if and only if $\mathbf{I} = 0$. Therefore, the following conditions are equivalent

(a) $\tau(y) = constant$;
(b) $\mathbf{I} = 0$;
(c) $\mathbf{C} = 0$;
(d) F is Euclidean.

To measure the rate of changes of the distortion along geodesics, we define

$$\mathbf{S}(y) := \frac{d}{dt}\Big[\tau\Big(\dot{c}(t)\Big)\Big]_{t=0}, \tag{7.28}$$

where $c(t)$ is the geodesic with $\dot{c}(0) = y$. $\mathbf{S}$ is called the *S-curvature* [Sh9]. It is also called the *mean covariation* in [Sh2] and *mean tangent curvature* in [Sh4].

Let Y be a non-zero geodesic field on an open subset $\mathcal{U} \subset M$. By (7.28), we have

$$\mathbf{S}(Y) = Y\Big[\tau(Y)\Big]. \tag{7.29}$$

The S-curvature $\mathbf{S}$ satisfies the following homogeneity condition

$$\mathbf{S}(\lambda y) = \lambda \mathbf{S}(y), \qquad \lambda > 0.$$

In general, $\mathbf{S}(y)$ is not linear in y. Differentiating $\mathbf{S}(y)$ twice with respect to y gives new quantity. For a vector $y \in T_xM\backslash\{0\}$, define $\mathbf{E}_y : T_xM \times T_xM \to$ R by

$$\mathbf{E}_y(u,v) := \frac{1}{2}\frac{\partial^2}{\partial s \partial t}\Big[\mathbf{S}(y + su + tv)\Big]_{s=t=0}, \qquad u, v \in T_xM. \tag{7.30}$$

We call $\mathbf{E} = \{\mathbf{E}_y\}_{y \in TM\backslash\{0\}}$ the *E-curvature*. We will show that the E-curvature is independent of the volume form, although the S-curvature does.

We first derive local formulas for the S-curvature and E-curvature. In local coordinates, $\tau(y)$ is given by (7.27). Observe that in a standard local coordinate system (x^i, y^i) in TM,

$$\frac{\partial}{\partial x^k}\Big[\ln\sqrt{\det(g_{ij})}\Big] = \frac{1}{2}g^{ij}\frac{\partial g_{ij}}{\partial x^k}, \tag{7.31}$$

$$\frac{\partial}{\partial y^k}\Big[\ln\sqrt{\det(g_{ij})}\Big] = g^{ij}C_{ijk}. \tag{7.32}$$

It follows from (5.27) that

$$\frac{1}{2}g^{ij}\frac{\partial g_{ij}}{\partial x^k}y^k - 2g^{ij}C_{ijk}G^k = N^m_m. \tag{7.33}$$

Let $c(t)$ be the geodesic with $\dot{c}(0) = y$. By (7.31)-(7.33), we obtain

$$\begin{aligned}
\mathbf{S}(\dot{c}) : &= \frac{d}{dt}\Big[\tau(\dot{c}(t))\Big] \\
&= \frac{d}{dt}\Big(\ln\sqrt{\det(g_{ij}(\dot{c}))}\Big) - \frac{d}{dt}\Big(\ln\sigma(c)\Big) \\
&= \frac{1}{2}g^{ij}(\dot{c})\frac{\partial g_{ij}}{\partial x^k}(\dot{c})\dot{c}^k + g^{ij}(\dot{c})C_{ijk}(\dot{c})\ddot{c}^k - \frac{\dot{c}^m}{\sigma(c)}\frac{\partial\sigma}{\partial x^m}(c) \\
&= \frac{1}{2}g^{ij}(\dot{c})\frac{\partial g_{ij}}{\partial x^k}(\dot{c})\dot{c}^k - 2g^{ij}(\dot{c})C_{ijk}(\dot{c})G^k(\dot{c}) - \frac{\dot{c}^m}{\sigma(c)}\frac{\partial\sigma}{\partial x^m}(c) \\
&= N^m_m(\dot{c}) - \frac{\dot{c}^m}{\sigma(c)}\frac{\partial\sigma}{\partial x^m}(c).
\end{aligned}$$

This gives

$$\mathbf{S}(y) = N^m_m(y) - \frac{y^m}{\sigma(x)}\frac{\partial\sigma}{\partial x^m}(x). \tag{7.34}$$

Thus

$$\mathbf{S}_{y^iy^j}(y)u^iv^j = \frac{\partial N^m_m}{\partial y^i\partial y^j}(y)u^iv^j = \frac{\partial^3 G^m}{\partial y^i\partial y^j\partial y^m} = 2E_{ij}(y)u^iv^j.$$

We obtain a local formula for the E-curvature

$$\mathbf{E}_y(u,v) = \frac{1}{2}\mathbf{S}_{y^iy^j}(y)u^iv^j = \frac{1}{2}\frac{\partial^3 G^m}{\partial y^i\partial y^j\partial y^m}(y)u^iv^j, \tag{7.35}$$

where $u = u^i\frac{\partial}{\partial x^i}|_x, v = v^j\frac{\partial}{\partial x^j}|_x \in T_xM$.

From (7.35), we see that the E-curvature is independent of the volume form. It is purely a geometric quantity of Finsler metrics.

By (7.30), we see that $\mathbf{S}(y)$ is linear in $y \in T_xM$ if and only if $\mathbf{E} = 0$ on $T_xM \setminus \{0\}$. In particular, if F is a Berwald metric, then $\mathbf{S}(y)$ is linear in $y \in T_xM$ for all x [Sh2]. In fact, $\mathbf{S} = 0$ for Berwald metrics if we consider the S-curvature of the Busemann-Hausdorff measure Vol_F. More precisely, we have

Proposition 7.3.1 *For any Berwald space (M, F), $\mathbf{S} = 0$ with respect to the Busemann-Hausdorff volume form dV_F.*

Proof. Let $c : (-\varepsilon, \varepsilon) \to M$ be an arbitrary geodesic. Take an arbitrary parallel frame $\{\mathbf{b}_i(t)\}_{i=1}^n$ along c. Let $\{\theta^i(t)\}_{i=1}^n$ denote the dual coframe along c. By (2.7), the Busemann-Hausdorff volume form dV_F of F is given by

$$dV_F|_{c(t)} = \sigma_F(t)\theta^1(t) \wedge \cdots \wedge \theta^n(t),$$

where

$$\sigma_F(t) := \frac{\omega_n}{\mathrm{Vol}\Big\{(y^i) \in \mathrm{R}^n \ \Big|\ F(y^i\mathbf{b}_i(t)) < 1\Big\}}.$$

By (7.28), the S-curvature of F in the direction $\dot{c}(t)$ is given by

$$\mathbf{S}(\dot{c}(t)) = \frac{d}{dt}\Big[\ln\Big(\frac{\sqrt{\det\big[\mathbf{g}_{\dot{c}(t)}(\mathbf{b}_i(t), \mathbf{b}_j(t))\big]}}{\sigma_F(t)}\Big)\Big].$$

By Lemmas 5.3.1 and 5.3.2,

$$\mathbf{g}_{\dot{c}(t)}\Big(\mathbf{b}_i(t), \mathbf{b}_j(t)\Big) = \mathbf{g}_{\dot{c}(0)}\Big(\mathbf{b}_i(0), \mathbf{b}_j(0)\Big) = constant,$$

$$F\Big(y^i\mathbf{b}_i(t)\Big) = F\Big(y^i\mathbf{b}_i(0)\Big) = constant.$$

Thus

$$\det\Big[\mathbf{g}_{\dot{c}(t)}\Big(\mathbf{b}_i(t), \mathbf{b}_j(t)\Big)\Big] = \det\Big[\mathbf{g}_y\Big(\mathbf{b}_i, \mathbf{b}_j\Big)\Big],$$

$$\mathrm{Vol}\Big\{(y^i) \in \mathrm{R}^n \ \Big|\ F\Big(y^i\mathbf{b}_i(t)\Big) < 1\Big\} = \mathrm{Vol}\Big\{(y^i) \in \mathrm{R}^n, \ \Big|\ F\Big(y^i\mathbf{b}_i\Big) < 1\Big\}.$$

The last equality implies that $\sigma_F(t) = constant$. Thus $\mathbf{S}(\dot{c}(t)) = 0$. Since c is arbitrary, we conclude that $\mathbf{S} = 0$. Q.E.D.

Now we use (7.34) to prove the following

Example 7.3.1 Let $F = \alpha + \beta$ be a Randers metric on a manifold M, where $\alpha(y) = \sqrt{a_{ij}(x)y^iy^j}$ and $\beta(y) = b_i(x)y^i$ with $\|\beta\| := \sup_{\alpha(y)=1}\beta(y) < 1$. WE will find a sufficient and necessary condition on α and β for $\mathbf{S} = 0$. In particular, we will show that if β is a Killing form of constant length, then $\mathbf{S} = 0$.

In a standard local coordinate system (x^i, y^i) in TM, define $b_{i|j}$ by

$$b_{i|j}\theta^j := db_i - b_j\theta_i^{\ j},$$

where $\theta^i := dx^i$ and $\theta_i^{\ j} := \bar{\Gamma}^j_{ik} dx^k$ denote the Levi-Civita connection forms of α. Let

$$r_{ij} := \frac{1}{2}\Big(b_{i|j} + b_{j|i}\Big), \qquad s_{ij} := \frac{1}{2}\Big(b_{i|j} - b_{j|i}\Big).$$

The geodesic coefficients G^i of F are related to the coefficients $\bar{G}^i$ of α by

$$G^i = \bar{G}^i + Py^i + Q^i,$$

where

$$\begin{aligned} P(y): &= \frac{1}{2F(y)}\Big\{r_{ij}y^iy^j - 2\alpha(y)b_r a^{rp}s_{pl}y^l\Big\} \\ Q^i(y): &= \alpha a^{ir}s_{rl}y^l. \end{aligned}$$

Observe that

$$\frac{\partial Q^i}{\partial y^i} = \alpha^{-1}a_{ij}y^j a^{ir}s_{rl}y^l + \alpha a^{ir}s_{ri} = \alpha^{-1}s_{jl}y^jy^l + \alpha a^{ir}s_{ri} = 0.$$

Thus

$$N^i_i = \bar{N}^i_i + (n+1)P.$$

Put

$$dV_F = \sigma(x)dx^1\cdots dx^n, \qquad dV_\alpha = \sigma_\alpha(x)dx^1\cdots dx^n.$$

We have

$$\sigma_F = \Big(1 - \|\beta\|^2\Big)^{\frac{n+1}{2}}\sigma_\alpha.$$

Note that

$$d\Big[\ln\sigma_\alpha\Big] = \bar{N}^i_i.$$

By (7.34), we obtain a formula for $\mathbf{S}$,

$$\mathbf{S}(y) = (n+1)\Big\{P(y) - d\Big[\ln\sqrt{1-\|\beta\|^2}\Big](y)\Big\}, \tag{7.36}$$

where

$$P(y) := \frac{1}{2F(y)}\Big(r_{ij}y^iy^j - 2\alpha(y)b_r a^{rp}s_{pl}y^l\Big).$$

Now we are going to find a sufficient and necessary condition on β under which $\mathbf{S} = 0$. For the sake of simplicity, we choose an orthonormal basis for T_xM such that $a_{ij} = \delta_{ij}$. Let

$$A_l := \Big[\ln\sqrt{1-\|\beta\|^2}\Big]_{x^l} = -\frac{b_k b_{k|l}}{1-\|\beta\|^2}.$$

Observe that $\mathbf{S} = 0$ if and only if

$$r_{ij}y^iy^j - 2\alpha b_p s_{pl} y^l = 2(\alpha+\beta)A_l y^l.$$

This is equivalent to the following equations

$$r_{ij} = b_j A_i + b_i A_j \tag{7.37}$$

$$b_p s_{pl} + A_l = 0. \tag{7.38}$$

We claim that (7.37) implies (7.38). First, contracting (7.37) with b_i and b_j yields

$$b_i b_{i|j} b_j = -2\|\beta\|^2 \frac{b_i b_{i|j} b_j}{1-\|\beta\|^2}.$$

Thus

$$b_i A_i = -\frac{b_i b_{i|j} b_j}{1-\|\beta\|^2} = 0.$$

Using this fact, we obtain

$$\begin{aligned} b_i b_{j|i} &= 2b_i r_{ij} - b_i b_{i|j} \\ &= 2\|\beta\|^2 A_j + (1-\|\beta\|^2)A_j = (1+\|\beta\|^2)A_j. \end{aligned}$$

Finally, we obtain

$$\begin{aligned} b_p s_{pl} &= \frac{1}{2}(b_p b_{p|l} - b_p b_{l|p}) \\ &= \frac{1}{2}\Big\{-(1-\|\beta\|^2)A_l - (1-\|\beta\|^2)A_l\Big\} = -A_l. \end{aligned}$$

This gives (7.38). We conclude that $\mathbf{S} = 0$ if and only if (7.37) holds. We can rewrite (7.37) as follows.

$$(1-\|\beta\|^2)\Big(b_{i|j} + b_{j|i}\Big) + 2b_k b_{k|i} b_j + 2b_k b_{k|j} b_i = 0. \tag{7.39}$$

Note that if β is a Killing form ($r_{ij} = 0$) with constant length ($b_i b_{i|j} = 0$), then (7.39) holds. Hence $\mathbf{S} = 0$ and $\mathbf{E} = 0$. ♮

Below is a specific family of Randers metrics on S^3 with $\mathbf{S} = 0$.

Example 7.3.2 Let $\{\zeta^1, \zeta^2, \zeta^3\}$ be the canonical left-invariant co-frame on the Lie group $Sp(1) = S^3$ satisfying

$$d\zeta^1 = 2\zeta^2 \wedge \zeta^3, \quad d\zeta^2 = 2\zeta^3 \wedge \zeta^1, \quad d\zeta^3 = 2\zeta^1 \wedge \zeta^2. \tag{7.40}$$

Consider $F = \alpha + \beta$, where

$$\alpha(y) := \sqrt{\kappa^2[\zeta^1(y)]^2 + \lambda^2[\zeta^2(y)]^2 + \lambda^2[\zeta^3(y)]^2}, \qquad \beta(y) := \varepsilon\, \zeta^1(y),$$

where $\kappa > 0$, $\lambda > 0$ and $\varepsilon > 0$. Assume that

$$\|\beta\|_\alpha = \left|\frac{\varepsilon}{\kappa}\right| < 1.$$

Then F is a special Randers metric. Let

$$\theta^1 := \kappa\zeta^1, \quad \theta^2 := \lambda\zeta^2, \quad \theta^3 := \lambda\zeta^3.$$

$\{\theta^1, \theta^2, \theta^3\}$ is an orthonormal co-frame for TS^3 with respect to α. Put

$$\beta = b_1\theta^1 + b_2\theta^2 + b_3\theta^3,$$

where

$$b_1 = \frac{\varepsilon}{\kappa}, \quad b_2 = 0, \quad b_3 = 0.$$

It follows from (7.40) that

$$d\theta^i = \theta^j \wedge \theta_j{}^i,$$

where $\theta_j{}^i$ are the Levi-Civita connection forms of α, given by $\theta_j{}^i + \theta_i{}^j = 0$ and

$$\theta_2{}^1 = \frac{\kappa}{\lambda^2}\,\theta^3, \quad \theta_3{}^1 = -\frac{\kappa}{\lambda^2}\,\theta^2, \quad \theta_3{}^2 = \left(\frac{2}{\kappa} - \frac{\kappa}{\lambda^2}\right)\theta^1.$$

By definition, $b_{i|j}$ are defined by

$$db_i - b_j\theta_i{}^j =: b_{i|j}\theta^i.$$

A direct computation gives

$$b_{2|3} = -\frac{\varepsilon}{\lambda^2}, \quad b_{3|2} = \frac{\varepsilon}{\lambda^2}$$

and all other components $b_{i|j} = 0$. Thus β is a Killing form of constant length. By Example 7.3.1, we conclude that $\mathbf{S} = 0$, hence $\mathbf{E} = 0$. ♯

Example 7.3.3 Let F denote the Funk metric on a strongly convex domain Ω in R^n. The geodesic coefficients are given by

$$G^i = \frac{1}{2}F\, y^i.$$

Observe that

$$N^i_i = \frac{1}{2}(Fy^i)_{y^i} = \frac{1}{2}\Big(F_{y^i}y^i + F\delta^i_i\Big) = \frac{n+1}{2}F.$$

According to Example 2.2.4, the Busemann-Hausdorff volume form $dV_F = \sigma(x)dx^1 \cdots dx^n$ has constant coefficient,

$$\sigma_F(x) = constant.$$

By (7.34), we obtain

$$\mathbf{S}(y) = \frac{n+1}{2}F(y). \tag{7.41}$$

The angular form $\mathbf{h}_y(u,v) := h_{ij}(y)u^iv^j$ on T_xM is given by

$$h_{ij} := F\, F_{y^iy^j} = g_{ij} - g_{is}\frac{y^s}{F(y)}g_{jt}\frac{y^t}{F(y)}.$$

$\mathbf{h} = \{\mathbf{h}_y\}$ is called the *angular metric*. By (7.35), we obtain

$$E_{ij} = \frac{1}{2}\mathbf{S}_{y^iy^j} = \frac{n+1}{4}F_{y^iy^j}.$$

In index-free form,

$$\mathbf{E}_y(u,v) = \frac{n+1}{4F(y)}\mathbf{h}_y(u,v). \tag{7.42}$$

♮

Chapter 8

Structure Equations

In previous chapters, we introduced the Riemann curvature and many non-Riemannian quantities. We discovered some relationship among these quantities. In this chapter, we are going to use the exterior differential methods to find more relationships among these quantities. At the end, we compute the Riemann curvature of a special class of Randers metrics.

8.1 Structure Equations of Finsler Spaces

In this section, we first introduce the curvature forms by differentiating the Chern connection forms. Then we derive some identities for the curvature coefficients by differentiating the fundamental tensor.

Let (M, F) be a Finsler space. In Section 5.2, we introduced the fundamental tensor g, the Cartan tensor $\mathcal{C}$ and the Landsberg tensor $\mathcal{L}$ on π^*TM. See (5.12), (5.13) and (5.32) for definitions.

Let $\{e_i\}_{i=1}^n$ be an arbitrary local frame for π^*TM and $\{\omega^i, \omega^{n+i}\}_{i=1}^n$ the corresponding local coframe for $T^*(TM \setminus \{0\})$. See (5.18) for the correspondence. Let

$$(y, y) = y^i e_i.$$

and

$$g_{ij} := g(e_i, e_j), \qquad C_{ijk} := \mathcal{C}(e_i, e_j, e_k), \qquad L_{ijk} := \mathcal{L}(e_i, e_j, e_k).$$

y^i, g_{ij}, C_{ijk} and L_{ijk} are local functions on TM. By Theorem 5.2.2, there is a set of 1-forms $\{\omega_j{}^i\}_{i,j=1}^n$ on $TM \setminus \{0\}$. The Chern connection ∇ on

π^*TM is expressed by

$$\nabla U = \left\{ dU^i + U^j \omega_j^{\ i} \right\} \otimes e_i,$$

where $U = U^i e_i \in C^\infty(\pi^*TM)$.

The Chern connection is uniquely determined by (5.21) and (5.22). For convenience, we rewrite (5.21) and (5.22) again.

$$d\omega^i = \omega^j \wedge \omega_j^{\ i} \tag{8.1}$$

$$dg_{ij} = g_{kj}\omega_i^{\ k} + g_{ik}\omega_j^{\ k} + 2C_{ijk}\omega^{n+k}. \tag{8.2}$$

Let

$$\Omega_j^{\ i} := d\omega_j^{\ i} - \omega_j^{\ k} \wedge \omega_k^{\ i}. \tag{8.3}$$

$\{\Omega_j^{\ i}\}_{ij=1}^n$ is a set of local 2-forms on $TM \setminus \{0\}$. Thus we can express $\Omega_j^{\ i}$ in the following form

$$\Omega_j^{\ i} = \frac{1}{2} R_j^{\ i}{}_{kl}\omega^k \wedge \omega^l + P_j^{\ i}{}_{kl}\omega^k \wedge \omega^{n+l} + \frac{1}{2} Q_j^{\ i}{}_{kl}\omega^{n+k} \wedge \omega^{n+l}, \tag{8.4}$$

where

$$R_j^{\ i}{}_{kl} + R_j^{\ i}{}_{lk} = 0, \tag{8.5}$$

$$Q_j^{\ i}{}_{kl} + Q_j^{\ i}{}_{lk} = 0.$$

Differentiating (8.1), we obtain

$$\omega^j \wedge \Omega_j^{\ i} = 0.$$

This gives

$$R_j^{\ i}{}_{kl} + R_k^{\ i}{}_{lj} + R_l^{\ i}{}_{jk} = 0, \tag{8.6}$$

$$P_j^{\ i}{}_{kl} = P_k^{\ i}{}_{jl}, \tag{8.7}$$

$$Q_j^{\ i}{}_{kl} = 0.$$

Equation (8.4) simplifies to

$$\Omega_j^{\ i} = \frac{1}{2} R_j^{\ i}{}_{kl}\omega^k \wedge \omega^l + P_j^{\ i}{}_{kl}\omega^k \wedge \omega^{n+l}. \tag{8.8}$$

Define

$$\mathcal{R}(U,V)W := R_j{}^i{}_{kl}(y)U^kV^lW^je_i,$$
$$\mathcal{P}(U,V)W := P_j{}^i{}_{kl}(y)U^kV^lW^j\, e_i,$$

where $U = U^ke_k, V = V^le_l, W = W^je_j \in \pi^*TM$. We call $\mathcal{R}$ and $\mathcal{P}$ the *Riemannian curvature tensor* and the *Chern curvature tensor* respectively.

Let

$$\Omega^i := d\omega^{n+i} - \omega^{n+j}\wedge\omega_j{}^i. \tag{8.9}$$

According to Theorem 5.2.2, ω^{n+i} are given by

$$\omega^{n+i} = dy^i + y^j\omega_j{}^i. \tag{8.10}$$

Differentiating (8.10) yields

$$\Omega^i := y^j\Omega_j{}^i.$$

By (8.8), Ω^i can be expressed by

$$\Omega^i = \frac{1}{2}R^i{}_{kl}\omega^k\wedge\omega^l + P^i{}_{kl}\omega^k\wedge\omega^{n+l},$$

where

$$R^i{}_{kl} := y^jR_j{}^i{}_{kl}, \qquad P^i{}_{kl} := y^jP_j{}^i{}_{kl}.$$

Put

$$R^i{}_k := R^i{}_{kl}y^l = y^jR_j{}^i{}_{kl}y^l. \tag{8.11}$$

We claim that $R^i{}_k$ are just the coefficients of the Riemann curvature in (6.4) and $P_j{}^i{}_{kl}$ are just the coefficients of the Chern curvature in (7.13). Hence $P^i{}_{kl} = -g^{ij}L_{jkl}$ are just the coefficients of the Landsberg curvature in (7.12).

Let (x^i, y^i) be a standard local coordinate system in TM. Take

$$\omega^i = dx^i, \qquad \omega^{n+i} = \delta y^i.$$

Then by (5.25),

$$\omega_j{}^i = \Gamma^i_{jk}\, dx^k = \left(\frac{\partial^2 G^i}{\partial y^j \partial y^k} - L^i{}_{jk}\right) dx^k.$$

Plugging them into (8.3), we obtain

$$\begin{aligned} R_j{}^i{}_{kl} &= \frac{\partial \Gamma^i_{jl}}{\partial x^k} - \frac{\partial \Gamma^i_{jk}}{\partial x^l} + \frac{\partial \Gamma^i_{jk}}{\partial y^m} N^m_l - \frac{\partial \Gamma^i_{jl}}{\partial y^m} N^m_k \\ &\quad + \Gamma^m_{jl}\Gamma^i_{mk} - \Gamma^m_{jk}\Gamma^i_{ml} \qquad (8.12) \\ P_j{}^i{}_{kl} &= -\frac{\partial \Gamma^i_{jk}}{\partial y^l} = -\frac{\partial^2 G^i}{\partial y^j \partial y^k \partial y^l} + \frac{\partial L^i{}_{jk}}{\partial y^l}. \qquad (8.13) \end{aligned}$$

Equation (8.13) is just (7.13). By the homogeneity of F, we obtain

$$\begin{aligned} R^i{}_k &= 2\frac{\partial G^i}{\partial x^k} - y^j \frac{\partial^2 G^i}{\partial x^j \partial y^k} + 2G^j \frac{\partial^2 G^i}{\partial y^j \partial y^k} - \frac{\partial G^i}{\partial y^j}\frac{\partial G^j}{\partial y^k}, \qquad (8.14) \\ P^i{}_{kl} &= -L^i{}_{kl}. \qquad (8.15) \end{aligned}$$

Equation (8.11) is just (6.4). Thus $R^i{}_k$ defined in this section are the coefficients of the Riemann curvature.

Now we go back to the general setting. Define $C_{ijk|l}$ and C_{ijkl} by

$$dC_{ijk} - C_{ljk}\omega_i{}^l - C_{ilk}\omega_j{}^l - C_{ijl}\omega_k{}^l =: C_{ijk|l}\omega^l + C_{ijkl}\omega^{n+l}.$$

In virtue of (5.24), we have

$$L_{ijk} = C_{ijk|l}\, y^l. \qquad (8.16)$$

Differentiating (8.2) and using (8.9), we obtain

$$g_{pj}\Omega_i{}^p + g_{ip}\Omega_j{}^p + 2C_{ijp}\Omega^p + 2\Big(C_{ijl|k}\omega^k + C_{ijl\cdot k}\omega^{n+k}\Big)\wedge\omega^{n+l} = 0. \quad (8.17)$$

From (8.17), we obtain

$$g_{pj}R_i{}^p{}_{kl} + g_{ip}R_j{}^p{}_{kl} + 2C_{ijp}R^{\ p}_{\ kl} = 0, \qquad (8.18)$$

$$g_{pj}P_i{}^p{}_{kl} + g_{ip}P_j^p{}_{kl} + 2C_{ijp}P^{\ p}_{\ kl} + 2C_{ijl|k} = 0, \qquad (8.19)$$

$$C_{ijl\cdot k} = C_{ijk\cdot l}. \qquad (8.20)$$

We first derive some important identities for $R_j{}^i{}_{kl}$. Let

$$R_j{}^i{}_k := R_j{}^i{}_{kl} y^l.$$

Contracting (8.18) with y^j yields

$$y^j g_{pj} R_i{}^p{}_{kl} = -g_{ip} R^p{}_{kl}. \tag{8.21}$$

By (8.21), we obtain

$$y^j g_{jp} R^p{}_{kl} = -y^i g_{ip} R^p{}_{kl}.$$

This gives

$$y^i g_{ip} R^p{}_{kl} = 0. \tag{8.22}$$

Contracting (8.18) with y^l yields

$$g_{pj} R_i{}^p{}_k + g_{ip} R_j{}^p{}_k + 2C_{ijp} R^p{}_k = 0. \tag{8.23}$$

Contracting (8.21) with y^l yields

$$g_{ip} R^p{}_k = -y^j g_{pj} R_i{}^p{}_k. \tag{8.24}$$

It follows from (8.5) and (8.6) that

$$R_i{}^p{}_k = R_i{}^p{}_{kl} y^l = -R_k{}^p{}_{li} y^l - R_l{}^p{}_{ik} y^l = R_k{}^p{}_i - R^p{}_{ik}.$$

It follows from (8.22) and (8.24) that

$$\begin{aligned} g_{ip} R^p{}_k &= -y^j g_{pj} R_i{}^p{}_k \\ &= y^j g_{pj} \left(R^p{}_{ik} - R_k{}^p{}_i \right) \\ &= -y^j g_{pj} R_k{}^p{}_i = g_{kp} R^p{}_i. \end{aligned}$$

We obtain

$$g_{ip} R^p{}_k = g_{kp} R^p{}_i. \tag{8.25}$$

Now we derive some important identities for $P_j{}^i{}_{kl}$. Applying (8.19) three times to the combination

$$(g_{pj} P_i{}^p{}_{kl} + g_{ip} P_j{}^p{}_{kl}) - (g_{pk} P_j{}^p{}_{il} + g_{pj} P_k{}^p{}_{il}) + (g_{pi} P_k{}^p{}_{jl} + g_{pk} P_i{}^p{}_{jl}),$$

we obtain

$$\begin{aligned} g_{ip}P_{j\ kl}^{\ p} &= -C_{ijp}P^{p}_{\ kl} - C_{jkp}P^{p}_{\ il} - C_{ikp}P^{p}_{\ jl} \\ & \quad -C_{ijl|k} - C_{jkl|i} - C_{ikl|j}. \end{aligned} \tag{8.26}$$

By (8.26), we can easily obtain the following

$$L_{ikl} = -g_{ip}P^{p}_{\ kl}, \tag{3.27}$$

$$L_{ikl} = y^j g_{jp}P_{i\ kl}^{\ p}, \tag{8.28}$$

and

$$y^j g_{jp}P^{p}_{\ kl} = L_{ikl}y^i = 0. \tag{8.29}$$

8.2 Structure Equations of Riemannian Metrics

In this section, we will deal with Riemannian metrics in a traditional way. All the quantities are defined on the base manifold. At the end, we view Riemannian metrics as special Finsler metrics and deal with them as in the previous section. We will give a link between these two different approaches.

Let (M, g) be an n-dimensional Riemannian space. In a local coordinate system (x^i) in M, $g = g_{ij}(x)dx^i \otimes dx^j$ and the Christoffel symbols

$$\Gamma^i_{jk}(x) := \frac{\partial^2 G^i}{\partial y^j \partial y^k}(y) = \frac{1}{2}g^{il}(x)\Big\{\frac{\partial g_{lk}}{\partial x^j}(x) + \frac{\partial g_{jl}}{\partial x^k}(x) - \frac{\partial g_{jk}}{\partial x^l}(x)\Big\}$$

are functions of x only, so that

$$N^i_j(y) = \Gamma^i_{jk}(x)y^k, \qquad y = y^k\frac{\partial}{\partial x^k}|_x.$$

The Levi-Civita connection is given by

$$\mathrm{D}_y U = \Big\{dU^i(y) + U^j\Gamma^i_{jk}y^k\Big\}\frac{\partial}{\partial x^i},$$

where $U = U^i \frac{\partial}{\partial x^i} \in C^\infty(TM)$ and $y = y^k \frac{\partial}{\partial x^k}|_x$. The Levi-Civita connection is uniquely determined by

$$\mathrm{D}_U V - \mathrm{D}_V U = [U, V], \tag{8.30}$$

$$W[g(U, V)] = g(\mathrm{D}_W U, V) + g(U, \mathrm{D}_W V). \tag{8.31}$$

where $U, V, W \in C^\infty(TM)$.

Let $\{\mathbf{b}_i\}_{i=1}^n$ be a local frame for TM and $\{\theta^i\}_{i=1}^n$ the dual coframe for T^*M. Express the Levi-Civita connection D by

$$\mathrm{D}\mathbf{b}_j = \theta_j^{\ i} \otimes \mathbf{b}_i.$$

The set of 2-forms $\{\theta_j^{\ i}\}$ are called the Levi-Civita connection forms. Let $g_{ij} := g(\mathbf{b}_i, \mathbf{b}_j)$. Equations (8.30) and (8.31) are equivalent to

$$d\theta^i = \theta^j \wedge \theta_j^{\ i}, \tag{8.32}$$

$$dg_{ij} = g_{kj}\theta_i^{\ k} + g_{ik}\theta_j^{\ k}. \tag{8.33}$$

Set

$$\Theta_j^{\ i} = d\theta_j^{\ i} - \theta_j^{\ k} \wedge \theta_k^{\ i} = \frac{1}{2} R_j^{\ i}{}_{kl}\theta^k \wedge \theta^l, \tag{8.34}$$

where

$$R_j^{\ i}{}_{kl} + R_j^{\ i}{}_{lk} = 0. \tag{8.35}$$

Differentiating (8.32) and (8.33) yields

$$\theta^j \wedge \Theta_j^{\ i} = 0,$$

and

$$g_{kj}\Theta_i^{\ k} + g_{ik}\Theta_j^{\ k} = 0.$$

These two identities imply

$$R_j^{\ i}{}_{kl} + R_k^{\ i}{}_{lj} + R_l^{\ i}{}_{jk} = 0, \tag{8.36}$$

$$g_{kj}R_i^{\ k}{}_{kl} + g_{ik}R_j^{\ k}{}_{kl} = 0. \tag{8.37}$$

Define

$$\mathrm{R}(u,v)w := R_j{}^i{}_{kl} u^k v^l w^j \, \mathbf{b}_i,$$

where $u = u^k \mathbf{b}_k, v = v^l \mathbf{b}_l$ and $w = w^j \mathbf{b}_j$. We call $\mathrm{R}(u,v)w$ the *Riemann curvature tensor*. From the definition, we have

$$\mathrm{R}(U,V)W = \mathrm{D}_U \mathrm{D}_V W - \mathrm{D}_V \mathrm{D}_U W - \mathrm{D}_{[U,V]} W,$$

where $U, V, W \in C^\infty$.

Equations (8.35), (8.36) and (8.37) can be written

$$\mathrm{R}(u,v)w + \mathrm{R}(v,u)w = 0, \tag{8.38}$$

$$\mathrm{R}(u,v)w + \mathrm{R}(v,w)u + \mathrm{R}(w,u)v = 0, \tag{8.39}$$

$$g\Big(\mathrm{R}(u,v)w, z\Big) + g\Big(w, \mathrm{R}(u,v)z\Big) = 0. \tag{8.40}$$

It follows from (8.38) and (8.40) that

$$g\Big(\mathrm{R}(u,y)y, u\Big) = -g\Big(y, \mathrm{R}(u,y)u\Big) = g\Big(y, \mathrm{R}(y,u)u\Big) = g\Big(\mathrm{R}(y,u)u, y\Big).$$

This gives

$$\frac{g\Big(\mathrm{R}(u,y)y,u\Big)}{g(u,u)g(y,y) - g(y,u)g(y,u)} = \frac{g\Big(\mathrm{R}(y,u)u,y\Big)}{g(u,u)g(y,y) - g(y,u)g(y,u)}.$$

Thus the flag curvature $\mathbf{K}(P,y)$ is independent of $y \in P$ for any tangent plane $P \subset T_xM$. In this case, we denote by $\mathbf{K}(P) := \mathbf{K}(P,y)$. We call $\mathbf{K}(P)$ the *sectional curvature* of the section $P \subset T_xM$.

In a local coordinate system (x^i) in M, $\theta_j{}^i = \Gamma^i_{jk} dx^k$. Plugging them into (8.34) yields

$$R_j{}^i{}_{kl} = \Big(\frac{\partial \Gamma^i_{jl}}{\partial x^k} + \Gamma^m_{jl} \Gamma^i_{mk}\Big) - \Big(\frac{\partial \Gamma^i_{jk}}{\partial x^l} + \Gamma^m_{jk} \Gamma^i_{ml}\Big).$$

Using $G^i = \frac{1}{2}\Gamma^i_{jk} y^j y^k$, we can express $R_j{}^i{}_{kl} y^j y^l$ as follows

$$R_j{}^i{}_{kl} y^j y^l = 2\frac{\partial G^i}{\partial x^k} - y^j \frac{\partial^2 G^i}{\partial x^j \partial y^k} + 2G^j \frac{\partial^2 G^i}{\partial y^j \partial y^k} - \frac{\partial G^i}{\partial y^j}\frac{\partial G^j}{\partial y^k}. \tag{8.41}$$

Comparing (8.41) with the coefficients $R^i{}_k(y)$ of the Riemann curvature in (6.4), we obtain

$$R^i{}_k(y) = R_j{}^i{}_{kl} y^j y^l, \tag{8.42}$$

that is,

$$\mathbf{R}_y(u) = \mathrm{R}(u, y)y. \qquad y, u \in T_xM.$$

Observe that

$$\begin{aligned}
\frac{\partial^2 R^i{}_k}{\partial y^j \partial y^l} - \frac{\partial^2 R^i{}_l}{\partial y^j \partial y^k} &= \frac{\partial}{\partial y^j}\Big\{ R_l{}^i{}_{kq} y^q + R_p{}^i{}_{kl} y^p - R_k{}^i{}_{lq} y^q - R_p{}^i{}_{lk} y^p \Big\} \\
&= R_l{}^i{}_{kj} + R_j{}^i{}_{kl} - R_k{}^i{}_{lj} - R_j{}^i{}_{lk} \\
&= 3R_j{}^i{}_{kl}.
\end{aligned}$$

We obtain

$$R_j{}^i{}_{kl} = \frac{1}{3}\Big\{ \frac{\partial^2 R^i{}_k}{\partial y^j \partial y^l} - \frac{\partial^2 R^i{}_l}{\partial y^j \partial y^k} \Big\}. \tag{8.43}$$

We view a Riemann metric as a special Finsler metric. A natural question arises: What are the relationship between the Riemannian curvature tensor on the base manifold and that on the slit tangent bundle?

We lift $\{\mathbf{b}_i\}_{i=1}^n$ to a local frame $\{e_i\}_{i=1}^n$ for π^*TM, where

$$e_i|_y := (y, \mathbf{b}_i).$$

Let y^i be defined by

$$(y, y) = y^i e_i.$$

Let

$$\omega^i := \pi^*\theta^i, \qquad \omega^{n+i} := dy^i + y^j \pi^*\theta_j{}^i.$$

Then

$$\omega_j{}^i := \pi^*\theta_j{}^i$$

are the Chern connection forms of F with respect to $\{e_i\}_{i=1}^n$. We have

$$\begin{aligned}
\Omega_j{}^i : &= d\omega_j{}^i - \omega_j{}^k \wedge \omega_k{}^i \\
&= \pi^*\Big(d\theta_j{}^i - \theta_j{}^k \wedge \theta_k{}^i \Big)
\end{aligned}$$

$$= \pi^*\left(\frac{1}{2}R_j{}^i{}_{kl}\theta^j \wedge \theta^l\right)$$

$$= \frac{1}{2}R_j{}^i{}_{kl}\omega^k \wedge \omega^l.$$

Thus the coefficients of the Riemannian curvature tensor are functions of $x \in M$ only and the Chern curvature vanishes. However, if we take an arbitrary local frame $\{e_i\}_{i=1}^n$ for π^*TM, then the coefficients of the Riemannian metric curvature tensor might depend on $y \in T_xM$.

8.3 Riemann Curvature of Randers Metrics

In general, it is much more difficult to compute the Riemann curvature. In what follows, we are going to give a formula for the Riemann curvature of a Randers metric $F = \alpha + \beta$, where β is a Killing form with constant length.

Let $F = \alpha + \beta$ be a Randers metric on a manifold M. In a standard local coordinate system (x^i, y^i) in TM, α and β are expressed by

$$\alpha(y) = \sqrt{a_{ij}(x)y^iy^j} \qquad \beta(y) = b_i(x)y^i.$$

Define $b_{i|j}$ by

$$db_i - b_j\theta_i{}^j =: b_{i|j}\theta^j,$$

where $\theta^i := dx^i$ and $\theta_i{}^j := \bar{\Gamma}^j_{ik}dx^k$ denote the Levi-Civita connection forms. Let

$$r_{ij} := \frac{1}{2}\left(b_{i|j} + b_{j|i}\right), \qquad s_{ij} := \frac{1}{2}\left(b_{i|j} - b_{j|i}\right).$$

The geodesic coefficients G^i of F are related to the coefficients $\bar{G}^i$ of α by

$$G^i = \bar{G}^i + Py^i + Q^i,$$

where

$$P(y) : = \frac{1}{2F(y)}\left\{r_{ij}y^iy^j - 2\alpha(y)b_ra^{rp}s_{pl}y^l\right\}$$
$$Q^i(y) : = \alpha a^{ir}s_{rl}y^l.$$

From now on, we assume that β is a Killing form ($r_{ij} = 0$) with constant length ($b_i b_{i|j} = 0$), that is, β satisfies

$$r_{ij} = \frac{1}{2}\Big(b_{i|j} + b_{j|i}\Big) = 0, \qquad b_i a^{ij} b_{j|k} = 0. \tag{8.44}$$

Equations (8.44) implies that

$$b_r a^{rp} s_{pl} = b_r a^{rp} b_{p|l} = 0.$$

Thus $P = 0$ and G^i's are simplified to

$$G^i = \bar{G}^i + Q^i \tag{8.45}$$

with

$$Q^i = \alpha a^{ir} b_{r|l} y^l.$$

Note that Q^i's define a tensor $Q = Q^i \partial_i$ on $TM \setminus \{0\}$.

Let $\bar{R}^i{}_k$ and $R^i{}_k$ denote the coefficients of the Riemann curvature of α and F respectively. Plugging (8.45) into (6.4) yields

$$R^i{}_k = \bar{R}^i{}_k + \Big\{2Q^i{}_{|k} - y^j (Q^i{}_{|j})_{y^k} - (Q^i)_{y^j} (Q^j)_{y^k} + 2Q^j (Q^i)_{y^j y^k}\Big\}, \tag{8.46}$$

where $Q^i{}_{|j}$ denote the covariant derivatives of Q^i with respect to α. See (5.38). By Lemma 5.2.1), we know that $\alpha_{|j} = 0$. According to (5.39), we have

$$Q^i{}_{|j} = \alpha\Big(a^{ir} b_{r|l} y^l\Big)_{|j} = \alpha a^{ir} b_{r|l|j} y^l.$$

For the sake of simplicity, we take a local orthonormal frame $\{\mathbf{b}_i\}_{i=1}^n$ on M with respect to α so that

$$\alpha(y) = \sqrt{\sum_{i=1}^n (y^i)^i}, \qquad \beta(y) = \sum_{i=1}^n b_i y^i, \qquad y = y^i \mathbf{b}_i.$$

The local frame $\{\mathbf{b}_i\}_{i=1}^n$ determines a local frame $\{e_i = (y, \mathbf{b}_i)\}_{i=1}^n$ for $\pi^* TM$ and the corresponding local coframe $\{\omega^i, \omega^{n+i}\}_{i=1}^n$ for $T^*(TM \setminus \{0\})$. The formula (8.46) still holds the coefficients of the Riemann curvature with respect to $\{e_i\}_{i=1}^n$, although it is stated originally in a standard local coordinate system in TM. We have

$$Q^i = \alpha b_{i|p} y^p$$

$$Q^i{}_{|k} = \alpha b_{i|p|k} y^p$$
$$y^j (Q^i{}_{|j})_{y^k} = y^j (\alpha b_{i|p|j} y^p)_{y^k} = \alpha^{-1} (b_{i|p|j} y^j y^p)\, y^k + \alpha b_{i|k|p} y^p$$
$$(Q^i)_{y^j} = \alpha^{-1} (b_{i|p} y^p)\, y^j + \alpha b_{i|j}$$
$$(Q^i)_{y^j y^k} = \alpha^{-1}(b_{i|p}y^p)\,\delta_{jk} - \alpha^{-3}(b_{i|p}y^p)\, y^j y^k + \alpha^{-1} b_{i|j}\, y^k + \alpha^{-1} b_{i|k}\, y^j$$

Plugging them into (8.46) yields

$$\begin{aligned} R^i{}_k \quad = \quad & \bar{R}^i{}_k + 2\alpha b_{i|p|k} y^p - \alpha b_{i|k|p} y^p - \alpha^{-1} (b_{i|p|q} y^p y^q)\, y^k \\ & -\alpha^2 b_{i|m} b_{m|k} + 3(b_{i|p} y^p)\,(b_{k|q} y^q) + (b_{i|m} b_{m|p} y^p)\, y^k. \end{aligned} \tag{8.47}$$

We have the following Ricci identity for the covariant derivatives of β,

$$b_{i|j|k} - b_{i|k|j} = b_m \bar{R}_i{}^m{}_{jk}.$$

Using the Ricci identity, we obtain

$$b_{i|p|k} y^p = \Big(b_{i|k|p} + b_m \bar{R}_i{}^m{}_{pk} \Big) y^p = b_{i|k|p} y^p - b_m \bar{R}_i{}^m{}_{kp} y^p,$$

$$\begin{aligned} b_{i|p|q} y^p y^q \quad & = \quad -b_{p|i|q} y^p y^q \\ & = \quad -\Big(b_{p|q|i} y^p y^q + b_m \bar{R}_p{}^m{}_{iq} \Big) y^p y^q \\ & = \quad -b_m \bar{R}^m_i. \end{aligned}$$

Plugging them into (8.47), we obtain

$$\begin{aligned} R^i{}_k \quad = \quad & \bar{R}^i{}_k + \alpha b_{i|k|p} y^p - 2\alpha b_m \bar{R}_i{}^m{}_{kp} y^p + \alpha^{-1} b_m \bar{R}^m_i\, y^k \\ & -\alpha^2 b_{i|m} b_{m|k} + 3(b_{i|p} y^p)(b_{k|q} y^q) + (b_{i|m} b_{m|p} y^p)\, y^k. \end{aligned} \tag{8.48}$$

Let R and $\bar{R}$ denote the Ricci scalar of F and α respectively. Since α is a Riemannian metric,

$$\bar{R}_i{}^m{}_{ip} = -\bar{R}_m{}^i{}_{ip} = -\frac{n-1}{2} \bar{R}_{y^m y^p}.$$

From (8.48), we immediately obtain an equation for the Ricci curvature.

$$R = \bar{R} + \alpha b_m \bar{R}_{y^m y^p} y^p + \frac{1}{n-1} \Big\{ \alpha^2 (b_{m|p})^2 + 2(b_{m|p} y^p)^2 \Big\}. \tag{8.49}$$

Chapter 9

Finsler Spaces of Constant Curvature

As we know, every Riemannian metric of constant curvature λ is locally isometric to a canonical Riemannian metric of constant curvature λ. However, for Finsler metrics, this is no longer true. For each λ, there are infinitely many non-isometric Finsler metrics of constant curvature λ. In this chapter, we will discuss some basic properties and well-known examples of Finsler metrics with constant curvature.

9.1 Finsler Metrics of Constant Curvature

The purpose of this section is to derive some equations for Finsler metrics of constant curvature λ. We will show that Landsberg metrics of non-zero constant curvature must be Riemannian.

Let F be a Finsler metric on a manifold M. Differentiating (8.1) yields the following Bianchi identities:

$$d\Omega_j{}^i = -\Omega_j{}^k \wedge \omega_k{}^i + \omega_j{}^k \wedge \Omega_k{}^i.$$

By (8.27), we obtain

$$R_j{}^i{}_{kl\cdot m} = P_j{}^i{}_{km|l} - P_j{}^i{}_{lm|k} + P_j{}^i{}_{kt}L^t{}_{lm} - P_j{}^i{}_{lt}L^t{}_{km}. \tag{9.1}$$

Contracting (9.1) with y^j yields

$$R_m^{\ i}{}_{kl} = R^i{}_{kl\cdot m} + L^i{}_{km|l} - L^i{}_{lm|k} + L^i{}_{kt}L^t{}_{lm} - L^i{}_{lt}L^t{}_{km}. \tag{9.2}$$

Contracting (9.2) with y^l yields

$$R_m^{\ i}{}_{kl}y^l = R^i{}_{k\cdot m} - R^i{}_{km} + L^i{}_{km|l}y^l. \tag{9.3}$$

By (8.6) and (9.3), we obtain

$$\begin{aligned} R^i{}_{km} &= R_m{}^i{}_{kl} y^l - R_k{}^i{}_{ml} y^l \\ &= R^i{}_{k\cdot m} - R^i{}_{m\cdot k} - 2R^i{}_{km}. \end{aligned}$$

This gives

$$R^i{}_{kl} = \frac{1}{3}\Big(R^i{}_{k\cdot l} - R^i{}_{l\cdot k}\Big). \tag{9.4}$$

Theorem 9.1.1 *Let (M, F) be a Finsler space of constant curvature $\lambda \neq 0$. Suppose that $\mathbf{J} = 0$. Then F is Riemannian.*

Proof. By assumption,

$$R^i{}_k = \lambda\Big(F^2 \delta^i_k - g_{kp} y^p y^i\Big). \tag{9.5}$$

Plugging (9.5) into (9.4) yields

$$R^i{}_{kl} = \lambda\Big\{g_{jl}\delta^i_k - g_{jk}\delta^i_l\Big\}. \tag{9.6}$$

Plugging (9.6) into (9.2), we obtain

$$\begin{aligned} R_j{}^i{}_{kl} &= \lambda\Big(g_{jl}\delta^i_k - g_{jk}\delta^i_l\Big) + L^i{}_{jk|l} - L^i{}_{jl|k} \\ &\quad + L^i{}_{kt} L^t{}_{jl} - L^i{}_{lt} L^t{}_{jk}. \end{aligned} \tag{9.7}$$

Using (9.7), we can rewrite (9.1) as follows

$$\begin{aligned} 2\lambda\Big(C_{jlm}\delta^i_k - C_{jkm}\delta^i_l\Big) &= P_j{}^i{}_{km|l} - P_j{}^i{}_{lm|k} + P_j{}^i{}_{kt} L^t{}_{lm} - P_j{}^i{}_{lt} L^t{}_{km} \\ &\quad + L^i{}_{lt\cdot m} L^t{}_{jk} - L^i{}_{kt\cdot m} L^t{}_{jl} + L^i{}_{lt} L^t{}_{jk\cdot m} - L^i{}_{kt} L^t{}_{jl\cdot m} \\ &\quad + L^i{}_{jl|k\cdot m} - L^i{}_{jk|l\cdot m}. \end{aligned} \tag{9.8}$$

Thus, for Finsler metrics of constant curvature λ, the Cartan torsion, the Chern curvature and the Landsberg curvature satisfy (9.8).

Contracting (9.8) with $y^s g_{is}$ yields

$$\lambda C_{jlm} g_{ks} y^s - \lambda C_{jkm} g_{ls} y^s = L_{jkm|l} - L_{jlm|k}. \tag{9.9}$$

Then contracting (9.9) with y^l, we obtain

$$L_{jkm|l} y^l + \lambda F^2 C_{jkm} = 0. \tag{9.10}$$

Contracting (9.10) with g^{jm} yields

$$J_{k|l}y^l + \lambda F^2 I_k = 0. \tag{9.11}$$

Thus if $J_k = 0$, then $I_k = 0$. By Deike's theorem, we conclude that F is Riemannian.

Q.E.D.

Theorem 9.1.1 improves a result by S. Numata in [Nu] where he that $\mathbf{L} = 0$ instead.

Let $c(t)$ be a unit speed geodesic in a Finsler space (M, F). Let $V = V(t)$ be an arbitrary parallel vector field along c. Consider

$$\mathbf{C}(t) := \mathbf{C}_{\dot c(t)}(V(t), V(t), V(t)), \qquad \mathbf{L}(t) := \mathbf{L}_{\dot c(t)}(V(t), V(t), V(t)).$$

$$\mathbf{I}(t) := \mathbf{I}_{\dot c(t)}(V(t)), \qquad \mathbf{J}(t) := \mathbf{J}_{\dot c(t)}(V(t)).$$

It follows from (7.16) and (7.24) that

$$\mathbf{L}(t) = \mathbf{C}'(t), \quad \mathbf{J}(t) = \mathbf{I}'(t).$$

By (9.10) and (9.11), we obtain the following ODE:

$$\mathbf{L}'(t) + \lambda \mathbf{C}(t) = 0$$
$$\mathbf{J}'(t) + \lambda \mathbf{I}(t) = 0.$$

Thus $\mathbf{C}(t)$ and $\mathbf{I}(t)$ satisfy

$$\mathbf{C}''(t) + \lambda \mathbf{C}(t) = 0 \tag{9.12}$$
$$\mathbf{I}''(t) + \lambda \mathbf{I}(t) = 0. \tag{9.13}$$

The general solutions of (9.12) and (9.13) are given by

$$\mathbf{C}(t) \;=\; \mathbf{s}_\lambda(t)\mathbf{L}(0) + \mathbf{s}'_\lambda(t)\mathbf{C}(0), \tag{9.14}$$
$$\mathbf{I}(t) \;=\; \mathbf{s}_\lambda(t)\mathbf{J}(0) + \mathbf{s}'_\lambda(t)\mathbf{I}(0), \tag{9.15}$$

where $\mathbf{s}_\lambda(t)$ is the unique solution of the following equation

$$y''(t) + \lambda y(t) = 0, \quad y(0) = 0, \quad y'(0) = 1.$$

Assume that F is complete, i.e., every geodesic is defined on the whole line $(-\infty, \infty)$. Let us take a look at the following cases:

Case 1: $\lambda = -1$. In this case, (9.14) gives

$$\mathbf{C}(t) = \sinh(t)\mathbf{L}(0) + \cosh(t)\mathbf{C}(0).$$

Assume that $\mathbf{C}$ is bounded. Then

$$\mathbf{L}(0) = 0 = \mathbf{C}(0).$$

Since c is arbitrary, we conclude that $\mathbf{C} = 0$ and F is Riemannian.

By a similar argument using (9.15), we can show that if $\mathbf{I}$ is bounded, then $\mathbf{I} = 0$. Hence F is Riemannian by Deike's theorem.

Case 2: $\lambda = 0$. In this case, (9.14) gives

$$\mathbf{C}(t) = t\mathbf{L}(0) + \mathbf{C}(0).$$

Assume that $\mathbf{C}$ is bounded. Then

$$\mathbf{L}(0) = 0.$$

Since c is arbitrary, we conclude that $\mathbf{L} = 0$ and F is a Landsberg metric.

By a similar argument using (9.15), we can show that if $\mathbf{I}$ is bounded, then $\mathbf{J} = 0$. Hence F is a weak Landsberg metric.

Theorem 9.1.2 ([AZ]) *Let (M, F) be a complete Finsler space of constant curvature $\mathbf{K} = \lambda$. Assume that the (mean) Cartan torsion is bounded.*

(a) If $\lambda = 0$, then F is a (weak) Landsberg metric;
(b) If $\lambda < 0$, then F is a Riemannian metric.

When M is compact, the Finsler metric is complete with bounded Cartan torsion. Thus the conclusions in Theorem 9.1.2 hold. When (M, F) is compact with vanishing flag curvature. In this case, all geometric quantities are bounded. By further argument, we can actually prove that the Chern curvature vanishes. It is known that every Finsler metric with vanishing Chern curvature and flag curvature must be locally Minkowskian [BCS1]. We conclude that every Finsler metric on a compact manifold with vanishing flag curvature must be locally Minkowskian. See also [BCS2][Pa] for some rigidity results on Finsler surfaces.

The geometric and topological structures of Finsler spaces of positive constant curvature will be discussed in Section 18.3 below.

9.2 Examples

In this section, we will discuss several important examples of Finsler metrics of constant curvature. These examples show that the classification problem seems to be not solved within my life time.

Example 9.2.1 Let F be the Funk metric on a strongly convex domain $\Omega \subset \mathrm{R}^n$. According to Example 7.2.1, the geodesic coefficients G^i of F are given by

$$G^i(y) = \frac{1}{2}F(y)y^i. \tag{9.16}$$

Thus geodesics are straight lines in Ω and F is projectively flat. Plugging (9.16) into (6.4) and using (7.19), we obtain

$$R^i{}_k = -\frac{1}{4}\Big\{F^2(y)\,\delta^i_k - g_{kl}(y)y^l\,y^i\Big\}.$$

In index-free form,

$$\mathbf{R}_y(u) = -\frac{1}{4}\Big\{\mathbf{g}_y(y,y)\,u - \mathbf{g}_y(y,u)\,y\Big\}, \qquad u \in T_xM.$$

Thus for any tangent plane $P = \mathrm{span}\{y,u\} \subset T_xM$, the flag curvature $\mathbf{K}$ of F satisfies

$$\mathbf{K}(P,y) = -\frac{1}{4}.$$

Now we take a look at a special case when $\Omega = \mathbb{B}^n$ is the unit ball in $\mathbb{R}^n$. From the definition, the Funk metric F is determined by

$$\left| x + \frac{y}{F(y)} \right| = 1.$$

We obtain

$$F(y) = \frac{\sqrt{|y|^2 - (|x|^2|y|^2 - \langle x,y\rangle^2)} + \langle x,y\rangle}{1-|x|^2}. \tag{9.17}$$

♯

Example 9.2.2 Let F denote the Funk metric on a strongly convex domain $\Omega \subset R^n$. The Klein metric $\tilde{F}$ on Ω is defined by

$$\tilde{F}(y) := \frac{1}{2}\Big\{F(y) + F(-y)\Big\}. \tag{9.18}$$

By (7.19), one can easily prove that $\tilde{F}$ satisfies the following equations

$$\tilde{F}_{x^k y^i} y^k = \tilde{F}_{x^i}$$

and the geodesic coefficients $\tilde{G}^i$ of $\tilde{F}$ are given by

$$\tilde{G}^i(y) = \frac{1}{2}\Big\{F(y) - \tilde{F}(-y)\Big\}\, y^i. \tag{9.19}$$

Thus geodesics are straight lines in Ω and $\tilde{F}$ is projectively flat. Plugging (9.16) into (6.4) and using (9.19), we obtain

$$R^i{}_k = -\Big\{\tilde{F}^2(y)\,\delta^i_k - \tilde{g}_{kl}(y)y^l\, y^i\Big\}.$$

In index-free form,

$$\tilde{\mathbf{R}}_y(u) = -\Big\{\tilde{\mathbf{g}}_y(y,y)\, u - \tilde{\mathbf{g}}_y(y,u)\, y\Big\}, \qquad u \in T_xM.$$

Thus for $P = \mathrm{span}\{y, u\}$, the flag curvature $\tilde{\mathbf{K}}$ of $\tilde{F}$ satisfies

$$\tilde{\mathbf{K}}(P, y) = -1.$$

By (9.17), we obtain the Klein metric $\tilde{F}(y)$ on the unit ball $\mathbb{B}^n \subset \mathbb{R}^n$:

$$\tilde{F}(y) = \frac{\sqrt{|y|^2 - (|x|^2|y|^2 - \langle x, y\rangle^2)}}{1 - |x|^2}.$$

$\sharp$

Example 9.2.3 ([Sh8]) Minkowski spaces are the most trivial Finsler spaces with $\mathbf{K} = 0$. We may construct non-trivial Finsler metrics with $\mathbf{K} = 0$ as follows.

Let F denote the Funk metric on a strongly convex domain Ω in R^n. Assume that a Finsler metric $\hat{F}$ on an open subset $\mathcal{U} \subset \Omega$ satisfies the following system

$$\hat{F}_{x^k} = (F\hat{F})_{y^k}. \tag{9.20}$$

Observe that

$$\hat{F}_{x^k y^i} y^k = (\hat{F}F)_{y^k y^i} y^k = (\hat{F}F)_{y^i} = \hat{F}_{x^i}.$$

This implies that

$$\frac{\hat{F}_{x^k} y^k}{2\hat{F}} = \frac{(\hat{F}F)_{y^k} y^k}{2\hat{F}} = \frac{2\hat{F}F}{2\hat{F}} = F.$$

By the above identities and (5.2), we obtain that

$$\hat{G}^i = F(y)\, y^i. \tag{9.21}$$

Thus, the geodesics of $\hat{F}$ are straight lines and $\hat{F}$ is projectively flat. Plugging (9.21) into (6.4) and using (7.19), we obtain that

$$R^i{}_k = 0.$$

Thus the flag curvature $\hat{\mathbf{K}}$ of $\hat{F}$ vanishes. When $\Omega = \mathbb{B}^n$ is the unit ball in $\mathbb{R}^n$, the Funk metric F can be expressed by elementary functions in (9.17). However, no explicit elementary expression for $\hat{F}$ has been found in this case.

♯

Now we give some examples of Finsler metrics on the n-sphere with positive constant curvature.

Example 9.2.4 ([Br1][Br2]) Let V^3 be a three-dimensional real vector space and $V^3 \otimes \mathbb{C}$ its complex vector space. Take a basis $\{\mathbf{b}_1, \mathbf{b}_2, \mathbf{b}_3\}$ for V^3 and define a quadratic Q on $V^3 \times \mathbb{C}$ by

$$Q(\mathbf{u}, \mathbf{v}) = e^{i\alpha} u^1 v^1 + e^{i\beta} u^2 v^2 + e^{-i\alpha} u^3 v^3,$$

where $\mathbf{u} = u^i \mathbf{b}_i, \mathbf{v} = v^i \mathbf{b}_i$ and $\alpha, \beta \in \mathrm{R}$. For $X \in V^3 \setminus \{0\}$, let $[X] := \{tX,\ t > 0\}$. Then $S^2 := \{[X],\ X \in V^3 \setminus \{0\}\}$ is diffeomorphic to the standard unit sphere $\mathbb{S}^2$ in the Euclidean space $\mathbb{R}^3$. For a vector $Y \in V^3$, denote by $[X, Y] \in T_{[X]}S^2$ the tangent vector to the curve $c(t) := [X + tY]$ at $t = 0$. Define $F : TS^2 \to \mathrm{R}$ by

$$F([X, Y]) := \mathcal{R}\Big[\sqrt{\frac{Q(X,X)Q(Y,Y) - Q(X,Y)^2}{Q(X,X)^2}} - i\frac{Q(X,Y)}{Q(X,X)}\Big], \tag{9.22}$$

where $\mathcal{R}[\cdot]$ denotes the real part of a complex number. Clearly, F is well-defined. Assume that $|\beta| \le \alpha < \frac{\pi}{2}$. Then F is a Finsler metric on S^2. Bryant has verified that F has constant curvature $\mathbf{K} = 1$.

Now we shall give an explicit expression for a special class of Bryant metrics on S^2 with $\beta = \alpha$. Take an arbitrary vector $\mathbf{y} \in T_{\mathbf{x}}R^2$, let

$$X := (\mathbf{x}, 1) \in R^3, \qquad Y := (\mathbf{y}, 0) \in R^3.$$

We have

$$\begin{aligned} Q(X,X) &= e^{i\alpha}|\mathbf{x}|^2 + e^{-i\alpha} \\ Q(X,Y) &= e^{i\alpha}\langle \mathbf{x}, \mathbf{y}\rangle \\ Q(Y,Y) &= e^{i\alpha}|\mathbf{y}|^2 \end{aligned}$$

Hence

$$Q(X,X)Q(Y,Y) - Q(X,Y)^2 = e^{2\alpha\, i}(|\mathbf{x}|^2|\mathbf{y}|^2 - \langle \mathbf{x}, \mathbf{y}\rangle^2) + |\mathbf{y}|^2.$$

The real part of $-i\, Q(X,Y)/Q(X,X)$ is

$$\mathcal{R}\Big[-i\,\frac{Q(X,Y)}{Q(X,X)}\Big] = \frac{\sin(2\alpha)\,\langle \mathbf{x}, \mathbf{y}\rangle}{|\mathbf{x}|^4 + 2\cos(2\alpha)|\mathbf{x}|^2 + 1}.$$

For a complex number z, the real part of $\sqrt{z}$ is given by

$$\mathcal{R}(\sqrt{z}) = \sqrt{\frac{\mathcal{R}(z) + |z|}{2}}.$$

For

$$Z = \frac{Q(\mathbf{x},\mathbf{x})Q(\mathbf{y},\mathbf{y}) - Q(\mathbf{x},\mathbf{y})^2}{Q(\mathbf{x},\mathbf{x})^2} = \frac{e^{2\alpha\, i}(|\mathbf{x}|^2|\mathbf{y}|^2 - \langle \mathbf{x}, \mathbf{y}\rangle^2) + |\mathbf{y}|^2}{2^{2\alpha\, i}|\mathbf{x}|^4 + 2|\mathbf{x}|^2 + e^{-2\alpha\, i}},$$

$$\begin{aligned} |Z| &= \frac{\sqrt{(|\mathbf{x}|^2|\mathbf{y}|^2 - \langle \mathbf{x}, \mathbf{y}\rangle^2)^2 + 2\cos(2\alpha)(|\mathbf{x}|^2|\mathbf{y}|^2 - \langle \mathbf{x}, \mathbf{y}\rangle^2)|\mathbf{y}|^2 + |\mathbf{y}|^4}}{|\mathbf{x}|^4 + 2\cos(2\alpha)|\mathbf{x}|^2 + 1} \\ \mathcal{R}(Z) &= \frac{|\mathbf{x}|^2|\mathbf{y}|^2 - \langle \mathbf{x}, \mathbf{y}\rangle^2 + \cos(2\alpha)|\mathbf{y}|^2}{|\mathbf{x}|^4 + 2\cos(2\alpha)|\mathbf{x}|^2 + 1} + 2\Big(\frac{\sin(2\alpha)\langle \mathbf{x}, \mathbf{y}\rangle}{|\mathbf{x}|^4 + 2\cos(2\alpha)|\mathbf{x}|^2 + 1}\Big)^2. \end{aligned}$$

We obtain

$$F([X,Y]) = \sqrt{\frac{\mathcal{R}(Z) + |Z|}{2}} + \frac{\sin(2\alpha)\langle \mathbf{x}, \mathbf{y}\rangle}{|\mathbf{x}|^4 + 2\cos(2\alpha)|\mathbf{x}|^2 + 1}.$$

Define

$$F_\alpha(y) := F([X,Y]), \qquad \mathbf{y} \in T_{\mathbf{x}}R^2,$$

where $X = (\mathbf{x}, 1), Y = (\mathbf{y}, 0) \in \mathrm{R}^3$. We obtain a Finsler metric F_α on R^2. This is the pull-back of the Finsler metric F on S^2 by

$$\psi(\mathbf{x}) := \Big(\frac{\mathbf{x}}{\sqrt{1+|\mathbf{x}|^2}}, \frac{1}{\sqrt{1+|\mathbf{x}|^2}}\Big).$$

Clearly, we can generalize F_α to a Finsler metric defined on R^n without any modification. One can show that the generalized metric F_α on R^n also has constant curvature $\mathbf{K} = 1$. Note that F_0 is just the spherical Riemannian metric on R^n,

$$F_0(\mathbf{y}) = \frac{\sqrt{|\mathbf{y}|^2 + (|\mathbf{x}|^2|\mathbf{y}|^2 - \langle \mathbf{x}, \mathbf{y}\rangle^2)}}{1+|\mathbf{x}|^2}. \qquad (9.23)$$

All F_α are pointwise projective to the Euclidean metric. ♯

9.3 Randers Metrics of Constant Curvature

The Funk metric F in (9.17) is a Randers metric. In fact the Funk metric on any shifted ellipsoid $\Omega \subset \mathrm{R}^n$ is a Randers metric. We would like to find more Randers metrics of constant curvature. According to Section 8.3, the Riemann curvature a Randers metric $F = \alpha + \beta$ with a Killing form β of constant length takes a relatively simple form (8.47). In this section we will investigate this type of Randers metrics.

Let $F = \alpha + \beta$ be a Randers metric on M, where β is a Killing form of constant length. At a point, take a local orthonormal frame $\{\mathbf{b}_i\}_{i=1}^n$ on M so that

$$\alpha(y) = \sqrt{\sum_{i=1}^n (y^i)^2}, \qquad \beta(y) = b_i y^i, \qquad y = y^i \mathbf{b}_i.$$

By assumption,

$$b_{i|j} + b_{j|i} = 0, \qquad b_i b_{i|j} = 0. \qquad (9.24)$$

Suppose that F is of constant curvature λ, i.e, the coefficients of Riemann curvature with respect to $\{\mathbf{b}_i\}_{i=1}^n$ are in the following form:

$$R^i{}_k = \lambda F^2\Big(\delta_{ik} - \frac{y^i}{F}F_{y^k}\Big)$$

$$\begin{aligned} &= \lambda\Big\{(\alpha^2+\beta^2)\delta_{ik} - y^i y^k - \beta b_k y^i\Big\} \\ &\quad +\lambda\alpha\Big(2\beta\delta_{ik} - b_k y^i\Big) - \lambda\alpha^{-1}\beta y^i y^k. \end{aligned}$$

Compare it with (8.47), we obtain

$$\begin{aligned} \bar{R}^i{}_k &= \lambda\Big\{(\alpha^2+\beta^2)\delta_{ik} - y^i y^k - \beta b_k y^i\Big\} \\ &\quad +\alpha^2 b_{i|m}b_{m|k} - 3(b_{i|p}y^p)(b_{k|q}y^q) - (b_{i|m}b_{m|p}y^p)y^k \end{aligned} \tag{9.25}$$

$$\begin{aligned} b_m\bar{R}^m_i y^k \quad &+ \quad \alpha^2\Big(b_{i|k|p}y^p - 2b_m\bar{R}_i{}^m{}_{kp}y^p\Big) \\ &= -\lambda\beta y^i y^k + \lambda\alpha^2\Big(2\beta\delta_{ik} - b_k y^i\Big). \end{aligned} \tag{9.26}$$

According to (8.42) and (8.43), $\bar{R}^i_k$ and $\bar{R}_j{}^i{}_{kl}$ are related by

$$\bar{R}^i{}_k = \bar{R}_j{}^i{}_{kl}y^j y^l, \qquad \bar{R}_j{}^i{}_{kl} = \frac{1}{3}\Big\{\frac{\partial^2\bar{R}^i{}_k}{\partial y^j\partial y^l} - \frac{\partial^2\bar{R}^i{}_l}{\partial y^j\partial y^k}\Big\}.$$

Differentiating $\bar{R}^i{}_k$ in (9.25) with respect to y^j and y^l, we obtain

$$\begin{aligned} \bar{R}_j{}^i{}_{kl} &= \lambda\Big(\delta_{ik}\delta_{jl} - \delta_{il}\delta_{jk}\Big) + \lambda\Big(b_j b_l\delta_{ik} - b_j b_k\delta_{il}\Big) \\ &\quad + b_{i|p}b_{p|k}\delta_{jl} - b_{i|p}b_{p|l}\delta_{jk} + b_{i|l}b_{j|k} - b_{i|k}b_{j|l} - 2b_{k|l}b_{i|j} \end{aligned} \tag{9.27}$$

Contracting (9.27) with b_i and y^l, and using (9.24), we obtain

$$b_m\bar{R}_j{}^m{}_{kl}y^l = \lambda\Big(b_k y^j - \beta\delta_{jk}\Big) \qquad b_m\bar{R}^m_k = \lambda\Big(\alpha^2 b_k - \beta y^k\Big).$$

Plugging them into (9.26) yields

$$b_{i|k|p} = \lambda\Big(b_k\delta_{ip} - b_i\delta_{kp}\Big). \tag{9.28}$$

By taking the Hessian of $\frac{1}{2}\|\beta\|^2 = \frac{1}{2}(b_i)^2$,

$$\begin{aligned} \frac{1}{2}\Big[(b_i)^2\Big]_{|k|l} &= b_{i|k}b_{i|l} + b_i b_{i|k|l} \\ &= b_{i|k}b_{i|l} + \lambda b_i\Big(b_k\delta_{il} - b_i\delta_{kl}\Big) \\ &= b_{i|k}b_{i|l} + \lambda\Big(b_k b_l - \|\beta\|^2\delta_{kl}\Big). \end{aligned}$$

we obtain

$$b_{p|k}b_{p|l} = \lambda\Big(\|\beta\|^2\delta_{kl} - b_k b_l\Big). \tag{9.29}$$

From (9.29), we obtain

$$(b_{p|k})^2 = \lambda(n-1)\|\beta\|^2. \tag{9.30}$$

Thus $\lambda \geq 0$. If $\lambda = 0$, then β is parallel and $\bar{R}^i{}_k = 0$.

Plugging (9.29) into (9.25) yields

$$\begin{aligned}\bar{R}^i{}_k &= \lambda\Big\{\Big[(1-\|\beta\|^2)\alpha^2 + \beta^2\Big]\delta_{ik} + \alpha^2 b_i b_k - (1-\|\beta\|^2)y^i y^k \\ &\quad -\beta(b_k y^i + b_i y^k)\Big\} - 3(b_{i|p}y^p)(b_{k|q}y^q).\end{aligned} \tag{9.31}$$

Conversely, one can easily verify that (9.28) and (9.31) imply (9.26). Thus under the assumption (9.24), F is of constant curvature λ if and only if (9.28) and (9.31) are satisfied.

Proposition 9.3.1 *Let $F = \alpha + \beta$ be a Randers metric where β is a Killing form of constant length. F is of constant flag curvature λ if and only if α and β satisfy (9.29) and (9.31). In this case, either $\lambda > 0$ or $\lambda = 0$ (hence α is flat and β is parallel).*

H. Yasuda and H. Shimada [YaSh] proved a better result than Proposition 9.3.1. They proved that a Randers metric $F = \alpha + \beta$ is of constant flag curvature $\lambda > 0$ if and only if

(a) β is a Killing form of constant length with respect to α,
(b) α and β satisfy (9.29) and (9.31).

However, their proof is much more complicated, so it is omitted here. See [Ma3] for another proof of Yasuda-Shimada's theorem.

Recently, D. Bao and Z. Shen have found a specific family of Randers metrics on S^3 with constant flag curvature $\mathbf{K} = 1$. The metrics on S^3 are described in the following example.

Example 9.3.1 ([BaSh2]) We view S^3 as a compact Lie group. Let $\zeta^1, \zeta^2, \zeta^3$ be the standard right invariant 1-forms on S^3 satisfying

$$d\zeta^1 = 2\zeta^2\wedge\zeta^3, \qquad d\zeta^2 = 2\zeta^3\wedge\zeta^1, \qquad d\zeta^3 = 2\zeta^1\wedge\zeta^2.$$

The metric

$$\alpha_1(y) := \sqrt{[\zeta^1(y)]^2 + [\zeta^2(y)]^2 + [\zeta^3(y)]^2}$$

is the standard Riemannian metric of constant curvature 1. For $k \geq 1$, define

$$\alpha_k(y) := \sqrt{k^2[\zeta^1(y)]^2 + k[\zeta^2(y)]^2 + k[\zeta^3(y)]^2}$$

and

$$\beta_k(y) := \sqrt{k^2 - k}\, \zeta^1.$$

Then

$$F_k := \alpha_k + \beta_k \tag{9.32}$$

is a Finsler metric on S^3. Bao-Shen show that F_k is of constant curvature $\mathbf{K} = 1$ for any $k \geq 1$. We give a brief argument below which is different from that in [BaSh2].

First, we show that β_k is a Killing form of constant length. Take an orthonormal coframe for T^*S^3,

$$\theta^1 := k\zeta^1, \quad \theta^2 := \sqrt{k}\zeta^2, \quad \theta^3 = \sqrt{k}\zeta^3,$$

so that

$$\begin{aligned}\alpha_k(y) &= \sqrt{[\theta^1(y)]^2 + [\theta^2(y)]^2 + [\theta^3(y)]^2},\\ \beta_k(y) &= b_1\theta^1(y) + b_2\theta^2(y) + b_3\theta^3(y),\end{aligned}$$

where

$$b_1 = \frac{\sqrt{k^2 - k}}{k}, \quad b_2 = 0, \quad b_3 = 0.$$

Clearly, β_k has constant length with respect to α_k, that is,

$$\|\beta_k\|_{\alpha_k} = \frac{\sqrt{k^2 - k}}{k}.$$

The Levi-Civita connection forms $\theta_j{}^i$ are given by $\theta_j{}^i + \theta_i{}^j = 0$ and

$$\theta_2{}^1 = \theta^3, \quad \theta_3{}^1 = -\theta^2, \quad \theta_3{}^2 = \Big(\frac{2}{k} - 1\Big)\theta^1.$$

By definition, $b_{i|j}$ are defined by

$$db_i - b_j\theta_j^{\ i} =: b_{i|j}\theta^j.$$

A direct computation gives

$$b_{1|1} = 0, \quad b_{1|2} = 0, \quad b_{1|3} = 0,$$

$$b_{2|1} = 0, \quad b_{2|2} = 0, \quad b_{2|3} = -\frac{\sqrt{k^2 - k}}{k},$$

$$b_{3|1} = 0, \quad b_{3|2} = \frac{\sqrt{k^2 - k}}{k}, \quad b_{3|3} = 0.$$

Thus

$$b_{i|j} + b_{j|i} = 0.$$

We conclude that β_k is a Killing form with respect to α_k.

By definition, $b_{i|j|k}$ are defined by

$$db_{i|j} - b_{k|j}\theta_i^{\ k} - b_{i|k}\theta_j^{\ k} =: b_{i|j|k}\theta^k.$$

A direct computation gives

$$b_{1|2|k} = -\frac{\sqrt{k^2 - k}}{k}\delta_{2k}, \quad b_{1|3|k} = -\frac{\sqrt{k^2 - k}}{k}\delta_{3k}, \quad b_{2|3|k} = 0.$$

Other components are determined by $b_{i|j|k} + b_{j|i|k} = 0$. One can easily verify that α_k and β_k satisfy (9.29) with $\lambda = 1$.

Now we compute the Riemann curvature of α_k. By definition, $\bar{R}_j^{\ i}{}_{kl}$ are defined by

$$d\theta_j^{\ i} - \theta_j^{\ k} \wedge \theta_k^{\ i} = \frac{1}{2}\bar{R}_j^{\ i}{}_{kl}\theta^k \wedge \theta^l,$$

where $\bar{R}_j^{\ i}{}_{kl} + \bar{R}_j^{\ i}{}_{lk} = 0$. A direct computation gives

$$\bar{R}_2^{\ 1}{}_{12} = 1, \quad \bar{R}_2^{\ 1}{}_{13} = 0, \quad \bar{R}_2^{\ 1}{}_{13} = 0,$$

$$\bar{R}_3^{\ 1}{}_{13} = 1, \quad \bar{R}_3^{\ 1}{}_{12} = 0, \quad \bar{R}_3^{\ 1}{}_{23} = 0,$$

$$\bar{R}_3^{\ 2}{}_{23} = \frac{4}{k} - 3, \quad \bar{R}_3^{\ 2}{}_{13} = 0, \quad \bar{R}_3^{\ 2}{}_{12} = 0.$$

Other components of $\bar{R}_j{}^i{}_{kl}$ are determined by

$$\bar{R}_j{}^i{}_{kl} + \bar{R}_i{}^j{}_{kl} = 0, \qquad \bar{R}_j{}^i{}_{kl} + \bar{R}_k{}^i{}_{lj} + \bar{R}_l{}^i{}_{jk} = 0.$$

One can easily verify that α_k and β_k satisfy (9.31) with $\lambda = 1$.

By Proposition 9.3.1, we conclude that $F_k = \alpha_k + \beta_k$ has constant flag curvature $\mathbf{K} = 1$. According to Example 7.3.2, F_k satisfies

$$\mathbf{S} = 0, \qquad \mathbf{E} = 0.$$

The volume form of α_k is given by

$$dV_{\alpha_k} = k^2 \zeta^1 \wedge \zeta^2 \wedge \zeta^3 = k^2 dV_{\alpha_1}.$$

By (2.10), the volume forms of F_k and α_k are related by

$$dV_{F_k} = \Big(1 - \|\beta_k\|^2_{\alpha_k}\Big)^2 dV_{\alpha_k} = \Big(1 - \Big(1 - \frac{1}{k}\Big)\Big)^2 k^2 dV_{\alpha_1} = dV_{\alpha_1}.$$

This gives

$$\mathrm{Vol}_{F_k}(\mathrm{S}^3) = \mathrm{Vol}_{\alpha_1}(\mathrm{S}^3) = \mathrm{Vol}(\mathbb{S}^3).$$

That is, the volume of (S^3, F_k) is a constant. This implies that for any point $p \in \mathrm{S}^3$, there is a unique point $q \in \mathrm{S}^3$ with $d(p,q) = \pi$ such that all geodesics $c(t)$, $0 \le t < \infty$, with $c(0) = p$ is minimizing on $[0, \pi]$ with $c(t) = q$. The details will be given in Section 18.3 later.

Let (x, y, z, u, v, w) be the standard coordinate system in $T\mathrm{R}^3 = \mathrm{R}^3 \times \mathrm{R}^3$. Let $\psi_\pm : \mathrm{R}^3 \to \mathrm{S}^3_\pm$ denote the diffeomorphisms defined by

$$\psi_\varepsilon(x, y, z) := \frac{1}{\sqrt{x^2 + y^2 + z^2 + 1}} \Big(x, y, z, \varepsilon\Big), \qquad (x, y, z) \in \mathrm{R}^3,$$

where $\varepsilon = \pm 1$. We can express F_κ on R^3. Pulling back ζ^i by ψ_ε onto R^3, we obtain

$$\begin{aligned}
\zeta^1 &= \frac{-\varepsilon dx - z dy + y dz}{x^2 + y^2 + z^2 + 1} \\
\zeta^2 &= \frac{z dx - \varepsilon dy - x dz}{x^2 + y^2 + z^2 + 1} \\
\zeta^3 &= \frac{-y dx + x dy - \varepsilon dz}{x^2 + y^2 + z^2 + 1}.
\end{aligned}$$

Plugging them into α_κ and β_κ, we obtain

$$\begin{aligned}\alpha_k &= \frac{\sqrt{k^2(\varepsilon u + zv - yw)^2 + k(zu - \varepsilon v - xw)^2 + k(yu - xv + \varepsilon w)^2}}{1 + x^2 + y^2 + z^2}\\ \beta_k &= \sqrt{k^2 - k}\,\frac{-\varepsilon u - zv + yw}{1 + x^2 + y^2 + z^2}.\end{aligned}$$

One can extend above construction to odd-dimensional spheres S^{2n+1}, because of the special fibration structure $S^{2n+1} \to \mathbb{C}P^n = S^{2n+1}/S^1$. ♯

A Finsler metric F on an n-dimensional manifold M is of constant Ricci curvature λ if

$$\mathbf{Ric}(y) = (n-1)\lambda\, F^2(y), \qquad y \in TM.$$

Finsler metrics of constant Ricci curvature are also called *Einstein metrics.*

Let us take a close look at Randers metrics of constant Ricci curvature λ. Let $F = \alpha + \beta$ be a Randers metric. We still assume that β is a Killing form of constant length, so that the Ricci curvature of F is in a relatively simple form (8.49). Take a local orthonormal frame $\{\mathbf{b}_i\}_{i=1}^n$ for TM so that

$$\alpha(y) = \sqrt{\sum_{i=1}^n (y^i)^2}, \qquad \beta(y) = b_i y^i, \qquad y = y^i \mathbf{b}_i \in T_xM.$$

By Assumption,

$$b_{i|j} + b_{j|i} = 0, \qquad b_i b_{i|j} = 0, \qquad b_i b_{j|i} = 0.$$

According to (8.49), the Ricci scalar R and $\bar{R}$ of F and α are related by

$$R = \bar{R} + \alpha b_m \bar{R}_{y^m y^p} y^p + \frac{1}{n-1}\Big\{\alpha^2 (b_{m|p})^2 + 2(b_{m|p} y^p)^2\Big\}. \tag{9.33}$$

Assume that F is of constant Ricci curvature λ. Then

$$\lambda(\alpha+\beta)^2 = \bar{R} + \alpha b_m \bar{R}_{y^m y^p} y^p + \frac{1}{n-1}\Big\{\alpha^2 (b_{m|p})^2 + 2(b_{m|p} y^p)^2\Big\}.$$

We obtain

$$b_m \bar{R}_{y^m y^p} y^p = 2\lambda\beta, \tag{9.34}$$

$$\bar{R} + \frac{1}{n-1}\Big\{\alpha^2 (b_{m|p})^2 + 2(b_{m|p}y^p)^2\Big\} = \lambda(\alpha^2 + \beta^2). \tag{9.35}$$

Equation (9.35) is equivalent to the following

$$\bar{R}_{y^i y^j} + \frac{2}{n-1}\Big\{\delta_{ij}(b_{m|p})^2 + b_{m|i}b_{m|j}\Big\} = 2\lambda\Big(\delta_{ij} + b_i b_j\Big). \tag{9.36}$$

Contracting (9.36) with b_i and b_j and using (9.34), we obtain

$$(b_{m|p})^2 = (n-1)\lambda\|\beta\|^2. \tag{9.37}$$

From (9.37), we see that $\lambda \geq 0$. Plugging (9.37) into (9.35) or (9.36), we obtain

$$\bar{R} = \lambda\Big\{(1 - \|\beta\|^2)\alpha^2 + \beta^2\Big\} - \frac{2}{n-1}(b_{m|p}y^p)^2. \tag{9.38}$$

or equivalently,

$$\bar{R}_{y^i y^j} = 2\lambda\Big\{(1 - \|\beta\|^2)\delta_{ij} + b_i b_j\Big\} - \frac{2}{n-1} b_{m|i} b_{m|j}. \tag{9.39}$$

Note that if $\lambda = 0$, then $b_{m|p} = 0$ by (9.37) and $\bar{R} = 0$ by (9.38). Thus α is Ricci-flat and β is parallel.

Conversely, assume that the Ricci scalar $\bar{R}$ of α satisfies (9.38), one can easily show that (9.34) and (9.35) hold.

We have proved the following

Proposition 9.3.2 *Let $F = \alpha + \beta$ be a Randers metric where β is a Killing form of constant length. F is of constant Ricci curvature λ if and only if the Ricci scalar $\bar{R}$ of α satisfies (9.38). In this case, either $\lambda > 0$ or $\lambda = 0$ (hence α is Ricci-flat and β is parallel).*

Chapter 10

Second Variation Formula

In previous chapters, we have introduced several important curvatures for Finsler metrics. We have also discussed some relationships among these quantities. In this chapter, we are going to introduce another non-Riemannian quantity — the T-curvature. This quantity has a close relationship with the distance function.

10.1 T-Curvature

Let (M, F) be a Finsler space. Given a vector $y \in T_xM \setminus \{0\}$. Extend y to a geodesic field Y in a neighborhood $\mathcal{U}$ of x. Let $\hat{\mathrm{D}}$ denote the Levi-Civita connection of the induced Riemannian metric $\hat{g} = \mathbf{g}_Y$ on $\mathcal{U}$. Define

$$\mathbf{T}_y(v) := \mathbf{g}_y\Big(\mathrm{D}_v V, y\Big) - \hat{g}_x\Big(\hat{\mathrm{D}}_v V, y\Big), \qquad v \in T_xM, \tag{10.1}$$

where V is a vector field with $V_x = v$. $\mathbf{T} = \{\mathbf{T}_y\}_{y \in TM\setminus\{0\}}$ is called the *T-curvature.* $\mathbf{T}$ is also called the *tangent curvature* in [Sh4].

From the definition of $\mathbf{T}$ we have

$$\mathbf{T}_y(\lambda v) = \lambda^2 \mathbf{T}_y(v), \qquad \lambda > 0$$

and

$$\mathbf{T}_y(y) = 0.$$

Now we are going to derive a formula for $\mathbf{T}_y(v)$ in a local coordinate system. First, contracting (5.28) with y^l gives

$$y^l g_{kl}\Gamma^k_{jm} = \frac{1}{2}\Big\{\frac{\partial g_{jl}}{\partial x^m} + \frac{\partial g_{ml}}{\partial x^j} - \frac{\partial g_{jm}}{\partial x^l}\Big\} + 2C_{jkm}G^k. \qquad (10.2)$$

Observe that

$$\frac{\partial \hat{g}_{ij}}{\partial x^k} = \frac{\partial g_{ij}}{\partial x^k}(Y) + 2C_{ijl}(Y)\frac{\partial Y^l}{\partial x^k}.$$

Since $Y = Y^i \frac{\partial}{\partial x^i}$ is geodesic, Y satisfies

$$Y^k \frac{\partial Y^i}{\partial x^k} + 2G^i(Y) = 0. \qquad (10.3)$$

We obtain

$$Y^i \frac{\partial \hat{g}_{ij}}{\partial x^k} = Y^i \frac{\partial g_{ij}}{\partial x^k}(Y), \qquad (10.4)$$

$$Y^k \frac{\partial \hat{g}_{ij}}{\partial x^k} = Y^k \frac{\partial g_{ij}}{\partial x^k}(Y) - 4C_{ijk}(Y)G^k(Y). \qquad (10.5)$$

Let $\hat{\Gamma}^i_{jk}$ denote the Christoffel symbols of $\hat{g}$. By (10.2), (10.4) and (10.5), we obtain

$$\begin{aligned} Y^l \hat{g}_{kl}\hat{\Gamma}^k_{jm} &= \frac{1}{2}\Big\{\frac{\partial \hat{g}_{jl}}{\partial x^m} + \frac{\partial \hat{g}_{ml}}{\partial x^j} - \frac{\partial \hat{g}_{jm}}{\partial x^l}\Big\}Y^l \\ &= \frac{1}{2}\Big\{\frac{\partial g_{jl}}{\partial x^m}(Y) + \frac{\partial g_{ml}}{\partial x^j}(Y) - \frac{\partial g_{jm}}{\partial x^l}(Y)\Big\}Y^l + 2C_{jkm}(Y)G^k(Y) \\ &= Y^l g_{kl}(Y)\Gamma^k_{jm}(Y) \end{aligned}$$

In particular, the following holds

$$y^l \hat{g}_{kl}(x)\hat{\Gamma}^k_{jm}(x) = y^l g_{kl}(y)\Gamma^k_{jm}(y). \qquad (10.6)$$

Thus

$$\begin{aligned} \hat{g}_x\Big(\hat{\mathrm{D}}_v V, y\Big) &= y^l \hat{g}_{kl}(x)\Big\{dV^k(v) + \hat{\Gamma}^k_{jm}(x)v^j v^m\Big\} \\ &= y^l g_{kl}(y)\Big\{dV^k(v) + \Gamma^k_{jm}(y)v^j v^m\Big\}. \end{aligned} \qquad (10.7)$$

On the other hand,

$$\mathbf{g}_y\Big(\mathrm{D}_v V, y\Big) = y^l g_{kl}(y)\Big\{dV^k(v) + \Gamma^k_{jm}(v)v^j v^m\Big\}. \qquad (10.8)$$

From (10.7) and (10.8), we obtain

$$\mathbf{T}_y(v) := y^l g_{kl}(y)\Big\{\Gamma^k_{jm}(v) - \Gamma^k_{jm}(y)\Big\}v^j v^m. \tag{10.9}$$

From (10.9), we see that $\mathbf{T}_y(v)$ is independent of the extensions of y and v.

Proposition 10.1.1 $\mathbf{P} = 0$ *if and only if* $\mathbf{T} = 0$.

Proof: Suppose $\mathbf{P} = 0$. By (7.13), we see that $\Gamma^k_{jm}(y)$ are independent of y. (10.9) immediately implies $\mathbf{T} = 0$.

Suppose that $\mathbf{T} = 0$. In local coordinates,

$$y^l g_{kl}(y)\Big\{\Gamma^k_{jm}(v) - \Gamma^k_{jm}(y)\Big\}v^j v^m = 0. \tag{10.10}$$

This implies that $\Gamma^i_{jk}(v)$ are independent of v. Hence $\mathbf{P} = 0$. The proof is left for the reader. Q.E.D.

The T-curvature is defined for Finsler metrics. There is another interesting quantity which depends only on the spray of the Finsler metric.

Let F b4 a Finsler metric and

$$\mathbf{G} = y^i \frac{\partial}{\partial x^i} - 2G^i(y)\frac{\partial}{\partial y^i}$$

be a spray of F. Let

$$N^i(u,v) := v^j N^i_j(u) - u^j N^i_j(v) = v^j \frac{\partial G^i}{\partial u^j}(u) - u^j \frac{\partial G^i}{\partial v^j}(v).$$

For a pair of vectors $(u,v) \in T_xM \times T_xM$, define

$$\mathbf{N}(u,v) := N^i(u,v)\frac{\partial}{\partial x^i}|_x.$$

One can verify that $\mathbf{N} : T_xM \times T_xM \to T_xM$ is a well-defined map. Further, it is anti-symmetric,

$$\mathbf{N}(u,v) + \mathbf{N}(v,u) = 0,$$

and satisfies the following homogeneity

$$\mathbf{N}(\lambda u, v) = \lambda\mathbf{N}(u,v), \qquad \mathbf{N}(u,\lambda v) = \lambda\mathbf{N}(u,v), \qquad \lambda > 0.$$

Proposition 10.1.2 $\mathbf{P} = 0$ *if and only if* $\mathbf{N} = 0$.

The proof is left for the reader.

10.2 Second Variation of Length

In this section, we will derive the second variation formula for a geodesic. The geometric meanings of the Riemann curvature and the T-curvature lie in this important formula.

Let (M, F) be a Finsler space. Take an arbitrary local frame $\{e_i\}_{i=1}^n$ for π^*TM. Let $\{\omega^i, \omega^{n+i}\}_{i=1}^n$ denote the local coframe for $T^*(TM \setminus \{0\})$, corresponding to $\{e_i\}$. The correspondence is given by (5.18). For vectors $\hat{X}, \hat{Y}$ on $TM \setminus \{0\}$ and a section $\tilde{U} = \tilde{U}^j e_j$ of π^*TM, define

$$\Omega(\hat{X}, \hat{Y})\tilde{U} := \tilde{U}^j \Omega_j{}^i(\hat{X}, \hat{Y})\, e_i.$$

Let ∇ denote the Chern connection on π^*TM. By (8.3), we have

$$\nabla_{\hat{X}} \nabla_{\hat{Y}} \tilde{U} - \nabla_{\hat{Y}} \nabla_{\hat{X}} \tilde{U} - \nabla_{[\hat{X}, \hat{Y}]} \tilde{U} = \Omega(\hat{X}, \hat{Y})\tilde{U},$$

where $\tilde{U} \in C^\infty(\pi^*TM)$ and $\hat{X}, \hat{Y} \in C^\infty(T(TM \setminus \{0\}))$.

Let $\{\bar{e}_i, \dot{e}_i\}_{i=1}^n$ be the local frame for $T^*(TM \setminus \{0\})$ that is dual to $\{\omega^i, \omega^{n+i}\}_{i=1}^n$. We have

$$\Omega(\bar{e}_k, \bar{e}_l)e_j = R_j{}^i{}_{kl}\, e_i \tag{10.11}$$

$$\Omega(\bar{e}_k, \dot{e}_l)e_j = P_j{}^i{}_{kl}\, e_i. \tag{10.12}$$

For a vector on $TM \setminus \{0\}$,

$$\hat{X} = X^i \bar{e}_i + Y^i \dot{e}_i,$$

we always denote

$$\tilde{X} = X^i e_i.$$

Let

$$\hat{V} = V^i \bar{e}_i + U^i \dot{e}_i, \qquad \hat{T} = T^i \bar{e}_i + S^i \dot{e}_i$$

be vector fields on $TM \setminus \{0\}$. Fix a vector $y \in T_xM \setminus \{0\}$. Assume that

$$T^i|_y = y^i, \qquad S^i|_y = 0,$$

where y^i are determined by $(y, y) = y^i e_i$.

By (8.24) and (8.29), we obtain

$$\begin{aligned} g\Big(\Omega(\hat{V},\hat{T})\tilde{V},\tilde{T}\Big)|_y &= V^jV^ky^iy^lg\Big(\Omega(\bar{e}_k,\bar{e}_l)e_j,e_i\Big) \\ &\quad +V^jU^ky^iy^lg\Big(\Omega(\dot{e}_k,\bar{e}_l)e_j,e_i\Big) \\ &= V^jV^k\, y^ig_{ip}R_j{}^p{}_{kl}y^l - V^jU^k\, y^ig_{ip}P_j{}^p{}_{lk}y^l \\ &= V^jV^k\, y^ig_{ip}R_j{}^p{}_k + V^jU^k\, y^ig_{ip}P^p{}_{jk} \\ &= -V^jV^k\, g_{jp}R^p{}_k. \end{aligned}$$

We obtain an important identity

$$g\Big(\Omega(\hat{V},\hat{T})\tilde{V},\tilde{T}\Big)|_y = -V^jV^k\, g_{jp}R^p{}_k. \tag{10.13}$$

Let $\{\mathbf{b}_i\}_{i=1}^n$ be the basis for T_xM determined by

$$e_i|_y = (y, \mathbf{b}_i).$$

Let $V := V^i\mathbf{b}_i \in T_xM$. We rewrite (10.13) as follows

$$g\Big(\Omega(\hat{V},\hat{T})\tilde{V},\tilde{T}\Big)|_y = -\mathbf{g}_y\Big(\mathbf{R}_y(V),V\Big) \tag{10.14}$$

Let $c : [a,b] \to M$ be a geodesic with $F(\dot{c}) = \lambda > 0$. Consider a piecewise C^∞ variation of c

$$H : (-\varepsilon,\varepsilon) \times [a,b] \to M$$

More precisely, there is a partition $a = t_0 < \cdots < t_k = b$ such that

(a) H is C^0 on $(-\varepsilon,\varepsilon)$;
(b) H is C^∞ each $(-\varepsilon,\varepsilon) \times [t_i,t_{i+1}]$ and
(c) $H(0,t) = c(t)$, $a \le t \le b$.

Let

$$\hat{H}(s,t) := \frac{\partial H}{\partial t}.$$

The map

$$\hat{H} : (-\varepsilon,\varepsilon) \times [a,b] \to TM$$

is not C^0 on $(-\varepsilon,\varepsilon)\times[a,b]$, but it is C^∞ on each $(-\varepsilon,\varepsilon)\times[t_i,t_{i+1}]$. $\hat{H}$ is not C^0 on $(\varepsilon,\varepsilon)\times[a,b]$ unless H is C^∞. Let

$$\hat{T}: \;=\; \frac{\partial\hat{H}}{\partial t}=\frac{\partial H^i}{\partial t}\frac{\partial}{\partial x^i}+\frac{\partial^2 H^i}{\partial t^2}\frac{\partial}{\partial y^i}$$

$$\hat{V}: \;=\; \frac{\partial\hat{H}}{\partial s}=\frac{\partial H^i}{\partial s}\frac{\partial}{\partial x^i}+\frac{\partial^2 H^i}{\partial s\partial t}\frac{\partial}{\partial y^i}.$$

Then

$$\tilde{T}:=\frac{\partial H^i}{\partial t}\partial_i,\qquad \tilde{V}:=\frac{\partial H^i}{\partial s}\partial_i.$$

Let

$$V(t):=V^i(t)\frac{\partial}{\partial x^i}|_{c(t)}=\frac{\partial H}{\partial s}(0,t).$$

Then

$$\tilde{T}(0,t)=\Big(\dot{c}(t),\dot{c}(t)\Big),\qquad \tilde{V}(0,t)=\Big(\dot{c}(t),V(t)\Big).$$

Let ∇ denote the Chern connection on π^*TM. The torsion-free condition (5.21) is equivalent to that

$$\nabla_{\hat{X}}\tilde{Y}-\nabla_{\hat{Y}}\tilde{X}=\widetilde{[\hat{X},\hat{Y}]}. \tag{10.15}$$

Without loss of generality, we may assume that $\hat{T}$ and $\hat{V}$ can extended to smooth vector fields on $TM\setminus\{0\}$. Then $[\hat{T},\hat{V}]$ is independent of any extensions. Observe

$$[\hat{T},\hat{V}]=\Big(\frac{\partial^2 H^i}{\partial t\partial s}-\frac{\partial^2 H^i}{\partial s\partial t}\Big)\frac{\partial}{\partial x^i}+\Big(\frac{\partial^3 H^i}{\partial t^2\partial s}-\frac{\partial^3 H^i}{\partial s\partial t^2}\Big)\frac{\partial}{\partial y^i}=0.$$

By (10.15),

$$\nabla_{\hat{V}}\tilde{T}=\nabla_{\hat{T}}\tilde{V}. \tag{10.16}$$

Now we consider the following length function

$$L(s):=\int_a^b F\Big(\frac{\partial H}{dt}\Big)dt=\int_a^b g(\tilde{T},\tilde{T})^{1/2}dt.$$

We obtain

$$\begin{aligned} L'(s) &= \int_a^b g(\tilde{T},\tilde{T})^{-1/2} g\Big(\nabla_{\hat{V}}\tilde{T},\tilde{T}\Big)dt \\ &= \int_a^b g(\tilde{T},\tilde{T})^{-1/2} g\Big(\nabla_{\hat{T}}\tilde{V},\tilde{T}\Big). \end{aligned} \tag{10.17}$$

Differentiating (10.17) and using (10.16) yield

$$\begin{aligned} L''(s) &= \int_a^b g(\tilde{T},\tilde{T})^{-1/2}\Big\{g(\nabla_{\hat{V}}\nabla_{\hat{T}}\tilde{V},\tilde{T}\Big) + g(\nabla_{\hat{T}}\tilde{V},\nabla_{\hat{T}}\tilde{V}\Big)\Big\}dt \\ &\quad - \int_a^b g(\tilde{T},\tilde{T})^{-3/2} g\Big(\nabla_{\hat{T}}\tilde{V},\tilde{T}\Big)^2 dt \\ &= \int_a^b g(\tilde{T},\tilde{T})^{-1/2} g\Big(\Omega(\hat{V},\hat{T})\tilde{V},\tilde{T}\Big)dt \\ &\quad + \int_a^b g(\tilde{T},\tilde{T})^{-1/2}\Big\{g(\nabla_{\hat{T}}\nabla_{\hat{V}}\tilde{V},\tilde{T}\Big) + g\Big(\nabla_{\hat{T}}\tilde{V},\nabla_{\hat{T}}\tilde{V}\Big)\Big\}dt \\ &\quad - \int_a^b g(\tilde{T},\tilde{T})^{-3/2} g\Big(\nabla_{\hat{T}}\tilde{V},\tilde{T}\Big)^2 dt \end{aligned}$$

Now we deal with the above identity along c term by term. Since c is a geodesic,

$$\begin{aligned} \hat{T}(0,t) &= \frac{\partial H^i}{\partial t}(0,t)\frac{\partial}{\partial x^i} + \frac{\partial^2 H^i}{\partial t^2}(0,t)\frac{\partial}{\partial y^i} \\ &= \dot{c}^i(t)\frac{\partial}{\partial x^i} - 2G^i(\dot{c}(t))\frac{\partial}{\partial y^i} = \dot{c}(t)\frac{\delta}{\delta x^i}. \end{aligned}$$

Thus $\hat{T}|_{s=0}$ is horizontal. Equation (10.14) gives

$$g\Big(\Omega(\hat{V},\hat{T})\tilde{V},\ \tilde{T}\Big)|_{s=0} = -\mathbf{g}_{\dot{c}(t)}\Big(\mathbf{R}_{\dot{c}(t)}(V(t)),V(t)\Big).$$

Observe that

$$g\Big(\nabla_{\hat{T}}\nabla_{\hat{V}}\tilde{V},\tilde{T}\Big) = \hat{T}\Big[g\Big(\nabla_{\hat{V}}\tilde{V},\tilde{T}\Big)\Big] - g\Big(\nabla_{\hat{V}}\tilde{V},\nabla_{\hat{T}}\tilde{T}\Big). \tag{10.18}$$

First, we study $\nabla_{\hat{T}}\tilde{T}$.

$$\nabla_{\hat{T}}\tilde{T} = \Big\{\frac{\partial^2 H^i}{\partial t^2} + 2G^i(\hat{H})\Big\}\partial_i.$$

We obtain

$$\nabla_{\hat{T}}\tilde{T}|_{s=0} = \Big\{\ddot{c} + 2G^i(\dot{c})\Big\}\partial_i|_{\dot{c}} = 0.$$

Now we fix t with $a \leq t \leq b$. Let

$$\sigma_t(s) := H(s,t).$$

Similarly,

$$\nabla_{\hat{V}}\tilde{V} = \Big\{\frac{\partial^2 H^i}{\partial s^2} + \frac{\partial H^j}{\partial s}\Gamma^i_{jk}(\hat{H})\frac{\partial H^k}{\partial s}\Big\}\partial_i.$$

We obtain

$$\begin{aligned}\nabla_{\hat{V}}\tilde{V}|_{s=0} &= \Big\{\ddot{\sigma}^i_t + \dot{\sigma}^j_t\Gamma_{jk}(\dot{c})\dot{\sigma}^k_t\Big\}\partial_i|_{\dot{c}}\\ &= \Big\{\ddot{\sigma}^i + 2G^i(\dot{\sigma}_t)\Big\}\partial_i|_{\dot{c}} + \Big\{\Gamma^i_{jk}(\dot{c}) - \Gamma^i_{jk}(\dot{\sigma}_t)\Big\}\dot{\sigma}^j_t\dot{\sigma}^k_t\ \partial_i|_{\dot{c}}.\end{aligned}$$

Assume that σ_t has constant speed with $F(\dot{\sigma}_t(s)) =: C_t$. Then the geodesic curvature κ_t of $\sigma_t(s)$ at $s = 0$ is given by

$$\kappa_t = \frac{1}{(C_t)^2}\Big\{\ddot{\sigma}_t(0) + 2G^i(\dot{\sigma}_t(0))\Big\}\frac{\partial}{\partial x^i}|_{\sigma_t(0)}.$$

This gives

$$g\Big(\nabla_{\hat{V}}\tilde{V},\tilde{T}\Big)|_{s=0} = (C_t)^2\mathbf{g}_{\dot{c}(t)}\Big(\kappa_t,\dot{c}(t)\Big) - \mathbf{T}_{\dot{c}(t)}\Big(\dot{\sigma}_t(0)\Big).$$

Note that the above function is piecewise C^∞. Thus

$$\begin{aligned}&\int_a^b g(\tilde{T},\tilde{T})^{-1/2}g\Big(\nabla_{\hat{T}}\nabla_{\hat{V}}\tilde{V},\tilde{T}\Big)|_{s=0}dt =\\ &\qquad \lambda^{-1}\Big[(C_b)^2\ \mathbf{g}_{\dot{c}(b)}\Big(\kappa_b,\dot{c}(b)\Big) - (C_a)^2\ \mathbf{g}_{\dot{c}(a)}\Big(\kappa_a,\dot{c}(a)\Big)\Big]\\ &\qquad -\lambda^{-1}\Big[\mathbf{T}_{\dot{c}(b)}\Big(\dot{\sigma}_b(0)\Big) - \mathbf{T}_{\dot{c}(a)}\Big(\dot{\sigma}_a(0)\Big)\Big].\end{aligned}$$

Observe that

$$\nabla_{\hat{T}}\tilde{V} = \{\frac{\partial^2 H^i}{\partial t^2} + V^jN^i_j(\hat{H})\Big\}\partial_i.$$

We obtain

$$\nabla_{\hat{T}}\tilde{V}|_{s=0} = \Big\{\frac{dV^i}{dt} + V^jN^i_j(\dot{c})\Big\}\partial_i|_{\dot{c}}.$$

This gives

$$g\Big(\nabla_{\hat{T}}\tilde{V}, \nabla_{\hat{T}}\tilde{V}\Big)|_{s=0} = \mathbf{g}_{\dot{c}(t)}\Big(\mathrm{D}_{\dot{c}}V(t), \mathrm{D}_{\dot{c}}V(t)\Big),$$

$$g\Big(\nabla_{\hat{T}}\tilde{V}, \tilde{T}\Big)|_{s=0} = \mathbf{g}_{\dot{c}(t)}\Big(\mathrm{D}_{\dot{c}}V(t), \dot{c}(t)\Big).$$

Let

$$V^{\perp}(t) := V(t) - \lambda^{-2}\mathbf{g}_{\dot{c}(t)}(V(t), \dot{c}(t))\dot{c}(t).$$

We have

$$\begin{aligned}
&\mathbf{g}_{\dot{c}(t)}\Big(\mathrm{D}_{\dot{c}(t)}V^{\perp}(t), \mathrm{D}_{\dot{c}(t)}V^{\perp}(t)\Big) \\
&\quad = \mathbf{g}_{\dot{c}(t)}\Big(\mathrm{D}_{\dot{c}(t)}V(t), \mathrm{D}_{\dot{c}(t)}V(t)\Big) - \lambda^{-2}\mathbf{g}_{\dot{c}(t)}\Big(\mathrm{D}_{\dot{c}(t)}V(t), \dot{c}(t)\Big).
\end{aligned}$$

We obtain the following

Second Variation Formula:

$$\begin{aligned}
L''(0) \;=\; & \lambda^{-1}\int_a^b \Big\{\mathbf{g}_{\dot{c}}\Big(\mathrm{D}_{\dot{c}}V^{\perp}, \mathrm{D}_{\dot{c}}V^{\perp}\Big) - \mathbf{g}_{\dot{c}}\Big(\mathbf{R}_{\dot{c}}(V^{\perp}), V^{\perp}\Big)\Big\}dt \\
& +\lambda^{-1}\Big[F^2(V(b))\mathbf{g}_{\dot{c}(b)}\Big(\kappa_b(0), \dot{c}(b)\Big) \\
& -F^2(V(a))\mathbf{g}_{\dot{c}(a)}\Big(\kappa_a(0), \dot{c}(a)\Big)\Big] \\
& +\lambda^{-1}\Big[\mathbf{T}_{\dot{c}(a)}\Big(V(a)\Big) - \mathbf{T}_{\dot{c}(b)}\Big(V(b)\Big)\Big]. \qquad (10.19)
\end{aligned}$$

The second variation formula is derived in [AbPa]. But the T-curvature is not clearly defined. The second variation formula for variations with fixed endpoints is derived in [BCS1] in a different way.

Let $c(t)$, $a \le t \le b$, be a unit speed geodesic in (M, F). Take a special geodesic variation $H(s,t)$ of a geodesic $c(t)$ such that $H(s,a) = c(a)$ and

$$\sigma(s) := H(s, b)$$

is a geodesic with $\sigma(0) = c(b)$. Assume that $V(t) := \frac{\partial H}{\partial s}(0,t)$ is linearly independent of $\dot{c}(t)$. Then the length function $L(s)$ satisfies

$$L''(0) \;=\; \int_a^b \Big\{\mathbf{g}_{\dot{c}}\Big(\mathrm{D}_{\dot{c}}V^{\perp}, \mathrm{D}_{\dot{c}}V^{\perp}\Big) - \mathbf{K}(P, \dot{c})\mathbf{g}_{\dot{c}}\Big(V^{\perp}, V^{\perp}\Big)\Big\}dt$$

$$-\mathbf{T}_{\dot{c}(b)}\Big(\dot{\sigma}(0)\Big), \tag{10.20}$$

where $P(t) = \text{span}\{\dot{c}(t), V(t)\}$ is a family of targent planes along c. Thus the behavior of the geodesics issuing from a point depends on the flag curvature and the T-curvature.

10.3 Synge Theorem

The second variation formula has many applications. In this section, we will apply this formula to prove that any even-dimensional compact Finsler space with $\mathbf{K} > 0$ has simple fundamental group $\pi_1 = \{0\}$ or $\mathbb{Z}_2$ (Synge Theorem). First we prove the following

Lemma 10.3.1 *Let (M, F) be an even-dimensional, oriented Finsler space and c be a closed geodesic. There is at least one parallel vector field W along c, which is not tangent to $\dot{c}$.*

Proof: Let $c : [0, r] \to M$ denote the closed geodesic with $c(0) = x = c(r)$ and P_c denote the parallel translation along c. By Lemma 5.3.1, $P_c : T_xM \to T_xM$ is a linear isometry with respect to $\mathbf{g}_y$, where $y = \dot{c}(0) = \dot{c}(r)$, i.e.,

$$\mathbf{g}_y\Big(P_c(u), P_c(v)\Big) = \mathbf{g}_y(u, v), \qquad u, v \in T_xM. \tag{10.21}$$

Let

$$y^{\perp} := \Big\{v \in T_xM, \ \mathbf{g}_y(y, v) = 0.\Big\}.$$

It follows from (10.21) that $P_c : y^{\perp} \to y^{\perp}$ is a linear isometry. Since $\dim y^{\perp} = n - 1$ is odd, there is at least one vector $w \in y^{\perp}$ such that $P_c(w) = w$. Let $W(t)$ be the parallel vector field along c with $W(0) = w = W(r)$. Then

$$W(0) = P_c(w) = W(r).$$

Thus W must be a smooth parallel vector field along c. Q.E.D.

Lemma 10.3.1 has important application.

Theorem 10.3.2 (Synge, 1926) *Let (M, F) be an even-dimensional closed oriented Finsler manifold with $\mathbf{K} > 0$. Then M is simply connected.*

Proof: Suppose that $\pi_1(M)$ contains a non-trivial element α. By Proposition 12.3.1 below, there is a closed geodesic $c \in \alpha$ with $L(c) = |\alpha|$. Assume that $c(t)$, $0 \leq t \leq |\alpha|$, is parametrized by arclength. It follows from Lemma 10.3.1 that there is a parallel vector field $W(t)$ along c. Take a variation $H : (-\varepsilon, \varepsilon) \times [0, |\alpha|] \to M$ of c such that

$$W(t) = \frac{\partial H}{\partial s}(0, t).$$

Applying the second variation formula to

$$L(s) := \int_0^{|\alpha|} F\Big(\frac{\partial H}{\partial t}(s,t)\Big)dt$$

and using the fact that $W_{c(0)} = W_{c(|\alpha|)}$, we obtain

$$L''(0) = -\int_0^{|\alpha|} \mathbf{g}_{\dot{c}(t)}\Big(\mathbf{R}_{\dot{c}(t)}\big(W(t)\big),\ W(t)\Big)dt < 0.$$

Thus $c_s(t) := H(s,t)$ is shorter than c for small $s > 0$. Since c_s is homotopy equivalent to c, we have

$$|\alpha| \leq L(c_s) < L(c) = |\alpha|.$$

It is a contradiction, since c is assumed to be the shortest closed curve in α.

Q.E.D.

Chapter 11

Geodesics and Exponential Map

Geodesics are the most important objects in a Finsler space. In this chapter, we will define the exponential map and use it to discuss some basic properties of geodesics. We will show that the geodesic completeness is equivalent to the metric completeness (Hopf-Rinow theorem).

11.1 Exponential Map

In this section, we will define the exponential map and discuss its regularity. The exponential map plays a very important role in comparison geometry.

Let (M, F) be a Finsler space and $\pi : TM \to M$ denote the natural projection. For a vector $y \in T_xM \setminus \{0\}$, let $\hat{c}_y(t)$ denote the integral curve of the induced spray $\mathbf{G}$ in $TM \setminus \{0\}$. By definition,

$$c_y(t) := \pi \circ \hat{c}_y(t)$$

is a geodesic in M with $\dot{c}_y(0) = y$. In a standard local coordinate system (x^i, y^i) in TM, geodesics are characterized by the following equations

$$\ddot{c}^i + 2G^i(\dot{c}) = 0. \tag{11.1}$$

By the O.D.E. theory, $\hat{c}_y$ smoothly depends on $y \in TM \setminus \{0\}$. The homogeneity of $\mathbf{G}$ implies that for any $\lambda > 0$ and $y \in TM \setminus \{0\}$,

$$c_{\lambda y}(t) = c_y(\lambda t), \qquad \forall t \geq 0.$$

Let $\mathcal{U}(M)$ denote the set of all tangent vectors $y \in TM$ such that c_y is defined on $[0, 1+\varepsilon)$ for some $\varepsilon > 0$. Define $\exp : \mathcal{U}(M) \to M$ by

$$\exp(y) := c_y(1).$$

We call exp the *exponential map*. Since $\hat{c}_y$ smoothly depends on $y \in TM \setminus \{0\}$, we conclude that exp is C^∞ on $\mathcal{U}(M) \setminus \{0\}$. For a point $x \in M$, let

$$\exp_x := \exp|_{\mathcal{U}_x M} : \mathcal{U}_x M \to M.$$

$\exp_x$ is C^∞ on $\mathcal{U}_x M - \{0\}$, where $\mathcal{U}_x M := \mathcal{U}M \cap T_x M$. We call $\exp_x$ the *exponential map* at x.

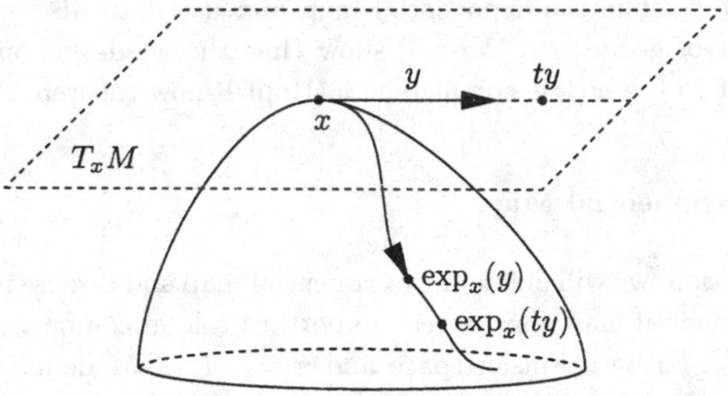

For any vector $y \in \mathcal{U}_x M$, we identify $T_y \mathcal{U}_x M$ with $T_x M$ in a natural way. Then the differential of $\exp_x$ at a vector $y \in \mathcal{U}_x M$ is a linear map

$$d(\exp_x)_{|y} : T_x M \to T_z M, \qquad z = \exp_x(y).$$

Thus,

$$d(\exp)_{|0} : T_x M \to T_x M$$

is an endomorphism, provided that $\exp_x$ is differentiable at the origin of $T_x M$. The problem is whether or not $\exp_x$ is C^∞ at the origin. This problem was answered by Whitehead.

Theorem 11.1.1 (Whitehead [Wh]) *Let (M, F) be a Finsler space. The exponential map* exp *is C^1 on the zero sections of $\pi : TM \to M$ and for* *any* $x \in M$, $d(\exp_x)_{|0} : T_xM \to T_xM$ *is the identity map at the origin* $0 \in T_xM$.

The proof is technical, so is omitted. See [BCS1] for details.

For any precompact open subset $A \subset M$, let $U_r(A)$ denote the r-neighborhood $U_r(A)$ of A in TM,

$$U_r(A) := \Big\{ y \in T_xM,\ F(y) < r,\ x \in A \Big\} \subset TM.$$

For a point $x \in M$, let $\mathrm{B}_r(x) := U_r(\{x\})$, i.e.,

$$\mathrm{B}_r(x) := \Big\{ y \in T_xM,\ F_x(y) < r \Big\} \subset T_xM.$$

By Theorem 11.1.1, we immediately obtain the following

Corollary 11.1.2 *Let* (M, F) *be a Finsler space. For any precompact open subset* $A \subset M$, *there is a positive number* $r_o = r_o(A) > 0$ *such that for any* $z \in A$, $\exp_z : T_zM \to M$ *is a* C^1*-diffeomorphism on* $\mathrm{B}_z(r) \subset T_zM$ *onto its image for all* $0 < r < r_o$.

In Section 11.3, we will show that $\exp_x[\mathrm{B}_x(r)]$ coincides with the metric ball $B(x, r)$.

11.2 Jacobi Fields

The Riemann curvature is defined via geodesic variations. Thus we will use geodesic variations to study the relationship between geodesics and the Riemann curvature.

Let (M, F) be a Finsler space. Let $c : [a, b] \to M$ be a geodesic and $H : (-\varepsilon, \varepsilon) \times [a, b] \to M$ a geodesic variation of c, namely, each $c_s(t) := H(s, t)$ is a geodesic. Let $J(t) := \frac{\partial H}{\partial s}(0, t)$. By Lemma 6.1.1, we know that $J(t)$ satisfies

$$\mathrm{D}_{\dot c}\mathrm{D}_{\dot c}J + \mathbf{R}_{\dot c}(J) = 0. \tag{11.2}$$

Equation (11.2) is called the *Jacobi equation* of c. Conversely, for every vector field $J(t)$ along c satisfying the Jacobi equation (11.2), there is a geodesic variation H of c whose variation field is equal to $J(t)$. Thus we call a vector field $J(t)$ along c satisfying (11.2) a *Jacobi field.*

Lemma 11.2.1 (Gauss) *Let $J(t)$ be a geodesic along a geodesic c with $F(\dot{c}) = \lambda$. Then*

$$\mathbf{g}_{\dot{c}(t)}\Big(J(t), \dot{c}(t)\Big) = \lambda^2(a + bt), \qquad \mathbf{g}_{\dot{c}(t)}\Big(\mathrm{D}_{\dot{c}}J(t), \dot{c}(t)\Big) = \lambda^2 b. \tag{11.3}$$

where

$$J^{\perp}(t) := J(t) - (a + bt)\dot{c}(t)$$

is also a Jacobi field along c such that $J^{\perp}(t)$ and $\mathrm{D}_{\dot{c}}J^{\perp}(t)$ are orthogonal to $\dot{c}(t)$ with respect to $\mathbf{g}_{\dot{c}(t)}$.

Proof: Observe that

$$\begin{aligned} \frac{d}{dt}\Big[\mathbf{g}_{\dot{c}(t)}\Big(J(t), \dot{c}(t)\Big)\Big] &= \mathbf{g}_{\dot{c}(t)}\Big(\mathrm{D}_{\dot{c}}J(t), \dot{c}(t)\Big) \\ \frac{d^2}{dt^2}\Big[\mathbf{g}_{\dot{c}(t)}\Big(J(t), \dot{c}(t)\Big)\Big] &= \mathbf{g}_{\dot{c}(t)}\Big(\mathrm{D}_{\dot{c}}\mathrm{D}_{\dot{c}}J(t), \dot{c}(t)\Big) \\ &= -\mathbf{g}_{\dot{c}(t)}\Big(\mathbf{R}_{\dot{c}(t)}(J(t)), \dot{c}(t)\Big) \\ &= -\mathbf{g}_{\dot{c}(t)}\Big(J(t), \mathbf{R}_{\dot{c}(t)}(\dot{c}(t))\Big) = 0. \end{aligned}$$

Thus (11.3) holds for some constants a, b. Plugging $J(t) = J^{\perp}(t) + (at + b)\dot{c}(t)$ into (11.2), we obtain

$$\mathrm{D}_{\dot{c}}\mathrm{D}_{\dot{c}}J^{\perp}(t) + \mathbf{R}_{\dot{c}}(J^{\perp}) = 0.$$

Thus $J^{\perp}(t)$ is also a Jacobi field along c. By (11.3),

$$\mathbf{g}_{\dot{c}(t)}\Big(J^{\perp}(t), \dot{c}(t)\Big) = 0, \quad \mathbf{g}_{\dot{c}(t)}\Big(\mathrm{D}_{\dot{c}}J^{\perp}(t), \dot{c}(t)\Big) = 0.$$

Q.E.D.

Fix a unit vector $y \in T_xM$ and let $c(t) := \exp_x(ty)$, $0 \le t < a$. Consider a special geodesic variation

$$H(s,t) := \exp_x[t(y + sv)]), \qquad 0 \le t < a, \ |s| < \varepsilon.$$

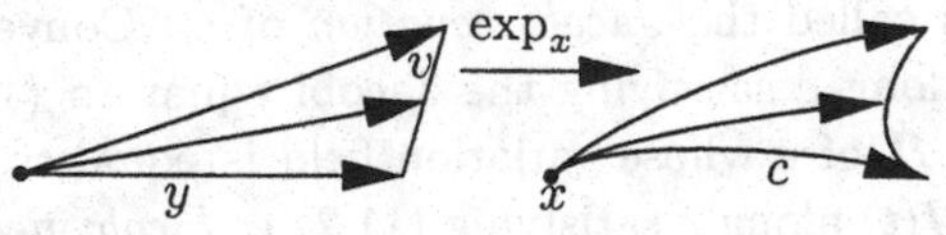

By Lemma 6.1.1,

$$J(t) := \frac{\partial H}{\partial s}(0,t) = d(\exp_x)|_{ty}(tv) \tag{11.4}$$

is a Jacobi field along c. Namely, $J(t)$ satisfies (11.2) on $(0, a)$. Since $\exp_x$ is C^1 at the origin, $J(t)$ is C^0 at $t = 0$. We can extend c to a geodesic $\tilde{c}$ on $(-\delta, a)$ for some $\delta > 0$. Take a number r close to 0 with $0 < r < a$. There is a unique Jacobi field $\tilde{J}$ along $\tilde{c}$ satisfying

$$\tilde{J}(r) = J(r), \qquad \mathrm{D}_{\dot{c}}\tilde{J}(r) = \mathrm{D}_{\dot{c}}J(r).$$

By the uniqueness,

$$\tilde{J}(t) = J(t), \qquad 0 < t < a.$$

Thus $J(t)$ can be extended to a Jacobi field $\tilde{J}(t)$ along $\tilde{c}$. By this observation, we can prove the following

Lemma 11.2.2 *The Jacobi field $J(t)$ in (11.4) is C^∞ along $c(t) = \exp_x(ty)$, $0 \leq t < a$. It satisfies the following initial conditions:*

$$J(0) = 0, \qquad \mathrm{D}_{\dot{c}}J(0) = v.$$

Proof. First observe that

$$J(0) = d(\exp_x)|_0(0) = 0.$$

By the above argument, we may extend J to a Jacobi field $\tilde{J}$ along a geodesic $\tilde{c}(t)$, $-\delta < t < a$. Write

$$\tilde{J}(t) = tW(t), \qquad -\delta < t < a$$

where $W(t)$ is a C^∞ vector field along $\tilde{c}$ and

$$W(t) = d(\exp_x)|_{ty}(v), \qquad 0 < t < a.$$

Observe that

$$\mathrm{D}_{\dot{c}}\tilde{J}(t) = W(t) + t\mathrm{D}_{\dot{c}}W(t).$$

Note that $W(t) = d(\exp_x)|_{ty}(v)$, $t \geq 0$, satisfies

$$W(0) = \lim_{t \to 0+} d(\exp_x)|_{ty}(v) = d(\exp_x)|_0(v) = v.$$

We obtain

$$\mathrm{D}_{\dot c}J(0) = W(0) = v.$$

Q.E.D.

Let $c : (-\varepsilon, \varepsilon) \to M$ be a unit speed geodesic with $\dot c(0) = y \in T_pM$. For a vector $v \in T_pM$, let J be the Jacobi field along c satisfying

$$J(0) = 0, \qquad \mathrm{D}_{\dot c}J(0) = v.$$

Consider the following function

$$f(t) := \mathbf{g}_{\dot c(t)}(J(t), Jt)).$$

Define $\dot{\mathbf{R}}_{\dot c(t)} : T_{c(t)}M \to T_{c(t)}M$ by

$$\dot{\mathbf{R}}_{\dot c(t)}(V(t)) := \mathrm{D}_{\dot c(t)}\Big[\mathbf{R}_{\dot c(t)}(V(t))\Big],$$

where $V(t)$ is a parallel vector field along c. Similarly, we can define $\ddot{\mathbf{R}}_{\dot c(t)}$. By the Jacobi equation (11.2), we obtain

$$\begin{aligned}
f' &= 2\mathbf{g}_{\dot c}\Big(\mathrm{D}_{\dot c}J, J\Big),\\
f'' &= 2\mathbf{g}_{\dot c}\Big(\mathrm{D}_{\dot c}J, \mathrm{D}_{\dot c}J\Big) - 2\mathbf{g}_{\dot c}\Big(\mathbf{R}_{\dot c}(J), J\Big),\\
f^{(3)} &= -8\mathbf{g}_{\dot c}\Big(\mathbf{R}_{\dot c}(J), \mathrm{D}_{\dot c}J\Big) - 2\mathbf{g}_{\dot c}\Big(\dot{\mathbf{R}}_{\dot c}(J), J\Big),\\
f^{(4)} &= -8\mathbf{g}_{\dot c}\Big(\mathbf{R}_{\dot c}(\mathrm{D}_{\dot c}J), \mathrm{D}_{\dot c}J\Big) + 8\mathbf{g}_{\dot c}\Big(\mathbf{R}_{\dot c}(J), \mathbf{R}_{\dot c}(J)\Big)\\
&\quad -12\mathbf{g}_{\dot c}\Big(\dot{\mathbf{R}}_{\dot c}(J), \mathrm{D}_{\dot c}J\Big) - 2\mathbf{g}_{\dot c}\Big(\ddot{\mathbf{R}}_{\dot c}(J), J\Big).
\end{aligned}$$

We obtain the following

Proposition 11.2.3 *The above Jacobi field $J(t)$ satisfies*

$$\mathbf{g}_{\dot c(t)}\Big(J(t), J(t)\Big) = \mathbf{g}_y(v, v)t^2 - \frac{1}{3}\mathbf{g}_y\Big(\mathbf{R}_y(v), v\Big)t^4 + o(t^4), \tag{11.5}$$

$$\mathbf{g}_{\dot c(t)}\Big(\mathrm{D}_{\dot c}J(t), J(t)\Big) = \mathbf{g}_y(v, v)t - \frac{2}{3}\mathbf{g}_y\Big(\mathbf{R}_y(v), v\Big)t^3 + o(t^3). \tag{11.6}$$

At the beginning, we have mentioned that exp is C^∞ on $\mathcal{U}(A) \setminus \{0\}$. Now we discuss the regularity of exp on the set of zero sections in TM.

We first consider the case when F is a Berwald metric, i.e., the geodesic coefficients $G^i(y) = \frac{1}{2}\Gamma^i_{jk}(x)y^j y^k$ are quadratic functions in $y \in T_xM$ for all $x \in M$. Geodesics are characterized by

$$\frac{d^2c^i}{dt^2} + \Gamma^i_{jk}(c)\frac{dc^j}{dt}\frac{dc^k}{dt} = 0.$$

By the O.D.E. theory, $c_y(t)$ smoothly depend on $y \in TM$ (for small t). Thus exp is C^∞ at $y = 0 \in TM$. In particular, $\exp_x$ is C^∞ at $y = 0$ for all $x \in A$.

Assume that $\exp_x$ is C^2 at $y = 0$ for all x. Let

$$f(y) := \exp(y).$$

By definition, for any $r > 0$ and any precompact open subset A, there is a number $\varepsilon > 0$ such that $c(t) = f(ty)$, $0 \le t \le 1 + \varepsilon$, is a geodesic for every $y \in T_xM$, where $x \in A$, with $F(y) < r$. Substituting c into the geodesic equation (11.1) and letting $t = 0$ yields

$$y^j y^k \frac{\partial^2 f^i}{\partial y^j \partial y^k}(0) + 2G^i(y) = 0.$$

In other words, $G^i(y)$ are quadratic in $y \in T_xM$. By definition, such a Finsler metric is a Berwald metric. We have proved the following

Theorem 11.2.4 (Akbar-Zadeh [AZ]) *Let (M, F) be a Finsler space. The exponential map* exp *is C^2 at zero sections if and only if F is a Berwald metric.*

11.3 Minimality of Geodesics

In Section 5.1, we define a geodesic to be a critical curve of the arc-length function of curves joining two fixed points. In this section, we shall prove that geodesics are locally minimizing curves. We have the following

Proposition 11.3.1 *Let (M, F) be a Finsler space. Suppose that* $\exp_x : T_xM \to M$ *is a C^1 diffeomorphism on $\mathrm{B}_x(r) \subset T_xM$. Then for any $y \in \mathrm{S}_xM$, the radial geodesic*

$$c(t) = \exp_x(ty), \qquad 0 \le t \le r,$$

has minimal length r among all piecewise C^1 curves in M joining x to $z = c(r)$. Moreover, for any $0 < t \le r$, the exponential map restricted to $\mathrm{B}_x(t) \subset T_xM$ and $\mathrm{S}_x(t) = \partial \mathrm{B}_x(t)$ are diffeomorphisms onto the metric t-ball $B(x,t) \subset M$ and $S(x,t) = \partial B(x,t)$ respectively,

$$\exp_x : \mathrm{B}_x(t) \to B(x,t), \quad \exp_x : \mathrm{S}_x(t) \to S(x,t). \tag{11.7}$$

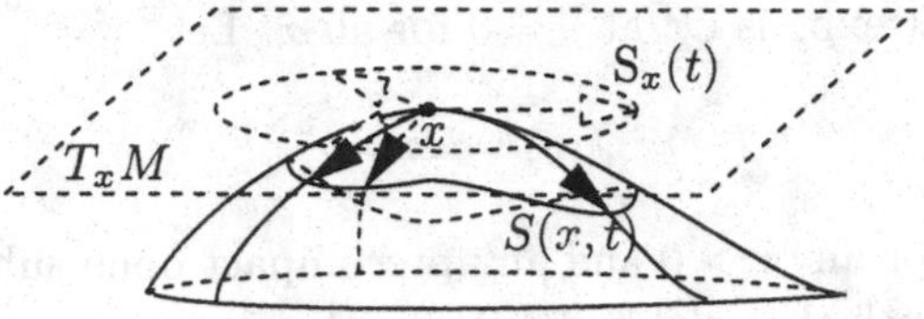

Proof. By Corollary 11.1.2, we may assume that for some r and $\epsilon > 0$, $\exp_x$ is a C^1-diffeomorphism from $\mathrm{B}_x(r+\epsilon) \subset T_xM$ onto its image. Take an arbitrary piecewise C^1 curve

$$\sigma : [0,1] \to M, \quad \sigma(0) = x = c(0),\ \sigma(1) = z = c(r).$$

First we assume that σ is contained in $\exp_x[\mathrm{B}_x(r+\varepsilon)]$. Then

$$\sigma(s) = \exp_x[t(s)y(s)] \quad y(s) \in \mathrm{S}_xM,$$

where $t = t(s)$ is a piecewise C^∞ function with $t(0) = 0,\ t(1) = r$ and

$$y : (0,1] \to \mathrm{S}_xM$$

is a piecewise C^∞ map with $y(1) = y$. It is easy to show that

$$\lim_{s \to 0^+} y(s) = \dot\sigma(0).$$

Hence, we may set $y(0) := \dot\sigma(0)$.

Consider the following geodesic variation,

$$H(s,t) := \exp_x(ty(s)), \quad 0 \le s \le 1,\ 0 \le t \le r,$$

such that

$$\sigma(s) = H(t(s), y(s)).$$

Let

$$Y := \frac{\partial H}{\partial t}, \quad V := \frac{\partial H}{\partial s}.$$

Then

$$\dot{\sigma} = Y \frac{dt}{ds} + V.$$

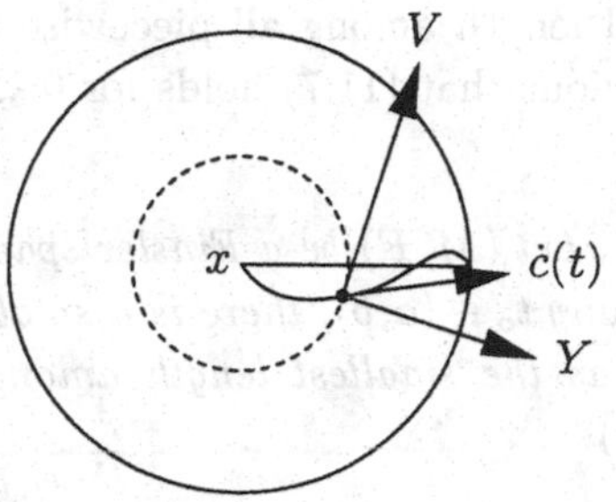

For each $s \in [0, 1]$, $V_s(t) := V(t, s)$ is a Jacobi field along

$$c_s(t) := \exp_x(ty(s)), \qquad 0 \le t \le r.$$

$V(t)$ satisfies

$$V_s(0) = 0, \qquad \mathrm{D}_{\dot{c}_s} V_s(0) = \dot{y}(s) \perp y(s).$$

By the Gauss Lemma (Lemma 11.2.1), $V_s(t) \perp \dot{c}_s(t)$ with respect to $\mathbf{g}_{\dot{c}_s(t)}$. Thus

$$\mathbf{g}_Y(Y, V) = 0. \tag{11.8}$$

Recall (1.14)

$$\mathbf{g}_y(y, w) \le F(y) F(w) \qquad \forall y \ne 0\ . \tag{11.9}$$

Taking $w = \dot{\sigma}(s)$ in (11.9) and using (11.8), we obtain

$$F(\dot{\sigma}) \ge \mathbf{g}_Y(Y, \dot{\sigma}) = \mathbf{g}_Y(Y, Y) \frac{dt}{ds} + \mathbf{g}_Y(Y, V) = \frac{dt}{ds}.$$

Hence,

$$L(\sigma) = \int_0^1 F(\dot{\sigma}) ds \ge \int_0^1 \frac{dt}{ds} ds = t(1) - t(0) = r = L(c).$$

This shows that if σ is contained in $\exp_x[\mathrm{B}_x(t+\varepsilon)]$, then its length is greater than or equal to that of c.

Now we suppose that σ is not completely contained in $\exp_x[\mathrm{B}_x(r+\varepsilon)]$. Let $0 < s_o < 1$ be the smallest number such that $\sigma(s_o) \in \exp_x[\mathrm{S}_x(r)]$. Then by the above argument, the length of $\sigma|_{[0,s_o]}$ is not less than r. We obtain

$$L(\sigma) = L(\sigma|_{[0,s_o]}) + L(\sigma|_{[s_o,1]}) > L(\sigma|_{[0,s_o]}) \geq r = L(c).$$

Thus c has minimal length among all piecewise C^1 curves joining x to z. Now it becomes obvious that (11.7) holds for $0 < t \leq r$. Q.E.D.

Corollary 11.3.2 *Let (M, F) be a Finsler space and $c(t)$, $a < t < b$ be a geodesic. Then for any $t_o \in (a, b)$, there is a small interval $[t_o - \varepsilon, t_o + \varepsilon] \subset (a, b)$ on which c has the smallest length among all curves issuing from $c(t_o - \varepsilon)$ to $c(t_o + \varepsilon)$.*

Proof: Take a sufficiently small $\varepsilon > 0$ such that $\exp_x$ is a diffeomorphism from $\mathrm{B}_x(r) \subset T_xM$ onto $B(x, r) \subset M$ for $x = c(t_o - \varepsilon)$ and $r \geq 2\varepsilon$. This is guaranteed by Corollary 11.1.2. Note that for $y = \dot{c}(t_o - \varepsilon)$,

$$\exp_x(ty) = c(t_o - \varepsilon + t), \qquad \forall t \in [0, 2\varepsilon].$$

By Proposition 11.3.1, we have

$$d\Big(c(t_o - \varepsilon), c(t_o - \varepsilon + t)\Big) = tF(y), \qquad \forall t \in [0, 2\varepsilon].$$

Thus c is minimizing on $[t_o - \varepsilon, t_o + \varepsilon]$. Q.E.D.

Now we study distance functions in a Finsler space (M, F). Consider the distance function from $x \in M$,

$$d_x(z) := d(x, z), \qquad z \in M.$$

According to Theorem 11.1.1, $\exp_x$ is C^∞ on $T_xM - \{x\}$ and C^1 at $y = 0 \in T_xM$. By the local Minimality of geodesics (Proposition 11.3.1), we see that for z nearby x,

$$d_x(z) = F_x\big(\exp_x^{-1}(z)\big).$$

Thus d_x^2 is C^∞ nearby x and C^1 at x. A natural question arises: Is d_x^2 always C^∞ at x? This question is answered in the following

Proposition 11.3.3 *Let (M, F) be a Finsler space. Suppose that d_x^2 is C^2 at x, then F_x is Euclidean at x. Thus d_x^2 is C^2 for all $x \in M$ if and only if F is Riemannian.*

Proof. Fix an arbitrary basis $\{\mathbf{b}_i\}_{i=1}^n$ for T_xM. Let $\varphi = u^i\mathbf{b}_i : \mathcal{U}_x \to T_xM$ be a local coordinate system at $x \in M$ with $\varphi(x) = 0$. Let

$$h(u) := d_x^2 \circ \varphi^{-1}(u).$$

Note that

$$h(u) \geq h(0) = 0.$$

By assumption, h is C^2 at 0. Thus

$$h(u) = \frac{1}{2}\frac{\partial^2 h}{\partial x^i \partial x^j}(0)u^i v^j + o(|u|^2).$$

On the other hand, for $u = u^i\mathbf{b}_i \in T_xM$, $\varphi^{-1}(tu) = c(t)$ is a C^2 curve with $\dot{c}(0) = u$.

$$d(x, c(t)) = \sqrt{h(tu)} = t\sqrt{\frac{1}{2}\frac{\partial^2 h}{\partial x^i \partial x^j}(0)u^i u^j} + o(t).$$

By (1.17), we obtain

$$F_x(y) = \lim_{r\to 0+}\frac{d(x, \varphi^{-1}(tu))}{t} = \lim_{r\to 0+}\frac{h(ru)}{r} = \sqrt{\frac{1}{2}\frac{\partial^2 h}{\partial x^i \partial x^j}(0)u^i u^j}.$$

Thus F_x is Euclidean at x. Q.E.D.

11.4 Completeness of Finsler Spaces

A Finsler metric F on a manifold M is said to be *positively (resp. negatively) complete* if every geodesic defined on $[0, a)$ (resp. $(-a, 0]$) can be extended to a geodesic defined on $[0, \infty)$ (resp. $(-\infty, 0]$). Thus F is positively complete if and only if exp is defined on the whole TM. F is positively complete if and only if $\bar{F}(y) := F(-y)$ is negatively complete. F is said to be *complete* if and only if F is both positively and negatively complete.

A sequence of points $\{x_i\}$ in a metric space (M, d) is called a *forward Cauchy sequence* if for any $\varepsilon > 0$, there exists a number N such that for any $j \geq i \geq N$,

$$d(x_i, x_j) < \varepsilon.$$

A sequence $\{x_i\}$ is said to be *forward convergent* (resp. *backward complete*) if there is a point $x \in M$ such that

$$\lim_{i\to\infty} d(x_i, x) = 0 \qquad \Big(\text{resp. } \lim_{i\to\infty} d(x, x_i) = 0\Big).$$

d is said to be *positively complete* (resp. *negatively compete*) if any forward Cauchy sequence (resp. backward Cauchy sequence) is forward convergent (resp. backward complete).

Lemma 11.4.1 *Let (M, F) be a connected Finsler space. Suppose that F is positively complete, then for any pair of points, $x, z \in M$, there exists a globally minimizing geodesic from x to z.*

Proof: Let $d = d_F$ be the metric induced by F. By Proposition 11.3.1, we know that for sufficiently small $r > 0$,

$$\exp_x[\mathrm{B}_x(r)] = B(x, r), \qquad \exp_x[\mathrm{S}_x(r)] = S(x, r). \tag{11.10}$$

Moreover, all geodesics from x to a point in $\mathrm{S}_x(r)$ are minimizing.

Let $r_o = d(x, z)$. If $z \in B(x, r)$, then we are done by the above fact. So we suppose that $z \notin B(x, r)$. There is a point $m \in S(x, r)$ such that

$$r + d(m, z) = d(x, z) = r_o.$$

Write $m = \exp_x(ry)$ for some $y \in \mathrm{S}_x M$.

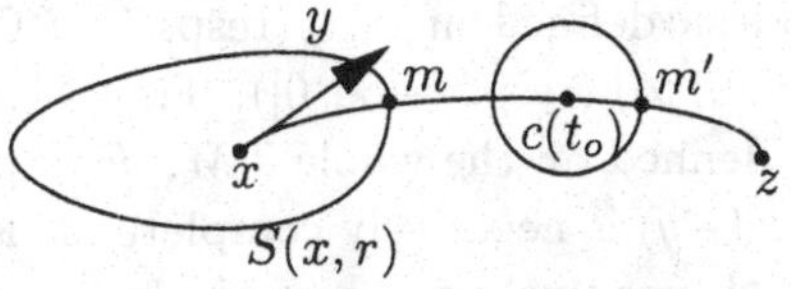

We claim that the geodesic $c(t) := \exp_x(ty)$, $0 \le t \le r_o$, is a minimizing geodesic. In this case, $c(r_o) = z$ since geodesics do not have branched points.

Let I be the set of $t \in [0, r_o]$ such that

$$t + d(c(t), z) = r_o.$$

In order to prove that c is minimizing, it suffices to prove that $I = [0, r_o]$. For any $0 < t \le r$,

$$r_o \le t + d(c(t), z) \le t + d(c(t), c(r)) + d(c(r), z) = r + d(c(r), z) = r_o.$$

This implies

$$t + d(c(t), z) = r_o.$$

Thus $[0, r] \subset I$. It is easy to see that I is closed. We claim that I is also open. Fix an arbitrary value $t_o \in I$ with $t_o \ge r$. For any t sufficiently close to t_o with $0 < t < t_o$,

$$r_o \le t + d(c(t), z) \le t + d(c(t), c(t_o)) + d(c(t_o), z) \le t_o + d(c(t_o), z) = r_o.$$

This implies that

$$t + d(c(t), z) = r_o.$$

Thus $[t_o - \varepsilon, t_o] \subset I$ for a sufficiently small $\varepsilon > 0$. Suppose that $t_o < r_o$. Take a sufficiently small $\delta > 0$. There is a point $m' \in S(c(t_o), \delta)$ such that

$$d(c(t_o), m') + d(m', z) = d(c(t_o), z).$$

Observe that

$$\begin{aligned} r_o &\le d(x, m') + d(m', z) \\ &\le d(x, c(t_o)) + d(c(t_o), m') + d(m', z) \\ &= t_o + d(c(t_o), z) = r_o. \end{aligned} \tag{11.11}$$

This implies that

$$d(x, m') = d(x, c(t_o)) + d(c(t_o), m').$$

We conclude that $m' = c(t_o + \delta)$ since geodesics do not have branched points. Plugging $m' = c(t_o + \delta)$ into (11.11), we obtain

$$t_o + \delta + d(c(t_o + \delta), z) = r_o.$$

Therefore $[t_o, t_o + \delta] \subset I$. Since I is close and open, we conclude that $I = [0, r_o]$. Q.E.D.

Using Lemma 11.4.1, one can easily prove the following Hopf-Rinow theorem.

Theorem 11.4.2 (Hopf-Rinow) *Let (M, F) be a Finsler space and d_F be the induced metric. The following two conditions are equivalent.*

(a) *The metric space (M, d_F) is positively complete,*

(b) *The Finsler space (M, F) is positively complete.*

The proof is omitted here. See [Ma1] [BCS1].

Chapter 12

Conjugate Radius and Injectivity Radius

For a positively complete Finsler space (M, F), the exponential map at any point $x \in M$,

$$\exp_x : T_xM \to M,$$

is an onto map. In this chapter, we discuss the singularity of $\exp_x$ and the largest radius r for which $\exp_x$ is a diffeomorphism on $\mathrm{B}_x(r) \subset T_xM$. Throughout this chapter, we always assume that Finsler spaces are positively complete.

12.1 Conjugate Radius

To study the singularity of the exponential map, we introduce the notion of conjugate value for unit tangent vectors.

Definition 12.1.1 *Let (M, F) be a positively complete Finsler space. For a unit vector $y \in \mathrm{S}_xM$, we define $\mathbf{c}_y > 0$ to be the first number $r > 0$ such that there is a Jacobi field $J(t)$ along $c(t) = \exp_x(ty)$, $0 \le t \le r$, satisfying*

$$J(0) = 0 = J(r).$$

$\mathbf{c}_y$ is called the conjugate value of y. Put

$$\mathbf{c}_x := \inf_{y \in \mathrm{S}_xM} \mathbf{c}_y, \qquad \mathbf{c}_M := \inf_{x \in M} \mathbf{c}_x.$$

$\mathbf{c}_x$ and $\mathbf{c}_M$ are called the conjugate radius of x and the conjugate radius of M respectively.

We will show that the function $\mathbf{c} : y \in SM \to \mathbf{c}_y \in (0, \infty]$ is lower-continuous. Let $y_i \in SM$ be a sequence of unit vectors converging to $y \in \mathrm{S}_x M$. Assume that $r := \liminf_{i\to\infty} \mathbf{c}_{y_i} < \infty$. Then $\mathbf{c}_{y_i} < \infty$ for large i. Since $\exp_x$ is singular at $\mathbf{c}_{y_i} y_i \in T_x M$, $\exp_x$ must be singular at ry by the continuity of $d(\exp_x)$ on $T_x M$. This implies that $\mathbf{c}_y \leq r$. Namely,

$$\mathbf{c}_y \leq \liminf_{y_i \to y} \mathbf{c}_{y_i}.$$

Thus $\mathbf{c}_y$ is lower-continuous in $y \in SM$.

To study the conjugate value of a tangent vector, we introduce the notion of index form. Let $c(t)$, $0 \leq t \leq r$, be a unit speed geodesic. For vector fields $U = U(t)$ and $V = V(t)$ along c, define

$$\mathcal{I}_c(U, V) := \int_0^r \Big\{ \mathbf{g}_{\dot c}(\mathrm{D}_{\dot c} U, \mathrm{D}_{\dot c} V) - \mathbf{g}_{\dot c}\Big(\mathbf{R}_{\dot c}(U), V\Big) \Big\} dt. \qquad (12.1)$$

$\mathcal{I}_c$ is called the *index form* along c. Since $\mathbf{R}_y$ is self-adjoint with respect to $\mathbf{g}_y$, $\mathcal{I}_c$ is symmetric, i.e.,

$$\mathcal{I}_c(U, V) = \mathcal{I}_c(V, U).$$

Lemma 12.1.2 (Index Lemma) *Let $c(t), 0 \leq t \leq r$, be a geodesic with $\dot c(0) = y \in T_x M$ and $J(t)$ be a Jacobi field along c with $J(0) = 0$. Suppose that $0 < r < \mathbf{c}_y$. For any piecewise C^∞ vector field $V(t), 0 \leq t \leq r$, with*

$$V(0) = 0, \qquad V(r) = J(r),$$

the following holds

$$\mathcal{I}_c(J, J) \leq \mathcal{I}_c(V, V).$$

The equality holds if and only if $V = J$.

Proof: First, define a map $\Phi : T_{c(0)} M \to T_{c(r)} M$ by

$$\Phi(v) := J_v(r),$$

where J_v denotes the Jacobi field along c with

$$J_v(0) = 0, \qquad \mathrm{D}_{\dot c} J_v(0) = v.$$

Φ is a linear map. By assumption, $\Phi(v) \neq 0$ for $v \in T_x M \backslash \{0\}$. We conclude that Φ is an isomorphism. Let $\{\mathbf{b}_i\}_{i=1}^n$ be a basis for $T_{c(0)} M$. Since Φ is non-singular, we can extend $\{\mathbf{b}_i\}_{i=1}^n$ to a Jacobi frame along c,

$$J_i(t) := J_{\mathbf{b}_i}(t), \qquad i = 1 \cdots n,$$

such that $J(t) = \sum_{i=1}^n a_i J_i(t)$. Since $V(0) = 0$, there are piecewise C^1 functions $f_i(t)$ on $[0, r]$ such that

$$V(t) = \sum_{i=1}^n f_i(t) J_i(t).$$

Since $J(r) = V(r)$, we get that $a_i = f_i(r)$ and

$$J(t) = \sum_{i=1}^n f_i(r) J_i(t).$$

Let

$$A(t) := \sum_{i=1}^n f_i'(t) J_i(t).$$

Integration by parts yields

$$\mathcal{I}_c(V, V) = \mathcal{I}_c(J, J) + \int_0^r \mathbf{g}_{\dot{c}}(A, A) dt.$$

Thus

$$\mathcal{I}_c(V, V) \geq \mathcal{I}_c(J, J).$$

Equality holds if and only if $A(t) = 0$, i.e.,

$$f_i(t) = constant, \qquad i = 1, \cdots, n,$$

This implies that $J(t) = V(t)$. Q.E.D.

Lemma 12.1.3 *Let $0 \leq r \leq \mathbf{c}_y$ and $c(t) = \exp_x(ty)$, $0 \leq t \leq r$. For any piecewise C^∞ vector field $V(t) \neq 0$ along c with $V(0) = 0 = V(r)$,*

$$\mathcal{I}_c(V, V) \geq 0$$

and the equality holds if and only if $r = \mathbf{c}_y$ and V is a Jacobi field along c.

Proof. First we assume that $r < \mathbf{c}_y$. The Jacobi field $J(t)$ along c with $J(0) = 0 = J(r)$ must vanish. By Lemma 12.1.2, we know that

$$\mathcal{I}_c(V,V) > \mathcal{I}_c(J,J) = \mathcal{I}_c(0,0) = 0.$$

Now we assume $r = \mathbf{c}_y$. Let $r_i := \mathbf{c}_y - 1/i$ and $V_i(t)$, $0 \le t \le r_i$, be a sequence of piecewise C^∞ vector fields along $c_i(t) = \exp_x(ty), 0 \le t \le r_i$ such that $V_i(0) = 0 = V_i(r_i)$ and $\lim_{i\to\infty} V_i = V$. By the above argument,

$$\mathcal{I}_{c_i}(V_i, V_i) > 0.$$

Letting $i \to \infty$ yields

$$\mathcal{I}_c(V,V) \ge 0.$$

Let $U(t)$ be an arbitrary piecewise C^∞ vector field along c with $U(0) = 0 = U(r)$. For any $\lambda \in \mathrm{R}$,

$$\mathcal{I}_c(V + \lambda U, V + \lambda U) = \mathcal{I}_c(V,V) + 2\lambda \mathcal{I}_c(U,V) + \lambda^2 \mathcal{I}_c(U,U) \ge 0.$$

This implies

$$\mathcal{I}_c(U,V) = 0.$$

Take a partition $0 = t_0 < t_1 < \cdots < t_{k-1} < t_k = r$ such that $V(t)$ is C^∞ on each $[t_{i-1}, t_i]$, $i = 1, \cdots, k$. Observe that for any piecewise C^∞ vector field U along c with the same regularity as V,

$$\mathcal{I}_c(U,V) = \sum_{i=0}^{k} \mathbf{g}_{\dot c}\Big(U, \mathrm{D}_{\dot c} V\Big)\Big|_{t_{i-1}}^{t_i} - \int_0^r \Big\{\mathbf{g}_{\dot c}\Big(U,\ \mathrm{D}_{\dot c}\mathrm{D}_{\dot c} V + \mathbf{R}_{\dot c}(V)\Big)\Big\} dt. \quad (12.2)$$

First we choose an arbitrary $U(t)$ with $U(t_i) = 0$, $i = 0, \cdots, k$. We obtain

$$\int_0^r \Big\{\mathbf{g}_{\dot c}\Big(U,\ \mathrm{D}_{\dot c}\mathrm{D}_{\dot c} V + \mathbf{R}_{\dot c}(V)\Big)\Big\} dt = 0.$$

We conclude that $V(t)$ is a Jacobi field on each $[t_{i-1}, t_i]$. Then for piecewise C^∞ vector field $U(t)$ along c with $U(0) = 0 = U(r)$, (12.2) simplifies to

$$\begin{aligned}\mathcal{I}_c(U,V) &= \sum_{i=0}^{k-1} \mathbf{g}_{\dot c}\Big(U, \mathrm{D}_{\dot c} V\Big)\Big|_{t_{i-1}}^{t_i} \\ &= \sum_{i=1}^{k} \mathbf{g}_{\dot c(t_i)}\Big(U(t_i), \mathrm{D}_{\dot c} V(t_i^-) - \mathrm{D}_{\dot c} V(t_i^+)\Big) = 0.\end{aligned}$$

Since $U(t_i)$ are arbitrary, we obtain

$$\mathrm{D}_{\dot{c}}V(t_i^-) = \mathrm{D}_{\dot{c}}V(t_i^+).$$

That is, $V(t)$ is C^1 at t_i, $i = 1, \cdots, k-1$. Since V is a Jacobi field on each $[t_{i-1}, t_i]$, we conclude that V is C^∞ along the whole c. Q.E.D.

12.2 Injectivity Radius

According to Proposition 11.3.1, for sufficiently small $r > 0$, the exponential map $\exp_x$ is a diffeomorphism from the tangent r-ball $\mathrm{B}_x(r) \subset T_xM$ onto the metric r-ball $B(x, r) \subset M$. In this section, we will discuss the largest possible domain around the origin in T_xM on which the exponential map $\exp_x$ is a diffeomorphism.

Definition 12.2.1 *Let (M, F) be a positively complete Finsler space and $x \in M$. For a vector $y \in \mathrm{S}_xM$, we define $\mathbf{i}_y$ to be the supremum of $r > 0$ such that $c(t) := \exp_x(ty)$ is minimizing on $[0, r]$. Set*

$$\mathbf{i}_x := \inf_{y \in \mathrm{S}_xM} \mathbf{i}_y, \qquad \mathbf{i}_M := \inf_{x \in M} \mathbf{i}_x.$$

We call $\mathbf{i}_x$ the injectivity radius at x and $\mathbf{i}_M$ the injectivity radius of M. Let

$$\mathcal{D}_x := \Big\{ \exp_x(ty), \ \ 0 \le t < \mathbf{i}_y, \ y \in \mathrm{S}_xM \Big\} \subset M$$

and

$$\mathcal{C}_x := M - \mathcal{D}_x$$

$\mathcal{C}_x$ and $\mathcal{D}_x$ are called the cut-locus and the cut-domain of x respectively. Let

$$\tilde{\mathcal{D}}_x := \Big\{ ty, \ \ 0 \le t < \mathbf{i}_y, \ y \in \mathrm{S}_xM \Big\} \subset T_xM.$$

$\tilde{\mathcal{D}}_x$ is called the tangent cut-domain at x. The exponential map

$$\exp_x : \tilde{\mathcal{D}}_x \to \mathcal{D}_x$$

is an onto diffeomorphism.

The function

$$\mathbf{i} : \mathrm{S}_x M \to [0, \infty]$$

is continuous. Thus for any compact subset $A \subset M$,

$$\mathbf{i}_A := \inf_{x \in A} \mathbf{i}_x > 0.$$

Moreover, the cut-locus $\mathcal{C}_x$ has null measure. See Section 8.4 in [BCS1] for the details.

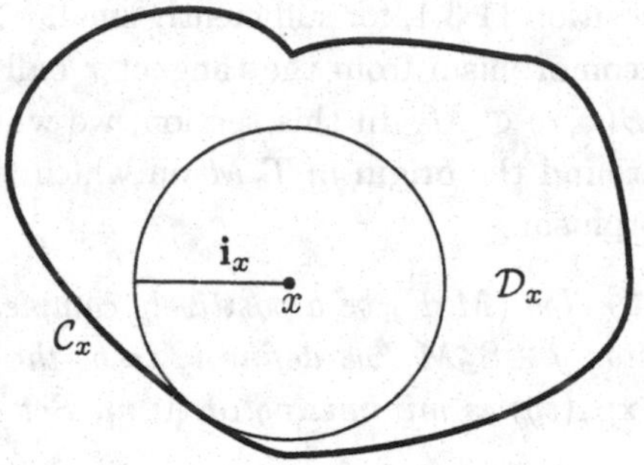

Lemma 12.2.2 *Let $c : [0, \infty) \to (M, F)$ be a unit speed geodesic with $\dot{c}(0) = y \in \mathrm{S}_x M$. If there is another geodesic σ issuing from x to $c(r)$ with $L(\sigma) \le r$, then $\mathbf{i}_y \le r$.*

Proof. Suppose that for some $r < \mathbf{i}_y$. Let $a = \mathrm{length}(\sigma)$. σ must be minimizing with the length of r. Since σ does not coincide with c, we have

$$v := \dot{\sigma}(a) \ne u := \dot{c}(r).$$

By Lemma 1.2.3

$$\mathbf{g}_v(v, u) < F(u)F(v) = 1. \tag{12.3}$$

Take a small $0 < \varepsilon < \mathbf{i}_y - r$, and consider the following distance function,

$$\rho(z) := d\Big(\sigma(a - \varepsilon), z\Big), \qquad z \in M.$$

The first variation formula gives

$$\frac{d}{dt}\Big[\rho(c(t))\Big]\Big|_{t=r} = \mathbf{g}_v(v, u) < 1.$$

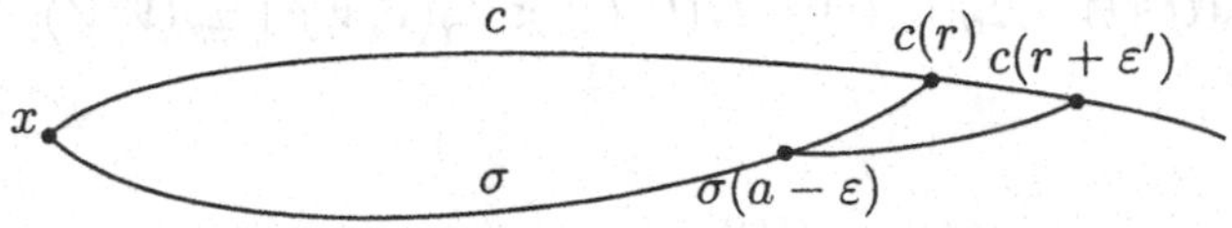

Thus, for sufficiently small $\varepsilon > 0$ and $\varepsilon' > 0$, there is a minimizing geodesic τ from $\sigma(a-\varepsilon)$ to $c(r+\varepsilon')$ with length $< \varepsilon + \varepsilon'$. This implies

$$d(x, c(r+\varepsilon')) \leq d(x, \sigma(a-\varepsilon)) + \text{length}(\tau) < r + \varepsilon'.$$

This is a contradiction. Q.E.D.

Lemma 12.2.3 *At each point $x \in M$, $\mathbf{i}_y \leq \mathbf{c}_y$ for all $y \in \mathrm{S}_x M$. Hence $\mathbf{i}_x \leq \mathbf{c}_x$.*

Proof. Let $c(t) := \exp_x(ty)$, $0 \leq t \leq \mathbf{c}_y + \varepsilon$. There is a nonzero Jacobi field $J(t)$, $0 \leq t \leq \mathbf{c}_y$, such that $J(0) = 0 = J(\mathbf{c}_y)$. Fix an arbitrary small $\varepsilon > 0$. Extend J to a piecewise C^∞ vector field $V(t)$, $0 \leq t \leq \mathbf{c}_y + \varepsilon$, by assigning $V(t) = 0$ for $\mathbf{c}_y \leq t \leq \mathbf{c}_y + \varepsilon$. Choosing a smaller $\varepsilon > 0$ if necessary, we know that there is a unique Jacobi field $J_\varepsilon(t)$, $\mathbf{c}_y - \varepsilon \leq t \leq \mathbf{c}_y + \varepsilon$ with the prescribed values

$$J_\varepsilon(\mathbf{c}_y - \varepsilon) = V(\mathbf{c}_y - \varepsilon) = J(\mathbf{c}_y - \varepsilon),$$

$$J_\varepsilon(\mathbf{c}_y + \varepsilon) = V(\mathbf{c}_y + \varepsilon) = 0.$$

Define a piecewise C^∞ vector field $U(t)$, $0 \leq t \leq \mathbf{c}_y + \varepsilon$ as follows.

$$U(t) := \begin{cases} V(t) = J(t) & \text{if } 0 \leq t \leq \mathbf{c}_y - \varepsilon \\ J_\varepsilon(t) & \text{if } \mathbf{c}_y - \varepsilon \leq t \leq \mathbf{c}_y + \varepsilon \,. \end{cases}$$

Let $\gamma := c|_{[0, \mathbf{c}_y - \varepsilon]}$ and $\tau := c|_{[\mathbf{c}_y - \varepsilon, \mathbf{c}_y + \varepsilon]}$. Note that

$$\mathcal{I}_\gamma(U, U) = \mathcal{I}_\gamma(V, V).$$

In view of Lemma 12.1.3,

$$\mathcal{I}_\tau(U, U) = \mathcal{I}_\tau(J_\varepsilon, J_\varepsilon) < \mathcal{I}_\tau(V, V).$$

We obtain

$$\mathcal{I}_c(U,U) = \mathcal{I}_\gamma(U,U) + \mathcal{I}_\tau(U,U) < \mathcal{I}_\gamma(V,V) + \mathcal{I}_\tau(V,V) = 0.$$

Let

$$U^\perp(t) := U(t) - \mathbf{g}_{\dot c(t)}\Big(\dot c(t), U(t)\Big)\, \dot c(t).$$

Then

$$\mathcal{I}_c(U^\perp, U^\perp) = \mathcal{I}(U,U) - \int_0^{\mathbf{c}_y+\varepsilon} \Big[\mathbf{g}_{\dot c(t)}\Big(\dot c(t), U(t)\Big)\Big]^2 dt < 0.$$

We can construct a piecewise C^∞ variation $H(s,t)$ of c such that the variation

$$H(s,0) = c(0), \qquad H(s, \mathbf{c}_y + \varepsilon) = c(\mathbf{c}_y + \varepsilon)$$

and

$$U(t) = \frac{\partial H}{\partial s}(0,t).$$

Consider the length function of $c_s(t) := H(s,t)$, $0 \le t \le \mathbf{c}_y + \varepsilon$,

$$L(s) := \int_0^{\mathbf{c}_y+\varepsilon} F\Big(\dot c_s(t)\Big) dt.$$

By the second variation formula,

$$L''(0) = \mathcal{I}_c(U^\perp, U^\perp) < 0.$$

Thus for small s,

$$d(x, c(\mathbf{c}_y + \varepsilon)) \le L(c_s) < L(c) = \mathbf{c}_y + \varepsilon.$$

Thus there is another minimizing geodesic σ issuing from $x = c(0)$ to $c(\mathbf{c}_y + \varepsilon)$ with

$$L(\sigma) < L(c) = \mathbf{c}_y + \varepsilon.$$

By Lemma 12.2.2, we conclude that

$$\mathbf{i}_y \le \mathbf{c}_y + \varepsilon.$$

Since ε is arbitrary, we conclude that $\mathbf{i}_y \le \mathbf{c}_y$. Q.E.D.

Lemma 12.2.4 *Let $c : [0, \infty) \to M$ be a unit speed geodesic with $\dot{c}(0) = y \in S_x M$. Suppose that $\mathbf{i}_y < \mathbf{c}_y$. Then there is another minimizing geodesic σ issuing from x to $c(\mathbf{i}_y)$.*

Proof. There is a sequence of minimizing unit speed geodesics σ_i issuing from x to $c(\mathbf{i}_y + \varepsilon_i)$ with $\lim_{i\to\infty} \varepsilon_i = 0$. Let $y_i := \dot{\sigma}_i(0) \in S_x M$. Without loss of generality, we may assume that $y_i \to y^*$ and

$$r_i := d(x, c(\mathbf{i}_y + \varepsilon_i)) \to \mathbf{i}_y.$$

We claim that $y^* \neq y$. Observe that

$$\exp_x(r_i y_i) \to \exp_x(\mathbf{i}_y y^*).$$

Since $d\exp_x$ is nonsingular at $\mathbf{i}_y y$, we see that $y^* \neq y$. Let $\sigma : [0, \mathbf{i}_y] \to M$ denote the geodesic with $\dot{\sigma}(0) = y^*$. We obtain the desired geodesic. Q.E.D.

Let (M, F) be a positively complete Finsler manifold. The injectivity radius $\mathbf{i}_x$ at x is also given by

$$\mathbf{i}_x = d(x, C_x) = \sup_{z \in C_x} d(x, z).$$

If $C_x = \emptyset$, then $\mathbf{i}_x = \infty$.

Lemma 12.2.5 *Let (M, F) be complete with $C_x \neq \emptyset$ at $x \in M$. Let $z \in C_x$ be a point which realizes the distance $d(x, z) = \mathbf{i}_x$. Suppose that $\mathbf{i}_x < \mathbf{c}_x$. Then there are exactly two unit speed minimizing geodesics c and σ from x to z such that*

$$\dot{c}(\mathbf{i}_x) + \dot{\sigma}(\mathbf{i}_x) = 0.$$

Proof. Take a unit speed minimizing geodesic $c(t) = \exp_x(ty)$ from x to $z = \exp_x(\mathbf{i}_x y)$. By Lemma 12.2.4, there is another unit speed minimizing geodesic segment $\sigma(t) = \exp_x(ty^*)$ from x to $z = \exp_x(\mathbf{i}_x y^*)$. We claim that c and σ form a smooth loop at x. Let $u := \dot{c}(\mathbf{i}_x)$ and $v := \dot{\sigma}(\mathbf{i}_x)$. It suffices to prove that

$$u + v = 0. \tag{12.4}$$

Suppose that (12.4) does not hold. That is,

$$w := -(u + v) \neq 0.$$

By the Schwartz inequality,

$$-\mathbf{g}_u(u,v) < F(u)F(v) = 1, \quad -\mathbf{g}_v(u,v) < F(u)F(v) = 1.$$

Thus

$$\begin{aligned}\mathbf{g}_u(u,w) &= -\mathbf{g}_u(u,u) - \mathbf{g}_u(u,v) = -1 - \mathbf{g}_u(u,v) < 0\\ \mathbf{g}_v(v,w) &= -\mathbf{g}_v(v,v) - \mathbf{g}_v(u,v) = -1 - \mathbf{g}_v(v,u) < 0.\end{aligned}$$

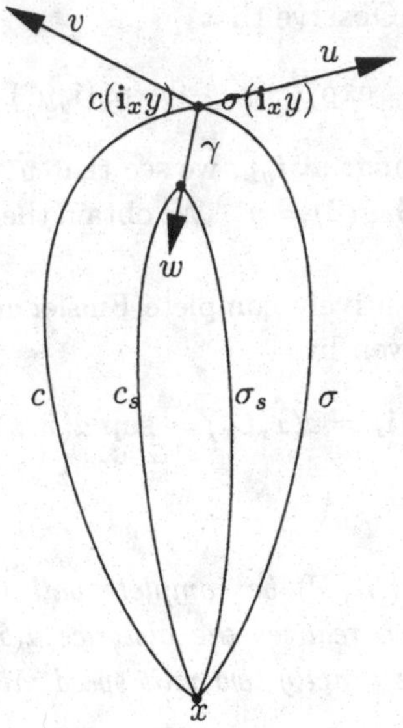

Take a smooth curve $\gamma : (-\varepsilon, \varepsilon) \to M$ with $\dot{c}(0) = w \in T_zM$. Since $\exp_x$ is a local diffeomorphism on $\mathrm{B}_{\mathbf{c}_x}(x) \subset T_xM$ and

$$\exp_x(\mathbf{i}_x y) = z = \exp_x(\mathbf{i}_x y^*),$$

there two curves $\tilde{\gamma}, \tilde{\gamma}^* : (-\varepsilon, \varepsilon) \to \mathrm{B}_{\mathbf{c}_x}(x)$ satisfying

$$\exp_x[\tilde{\gamma}(s)] = \exp_x[\tilde{\gamma}^*(s)] = c(s)$$

with $\tilde{\gamma}(0) = \mathbf{i}_x y$ and $\tilde{\gamma}^*(0) = \mathbf{i}_x y^*$. Consider the following variations c_s and c_s^*,

$$c_s(t) = \exp_x[t\tilde{\gamma}(s)], \qquad \sigma_s(t) = \exp_x[t\tilde{\gamma}^*(s)].$$

Let $L(s) := L(c_s)$ and $L^*(s) := L(\sigma_s)$. By the first variation formula,

$$\frac{dL}{ds}(0) = \mathbf{g}_u(u, w) < 0, \qquad \frac{dL^*}{ds}(0) = \mathbf{g}_v(v, w) < 0.$$

Thus both curves c_s and σ_s are shorter than c and σ, respectively, for small $s > 0$. We may assume that for some $s > 0$,

$$L(s) \leq L^*(s) < \mathbf{i}_x.$$

By Lemma 12.2.2, we know that $\mathbf{i}_{\bar{\gamma}^*(s)} \leq L^*(s) < \mathbf{i}_x$. It is a contradiction.

Q.E.D.

The smooth loop at x in Lemma 12.2.5 is not necessarily a geodesic loop, since the reverse $\bar{\sigma}(t) = \sigma(\mathbf{i}_x - t)$, $0 \leq t \leq \mathbf{i}_x$, might not be a geodesic unless F is reversible.

12.3 Geodesic Loops and Closed Geodesics

In this section, we show that geodesic loops exist in a Finsler space (M, F) with non-trivial fundamental group. In topology, the fundamental group $\pi_1(M, x)$ is the space of homotopy equivalent classes of continuous loops at x. Each class $\alpha \in \pi_1(M, x)$ contains at least one piecewise C^∞ loop at x. For a class $\alpha \in \pi_1(M, x)$, the *geometric length* $|\alpha|$ is defined by

$$|\alpha| = \inf_{c \in \alpha} L(c),$$

where the infimum is taken over all piecewise C^∞ loops c at x.

A piecewise C^∞ loop $c \in \alpha$ is said to be *shortest* if $L(c) = |\alpha|$. By the same argument as in §5.1, one can show that if c is a shortest piecewise C^∞ curve and parametrized by arc-length, then it must be a C^∞ geodesic loop at x.

By Theorem 11.1.1, we know that $\exp_x : \mathrm{B}_{\mathbf{i}_x}(x) \subset T_xM \to B(x, \mathbf{i}_x) \subset M$ is a diffeomorphism which is C^1 at the origin and C^∞ away from the origin. Thus for any non-trivial $\alpha \in \pi_1(M, x)$,

$$|\alpha| \geq 2\mathbf{i}_x.$$

Let $\pi_1(M)$ denote the space of free homotopy equivalent classes of closed curves in M. For a class $\alpha \in \pi_1(M)$, if a piecewise C^∞ closed curve $c \in \alpha$

satisfies $L(c) = |\alpha|$, then it must be closed C^∞ geodesic in α. Thus

$$|\alpha| \geq 2\mathbf{i}_M.$$

We have the following existence theorem for shortest geodesics.

Proposition 12.3.1 (Hilbert) *Let (M, F) be a Finsler space.*

(i) Suppose that (M, F) is positively complete. Then every non trivial class $\alpha \in \pi_1(M, x)$ contains a shortest geodesic loop at x;
(ii) Suppose that M is compact without boundary. Then every non-trivial class $\alpha \in \pi_1(M)$ contains a shortest closed geodesic.

Proof. We shall only prove (i). The proof of (ii) is similar. Let $c_i \in \alpha$ be a sequence of piecewise C^∞ loops such that

$$|\alpha| = \lim_{i\to\infty} L(c_i).$$

Clearly, c_i is contained in a compact subset K in M. Since K is compact, $\mathbf{i}_K > 0$. For any pair $x_1, x_2 \in K$ with $d(x^1, x^2) < \mathbf{i}_K$, there is exactly one minimizing geodesic segment c joining x_1 and x_2. Thus there is an integer $N > 0$ such that for each i, there are at most N points $x_i^s = c_i$ on c_i with

$$d(x_i^s, x_i^{s+1}) \leq L(c_i^s) \leq \frac{1}{2}\mathbf{i}_K, \qquad s = 1, \cdots, N,$$

where c_i^s denotes the portion of c_i from x_i^s to x_{i+1}^s. Let σ_i denote the piecewise C^∞ unit speed geodesic consisting of geodesic segments σ_i^s joining from x_i^s to x_i^{s+1}. From the construction we see that each σ_i is homotopic to c_i and

$$L(\sigma_i) \leq L(c_i).$$

Thus

$$|\alpha| = \lim_{i\to\infty} L(\sigma_i).$$

By taking a subsequence if necessary, one may assume that c_i consists of exactly N minimizing geodesic segments c_i^s, $s = 1, \cdots, N$, for all i. For a fixed $1 \leq s \leq N$, the endpoints x_i^s, x_i^{s+1} sub-converge to x^s, x^{s+1}. Let σ^s denote the minimizing geodesic segment from x^s to x^{s+1}. We must have

$$\lim_{i\to\infty} L(c_i^s) = L(\sigma^s).$$

Let $\sigma := \cup_{s=1}^{N} \sigma^s$. σ is a piecewise C^∞ geodesic loop at x and

$$L(\sigma) = \sum_{s=1}^{N} \lim_{i\to\infty} L(\sigma_i^s) = \lim_{i\to\infty} L(\sigma_i) = |\alpha|.$$

Thus σ is shortest. Q.E.D.

Lemma 12.3.2 *Let (M, F) be a closed reversible Finsler space. Suppose that $\mathbf{i}_M < \mathbf{c}_M$. Then there is a closed geodesic c with $L(c) = 2\mathbf{i}_M$.*

Proof. Assume that $\mathbf{i}_M < \mathbf{c}_M$. Since M is compact, there is a point $x \in M$ such that $\mathbf{i}_x = \mathbf{i}_M$. It follows from Lemma 12.2.5 that there is a geodesic loop at x,

$$c : [0, 2\mathbf{i}_x] \to M, \qquad c(0) = c(2\mathbf{i}_x).$$

Let $z = c(\mathbf{i}_x) \in C_x$. It is obvious that $\mathbf{i}_z = \mathbf{i}_x = \mathbf{i}_M$. By Lemma 12.2.5 again,

$$\dot{c}(0) + \dot{c}(2\mathbf{i}_x) = 0.$$

Thus c is a closed geodesic. Q.E.D.

There are many Finsler spaces satisfying the assumption in Lemma 12.3.2. We will show the following facts in Chapter 12.3.

(i) if $\mathbf{K} \le 0$, then $\mathbf{c}_M = \infty$. Hence $\mathbf{i}_M < \mathbf{c}_M = \infty$;
(ii) if $\mathbf{K} \ge 1$ and $\mathrm{Diam}(M) < \pi$, then $\mathbf{i}_M < \mathbf{c}_M$.

We know that for a generic Finsler metric on a closed manifold, the initial vectors to closed geodesics are dense in the unit tangent bundle. Finsler metrics with finitely many closed geodesics are rather exceptional.

The first interesting work was done by Katok [Kat] who found some non-Riemannian Finsler metrics on the n-sphere S^n with only finitely many closed geodesics. Later on, W. Ziller [Zi] made a close examination on Katok's examples. He actually generalized the construction of Katok and found some other interesting examples. In particular, he proves that there are Finsler metrics on S^n and S^{n-1} with only n closed geodesics, where n is even. These Finsler metrics exist in any neighborhood of the standard metric.

Katok's construction also shows that there is a big difference between non-reversible Finsler metrics and reversible ones. For example, on S^n, one can construct non-reversible Finsler metrics (close to the standard metric) with only n closed geodesics of lengths close to 2π and the length of all other closed geodesics larger than any prescribed number. (Note: any Finsler metric on S^n sufficiently C^2 close to the standard metric has at least n closed geodesic [Zi]).

However, for reversible Finsler metrics (including Riemannian metric s), Lusternik-Schnirelmann theory implies that for reversible Finsler metrics close to the standard metric, there are at least $g(n) = 2n - s - 1$ closed geodesics with lengths close to 2π, where

$$n = 2^k + s < 2^{k+1}.$$

Notice that

$$(3n-1)/2 \leq g(n) \leq 2n - 1.$$

Ziller constructed a reversible Finsler metric on S^n close to the standard metric, with only $2n - 1$ closed geodesics of length close to 2π and the length of all other geodesics larger than any prescribed number.

Chapter 13

Basic Comparison Theorems

In this chapter, we will discuss the geometric meanings of the Riemann curvature and its mean — the Ricci curvature. In particular, we will prove that the exponential map is nonsingular if the flag curvature is non-positive (Cartan-Hadamard Theorem) and the diameter must be bounded from above if the Ricci curvature is strictly positive (Bonnet-Myers Theorem).

13.1 Flag Curvature Bounded Above

Let (M, F) be a Finsler space. For a vector $y \in T_xM \setminus \{0\}$, the Riemann curvature $\mathbf{R}_y : T_xM \to T_xM$ is a self-adjoint linear transformation with respect to $\mathbf{g}_y$. For a tangent plane $P \subset T_xM$ with $y \in P$, the flag curvature $\mathbf{K}(P, y)$ is given by

$$\mathbf{K}(P,y) = \frac{\mathbf{g}_y\Big(\mathbf{R}_y(u), u\Big)}{\mathbf{g}_y(y,y)\mathbf{g}_y(u,u) - \mathbf{g}_y(y,u)\mathbf{g}_y(y,u)},$$

where $u \in P$ such that $P = \text{span}\{y, u\}$. We say that the flag curvature is bounded from above by λ, denoted by $\mathbf{K} \leq \lambda$, if for any flags (P, y),

$$\mathbf{K}(P,y) \leq \lambda.$$

This is equivalent to the following

$$\mathbf{g}_y\Big(\mathbf{R}_y(u), u\Big) \leq \lambda\Big\{\mathbf{g}_y(y,y)\mathbf{g}_y(u,u) - \mathbf{g}_y(u,y)\mathbf{g}_y(y,u)\Big\}.$$

Similarly, we define the lower bound $\mathbf{K} \geq \lambda$.

For a number $\lambda \in \mathrm{R}$, let

$$\mathbf{s}_\lambda(t) := \begin{cases} \frac{\sin(\sqrt{\lambda}t)}{\sqrt{\lambda}} & \text{if } \lambda > 0 \\ t & \text{if } \lambda = 0 \\ \frac{\sinh(\sqrt{-\lambda}t)}{\sqrt{-\lambda}} & \text{if } \lambda < 0, \end{cases}$$

and $\mathbf{c}_\lambda(t) := \mathbf{s}'_\lambda(t)$. Both $\mathbf{s}_\lambda(t)$ and $\mathbf{c}_\lambda(t)$ satisfy the following equation

$$y'' + \lambda y = 0.$$

Let $\mathbf{t}_\lambda > 0$ be the first positive zero of $\mathbf{s}_\lambda(t)$. More precisely,

$$\mathbf{t}_\lambda := \begin{cases} \pi/\sqrt{\lambda}, & \text{if } \lambda > 0 \\ \infty, & \text{if } \lambda \leq 0. \end{cases}$$

We have the following comparison result in one-dimension.

Lemma 13.1.1 *Let r be a number with $0 < r < \mathbf{t}_\lambda$. For any function $f \in C^1[0, r]$ with $f(0) = 0$,*

$$\int_0^r \Big\{(f')^2 - \lambda f^2\Big\} dt \geq \Big(\frac{f(r)}{\mathbf{s}_\lambda(r)}\Big)^2 \int_0^r \Big\{(\mathbf{s}'_\lambda)^2 - \lambda(\mathbf{s}_\lambda)^2\Big\} dt = f(r)^2 \frac{\mathbf{s}'_\lambda(r)}{\mathbf{s}_\lambda(r)}.$$

Equality holds if and only if

$$f(t) = \frac{f(r)}{\mathbf{s}_\lambda(r)} \mathbf{s}_\lambda(t), \qquad 0 \leq t \leq r.$$

The proof of Lemma 13.1.1 is elementary, so is omitted. Now we are ready to prove the following important theorem in Finsler geometry.

Theorem 13.1.2 (Cartan-Hadamard [Au]) *Let (M, F) be a positively complete Finsler space. Suppose that the flag curvature satisfies $\mathbf{K} \leq \lambda$. Then the conjugate radius satisfies $\mathbf{c}_y \geq \mathbf{t}_\lambda$ for any $y \in SM$. In particular, if $\mathbf{K} \leq 0$, then $\mathbf{c}_y = \infty$ for any $y \in SM$. Hence $\exp_x : T_xM \to M$ is non-singular for any $x \in M$.*

Proof. Fix a unit vector $y \in S_xM$. Let

$$c(t) := \exp_x(ty), \qquad t \geq 0.$$

Let $J(t)$ be a non-trivial Jacobi field along c with $J(0) = 0$ and $J(\mathbf{c}_y) = 0$. By Lemma 11.2.1, we have

$$J(t) = J^\perp(t) + at\,\dot{c}(t),$$

where $J^{\perp}(t)$ is a Jacobi field which is $\mathbf{g}_{\dot{c}(t)}$-orthogonal to $\dot{c}(t)$. We must have $a = 0$ and $J(t) = J^{\perp}(t)$. Namely, $J(t)$ is $\mathbf{g}_{\dot{c}(t)}$-orthogonal to $\dot{c}(t)$. Let

$$f(t) := \sqrt{\mathbf{g}_{\dot{c}(t)}(J(t), J(t))}, \qquad t \geq 0.$$

Observe that for $0 < t < \mathbf{c}_y$,

$$(f')^2 = \left[\frac{\mathbf{g}_{\dot{c}}(\mathrm{D}_{\dot{c}}J, J)}{\sqrt{\mathbf{g}_{\dot{c}}(J, J)}}\right]^2 \leq \mathbf{g}_{\dot{c}}\Big(\mathrm{D}_{\dot{c}}J, \mathrm{D}_{\dot{c}}J\Big).$$

Assume that $\mathbf{c}_y < \mathbf{t}_\lambda$. Integration by parts gives

$$\begin{aligned}
\mathbf{g}_{\dot{c}(t)}\Big(\mathrm{D}_{\dot{c}}J(t), J(t)\Big) &= \int_0^t \frac{d}{dt}\Big[\mathbf{g}_{\dot{c}}\Big(\mathrm{D}_{\dot{c}}J, J\Big)\Big] d\tau \\
&= \int_0^t \Big\{\mathbf{g}_{\dot{c}}\Big(\mathrm{D}_{\dot{c}}J, \mathrm{D}_{\dot{c}}J\Big) - \mathbf{g}_{\dot{c}}\Big(\mathbf{R}_{\dot{c}}(J), J\Big)\Big\} d\tau \\
&\geq \int_0^t \Big\{(f')^2 - \lambda f^2\Big\} d\tau \\
&\geq \left(\frac{f(t)}{\mathbf{s}_\lambda(t)}\right)^2 \int_0^t \Big\{(\mathbf{s}'_\lambda)^2 - \lambda(\mathbf{s}_\lambda)^2\Big\} d\tau \\
&= f(t)^2 \frac{\mathbf{s}'_\lambda(t)}{\mathbf{s}_\lambda(t)}.
\end{aligned}$$

This implies that for any $0 < t < \mathbf{c}_y$,

$$\frac{d}{dt}\left[\frac{f(t)}{\mathbf{s}_\lambda(t)}\right] = \frac{\mathbf{g}_{\dot{c}}\Big(\mathrm{D}_{\dot{c}}J(t), J(t)\Big)}{f(t)^2} - \frac{\mathbf{s}'_\lambda(t)}{\mathbf{s}_\lambda(t)} \geq 0.$$

Thus the quotient

$$\frac{\sqrt{\mathbf{g}_{\dot{c}(t)}(J(t), J(t))}}{\mathbf{s}_\lambda(t)} = \frac{f(t)}{\mathbf{s}_\lambda(t)}$$

is non-decreasing. Since $J(0) = 0$, we can express

$$J(t) = tW(t), \qquad 0 \leq t < \mathbf{c}_y,$$

where $W(t)$ is a C^∞ vector field along c with $W(0) = \mathrm{D}_{\dot{c}}J(0) = v$. This implies

$$\lim_{t \to 0^+} \frac{\sqrt{\mathbf{g}_{\dot{c}(t)}(J(t), J(t))}}{\mathbf{s}_\lambda(t)} = \sqrt{\mathbf{g}_y(v, v)}.$$

Therefore

$$\sqrt{\mathbf{g}_{\dot c(t)}(J(t),J(t))} \geq \mathbf{s}_\lambda(t)\sqrt{\mathbf{g}_y(v,v)} > 0, \qquad 0 < t \leq \mathbf{c}_y. \tag{13.1}$$

By assumption, $\mathbf{c}_y < \mathbf{t}_\lambda$, we conclude that $J(\mathbf{c}_y) \neq 0$. This is a contradiction.

Q.E.D.

Let $y \in \mathrm{S}_xM$ and $c(t) = \exp_x(ty)$, $t \geq 0$, be a unit speed geodesic. According to Lemma 11.2.1, for an arbitrary vector $v \in T_xM$, $J_v(t) := d(\exp_x)|_{ty}(tv)$ can be expressed in the form

$$J_v(t) = J_{v^\perp}(t) + \mathbf{g}_y(y,v)t\,\dot c(t),$$

where $v^\perp = v - \mathbf{g}_y(y,v)y$. Hence $J_{v^\perp}(t) \perp \dot c(t)$ for all $t > 0$. By the above argument,

$$g_{\dot c(t)}\Big(J_{v^\perp}(t), J_{v^\perp}(t)\Big) \geq [\mathbf{s}_\lambda(t)]^2\mathbf{g}_y(v^\perp,v^\perp), \qquad 0 < t \leq \mathbf{t}_\lambda.$$

Thus

$$\begin{aligned}
\mathbf{g}_{\dot c(t)}\Big(J_v(t),J_v(t)\Big) &= \mathbf{g}_{\dot c(t)}\Big(J_{v^\perp}(t),J_{v^\perp}(t)\Big) + [\mathbf{g}_y(y,v)]^2t^2 \\
&= [\mathbf{s}_\lambda(t)]^2\mathbf{g}_y(v^\perp,v^\perp) + [\mathbf{g}_y(y,v)]^2t^2 \\
&= [s_\lambda(t)]^2\mathbf{g}_y(v,v) + [\mathbf{g}_y(y,v)]^2\Big(t^2 - [\mathbf{s}_\lambda(t)]^2\Big).
\end{aligned}$$

Note that for $w = d(\exp_x)|_{ty}(v) \in T_{c(t)}M$,

$$d(\exp_x)|_{ty}(v) := \frac{1}{t}J_v(t).$$

Thus

$$\begin{aligned}
&\mathbf{g}_{\dot c(t)}\Big(d(\exp_x)|_{ty}(v),\ d(\exp_x)|_{ty}(v)\Big) \\
&\qquad \geq \Big(\frac{\mathbf{s}_\lambda(t)}{t}\Big)^2\mathbf{g}_y(v,v) + [\mathbf{g}_y(y,v)]^2\frac{t^2 - [\mathbf{s}_\lambda(t)]^2}{t^2}.
\end{aligned} \tag{13.2}$$

When $\mathbf{K} \leq \lambda = 0$, $\mathbf{s}_\lambda(t) = t$. Inequality (13.2) becomes

$$\mathbf{g}_{\dot c(t)}\Big(d(\exp_x)|_{ty}(v), d(\exp_x)|_{ty}(v)\Big) \geq \mathbf{g}_y(v,v), \qquad v \in T_xM. \tag{13.3}$$

Thus $\exp_x$ is non-singular at any point $ty \in T_xM$.

Theorem 13.1.3 (Cartan-Hadamard) *Let (M, F) be positively complete with $\mathbf{K} \leq 0$. For any $x \in M$, $\exp_x : T_xM \to M$ is a covering map. Moreover, for any $y \in S_xM$, (13.3) holds.*

By further argument, one can show that $\exp_x$ in Theorem 13.1.3 is a covering map. See [BCS1] for a proof.

A Finsler space (M, F) is called a *Hadamard* space if it is positively complete, simply connected with $\mathbf{K} \leq 0$. We have just proved that for a Hadamard space, the exponential map $\exp_x : T_xM \to M$ is non-singular for all $x \in M$. In [Eg1][Eg2], Egloff studied uniform Hadamard spaces and uniform negatively curved Finsler spaces. A Finsler space is said to be *uniform* with uniformity constant C if for any $x \in M$ and $u, v \in T_xM$,

$$C^{-1}\mathbf{g}_u \leq \mathbf{g}_v \leq C\mathbf{g}_u. \tag{13.4}$$

Hence

$$C^{-1}F^2(u) \leq \mathbf{g}_v(u, u) \leq CF^2(u).$$

We have the following

Theorem 13.1.4 *Let (M, F) be a complete reversible uniform Finsler space satisfying (13.4). Assume that M is simply connected and F satisfies $\mathbf{K} \leq 0$. Then $\exp_x : (T_xM, F_x) \to M$ is distance quasi-nondecreasing. More precisely,*

$$F_x(y_2 - y_1) \leq \sqrt{C}d\Big(\exp_x(y_1), \exp_x(y_2)\Big), \qquad \forall y_1, y_2 \in T_xM. \tag{13.5}$$

Proof: Let $\sigma : [0,1] \to M$ be a minimizing geodesic joining from $\exp_x(y_1)$ to $\exp_x(y_2)$. Since $\exp_x$ is a covering map and M is simply connected, $\exp_x$ is a diffeomorphism. There is a curve $\tilde{\sigma} : [0,1] \to T_xM$ joining from y_1 to y_2, such that

$$\exp_x(\tilde{\sigma}(t)) = \sigma(t), \qquad 0 \leq t \leq 1.$$

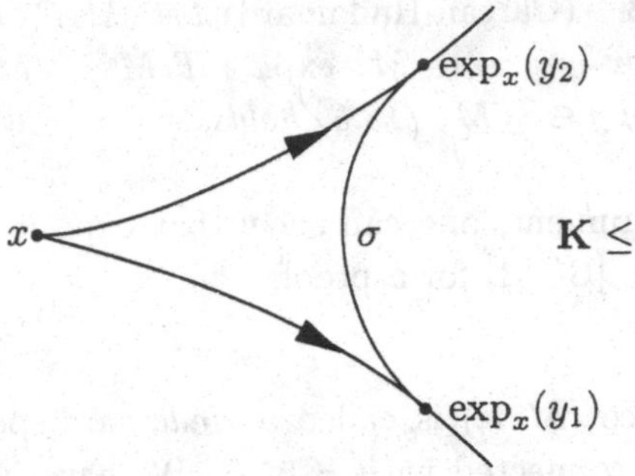

Observe that

$$\begin{aligned} F(\dot{\sigma}(t)) &= F\Big(d(\exp_x)(\dot{\bar{\sigma}}(t))\Big) \\ &\geq \sqrt{\mathbf{g}_{\dot{\sigma}(t)}\Big(\dot{\bar{\sigma}}(t), \dot{\bar{\sigma}}(t)\Big)} \\ &\geq \frac{1}{\sqrt{C}} F_x(\dot{\bar{\sigma}}(t)). \end{aligned}$$

Thus

$$\begin{aligned} F_x(y_2 - y_1) &\leq \int_0^1 F_x(\dot{\bar{\sigma}}(t))dt \\ &\leq \sqrt{C} \int_0^1 F(\dot{\sigma}(t))dt \\ &= \sqrt{C} d\Big(\exp_x(y_1), \exp_x(y_2)\Big). \end{aligned}$$

This proves (13.5). Q.E.D.

Speaking of Finsler spaces with negative curvature, we should mention Foulon's work on geodesic flows. In [Fo1], Foulon proved that the geodesic flow of a compact reversible Finsler space with $\mathbf{K} < 0$ is of Anosov type. See also [Eg1] for further studies on the geodesic flow of reversible Finsler spaces.

13.2 Positive Flag Curvature

Now we discuss the geometric properties of even-dimensional Finsler spaces with positive curvature. As we know, the conjugate radius is always less than or equal to the injectivity radius (Lemma 12.2.3). We will show that they are equal for even-dimensional oriented Finsler spaces with positive curvature.

Theorem 13.2.1 (Klingenberg) *Let (M, F) be an even-dimensional closed oriented Finsler space with $\mathbf{K} > 0$. Then $\mathbf{i}_M = \mathbf{c}_M$.*

Proof: We assume that $\mathbf{i}_M < \mathbf{c}_M$. By Lemma 12.3.2, there is a closed geodesic c with $L(c) = 2\mathbf{i}_M$. Let $c(t)$, $0 \le t \le 2\mathbf{i}_M$ be parametrized by arc-length. By Lemma 10.3.1, there is a parallel vector field $W(t)$ along c and not tangent to $\dot{c}(t)$. Choose a variation $H(s,t)$ of $c(t)$ as in the proof of the Synge Theorem. The length function

$$L(s) := \int_0^{2\mathbf{i}_M} F\Big(\frac{\partial H}{\partial t}(s,t)\Big)dt$$

satisfies

$$L''(0) = -\int_0^{2\mathbf{i}_M} \mathbf{g}_{\dot{c}}\Big(\mathbf{R}_{\dot{c}}(W), W\Big)dt < 0.$$

Thus for small $s > 0$, the curve $c_s(t) := H(s,t)$ is shorter than c,

$$L(c_s) < L(c).$$

Let $z_s \in c_s$ such that

$$d\Big(c_s(0), z_s\Big) = \sup_{z \in c_s} d\Big(c_s(0), z\Big) < \mathbf{i}_M.$$

There is a unique minimizing geodesic segment $\sigma_s : [0, r_s] \to M$ from $c_s(0)$ to $z_s = c_s(r_s)$. It follows from the first variation formula that

$$\mathbf{g}_{y_s}\Big(y_s, \dot{c}_s(r_s)\Big) = 0,$$

where $y_s = \dot{\sigma}_s(r_s)$.

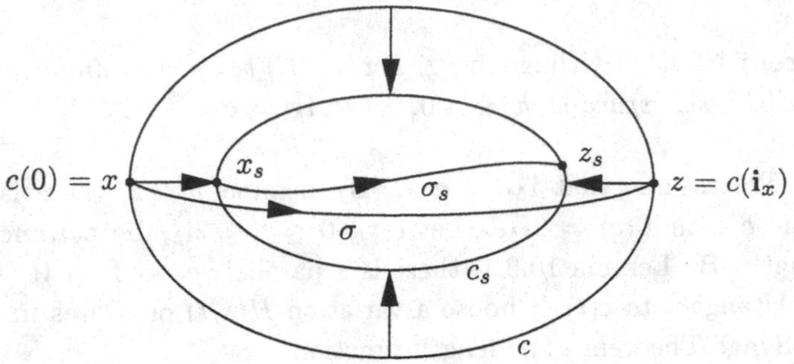

We may assume that z_{s_i} converges to $z = c(\mathbf{i}_M)$ for some sequence $s_i \to 0^+$. Further, we can assume that $\lim_{i\to\infty} \dot{\sigma}_{s_i}(r_{s_i}) \to v$. Thus σ_{s_i} converges to a minimizing geodesic segment σ from $c(0)$ to z. We still have

$$\mathbf{g}_{y_0}\Big(y_0, \dot{c}(\mathbf{i}_M)\Big) = 0,$$

where $y_o = \dot{\sigma}(r_o)$. We obtain two minimizing geodesic from $c(0)$ to z which do not form a smooth curve at z. This contradicts Lemma 12.2.5. Q.E.D.

Positively complete Finsler spaces with nonnegative flag curvature have not been completely understood. For Riemannian spaces with nonnegative sectional curvature, we have the following Cheeger-Gromoll's soul theorem.

Theorem 13.2.2 ([ChGr2]) *Let (M, g) be a complete Riemannian space with $\mathbf{K} \geq 0$. There is a totally geodesic closed submanifold $S \subset M$ such that M is diffeomorphic to the normal bundle of S in M.*

The proof is based on the Toponogov comparison theorem [To] and quite long, so is omitted. See also [ChEb].

13.3 Ricci Curvature Bounded Below

Let (M, F) be an n-dimensional Finsler space. The Riemann curvature in a direction $y \in T_xM$ is a linear transformation $\mathbf{R}_y : T_xM \to T_xM$. The Ricci curvature is defined as the trace of the Riemann curvature. Let $\{\mathbf{b}_i\}_{i=1}^n$ be an arbitrary basis for T_xM and express

$$\mathbf{R}_y(u) = R^i{}_k(y)u^k\, \mathbf{b}_i, \qquad u = u^k\mathbf{b}_k \in T_xM.$$

By definition,

$$\mathbf{Ric}(y) = \sum_{i=1}^n R^i{}_i(y).$$

Assume that $\{\mathbf{b}_i\}_{i=1}^n$ is an orthonormal basis with respect to $\mathbf{g}_y$ such that $\mathbf{b}_n = y/F(y)$. Let

$$P_i := \mathrm{span}\{\mathbf{b}_i, y\}, \qquad i = 1, \cdots, n-1.$$

Since $\mathbf{R}_y(y) = 0$, we have

$$\mathbf{K}(P_i, y) = \frac{1}{F^2(y)}\mathbf{g}_y\Big(\mathbf{R}_y(\mathbf{b}_i), \mathbf{b}_i\Big).$$

Thus

$$\mathbf{Ric}(y) = F^2(y)\sum_{i=1}^{n-1}\mathbf{K}(P_i, y).$$

By the above formula, we see that $\mathbf{K} \geq \lambda$ implies

$$\mathbf{Ric}(y) \geq (n-1)\lambda F^2(y), \qquad 0 \neq \forall y \in TM. \tag{13.6}$$

We say that $\mathbf{Ric} \geq (n-1)\lambda$ if (13.6) holds.

Theorem 13.3.1 (Bonnet-Myers[Au]) *Let (M, F) be an n-dimensional positively complete Finsler space. Suppose that* $\mathbf{Ric} \geq n-1$. *Then for any unit vector $y \in SM$, the conjugate value satisfies $\mathbf{c}_y \leq \pi$.*

Proof. Suppose that $\mathbf{c}_y > \pi$ for some $y \in SM$. Let $c(t) = \exp_x(ty)$, $0 \leq t \leq \pi$. Take a parallel $\mathbf{g}_{\dot c}$-orthonormal frame $\{E_i(t)\}_{0\leq t\leq\pi}$ along c with $E_n(t) = \dot c(t)$. For $1 \leq i \leq n-1$, let

$$V_i(t) = \sin(t)E_i(t), \qquad 0 \leq t \leq \pi.$$

By Lemma 12.1.3, we have

$$\sum_{i=1}^{n-1} \mathcal{I}_\pi(V_i, V_i) > 0.$$

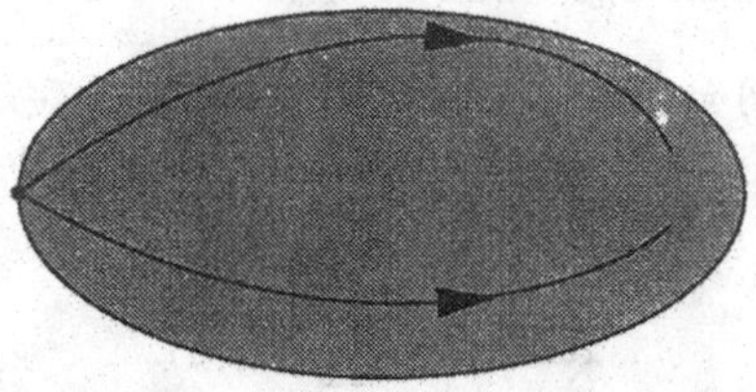

Ric $\geq n-1$

On the other hand,

$$\begin{aligned}\sum_{i=1}^{n-1} \mathcal{I}_\pi(V_i, V_i) &= \int_0^\pi \Big\{(n-1)\cos^2(t) - \mathbf{Ric}(\dot{c}(t))\sin^2(t)\Big\}dt \\ &\leq (n-1)\int_0^\pi \Big\{\cos^2(t) - \sin^2(t)\Big\}dt = 0.\end{aligned}$$

This is a contradiction. Thus $\mathbf{c}_y \leq \pi$ for any $y \in SM$. Q.E.D.

Let (M, F) be as in Theorem 13.3.1. By Lemma 12.2.3,

$$\mathbf{i}_y \leq \mathbf{c}_y \leq \pi, \qquad y \in \mathrm{S}_x M.$$

Note that the diameter at $x \in M$ satisfies

$$\mathrm{Diam}_x(M) := \sup_{z\in M} d(x, z) = \sup_{y\in \mathrm{S}_x M} \mathbf{i}_y.$$

Thus the diameter $\mathrm{Diam}_x(M)$ of M around x must be bounded by π, i.e.,

$$\mathrm{Diam}_x(M) \leq \pi.$$

Since the induced Finsler metric on the universal cover $\pi : \tilde{M} \to M$ also satisfies the same Ricci curvature bound, we still have

$$\mathrm{Diam}_{\tilde{x}}(\tilde{M}) \leq \pi,$$

where $\pi(\tilde{x}) = x$. Thus $\tilde{M}$ is compact and the fundamental group $\pi_1(M)$ is finite.

Positively complete Finsler spaces with nonnegative Ricci curvature are not completely understood yet. One of the most important theorems is the Cheeger-Gromoll splitting theorem for Riemannian spaces.

Theorem 13.3.2 ([ChGr1]) *Let (M, g) be a complete Riemannian space with* $\mathbf{Ric} \geq 0$. *Suppose that there is a line $c : (-\infty, \infty) \to M$ in M, i.e.,*

$$d\Big(c(t), c(t')\Big) = |t - t'|, \qquad t, t' \in \mathrm{R}.$$

Then M is isometric to the product space $N \times \mathrm{R}$.

Many results on Riemannian spaces with nonnegative Ricci curvature are obtained using the Cheeger-Gromoll splitting theorem. Hence they are not true for Finsler spaces.

13.4 Green-Dazord Theorem

By Proposition 5.4.3, we know that the geodesic flow ϕ_t preserves the Riemannian volume form $dV_{\dot{g}}$ on the unit tangent bundle SM. Using this property, we can prove the following

Theorem 13.4.1 (Green-Dazord [Gr][Da1]) *Let (M, F) be a closed Finsler space of dimension n. Suppose that the conjugate radius $\mathbf{c}_M \geq r > 0$. Then*

$$(n-1)\Big(\frac{\pi}{r}\Big)^2 \geq \frac{\int_{SM} \mathbf{Ric}(y) dV_{\dot{g}}}{\mathrm{Vol}_{\dot{g}}(SM)}. \tag{13.7}$$

The equality holds if and only if F has constant curvature $\mathbf{K} = \Big(\frac{\pi}{r}\Big)^2$.

Proof. Let $y \in S_xM$. The curve

$$c(t) := \phi_t(y) = \exp_x(ty), \qquad 0 \leq t \leq r,$$

is a geodesic with $\dot{c}(0) = y$. Take an arbitrary basis $\{\mathbf{b}_i\}_{i=1}^n$ with $\mathbf{b}_n = y$. Let $\{E_i(t)\}$ be a $\mathbf{g}_{\dot{c}}$-orthonormal frame along c with $E_n(t) = \dot{c}(t)$. According to Lemma 12.1.2, the following vector fields

$$V_i(t) := \sin\Big(\frac{\pi}{r}t\Big) E_i(t), \qquad 0 \leq t \leq r,$$

satisfy

$$\mathcal{I}_c(V_i, V_i) \geq 0,$$

and equality holds if and only if $V_i(t)$ are Jacobi fields along c. Observe that

$$\begin{aligned}\Phi(y) : &= \frac{1}{n-1}\sum_{i=1}^{n-1}\mathcal{I}_c(V_i, V_i) \\ &= \int_0^r \Big\{\Big(\frac{\pi}{r}\Big)^2\cos^2\Big(\frac{\pi}{r}t\Big) - \sin^2\Big(\frac{\pi}{r}t\Big)\frac{\mathbf{Ric}(\phi_t(y))}{n-1}\Big\}dt. \quad (13.8)\end{aligned}$$

Integrating Φ over SM yields

$$\int_{SM}\Phi(y)dV_{\dot g} = \frac{\pi^2}{2r}\mathrm{Vol}_{\dot g}(SM) - \int_0^r \sin^2\Big(\frac{\pi}{r}t\Big)\Big[\int_{SM}\frac{\mathbf{Ric}(\phi_t(y))}{n-1}dV_{\dot g}\Big]dt.$$

Since ϕ_t preserves the volume form $dV_{\dot g}$, we have

$$\int_{SM}\mathbf{Ric}(\phi_t(y))dV_{\dot g} = \int_{SM}\mathbf{Ric}(y)dV_{\dot g}.$$

Thus

$$\Big(\frac{\pi}{r}\Big)^2\mathrm{Vol}_{\dot g}(SM) - \int_{SM}\frac{\mathbf{Ric}(y)}{n-1}dV_{\dot g} = \frac{2}{r}\int_{SM}\Phi dV_{\dot g} \geq 0. \qquad (13.9)$$

We obtains (13.7).

Assume that the equality in (13.7) holds. (13.9) implies that $\Phi(y) = 0$ for any $y \in SM$. Thus for any set of vector fields $V_i(t) = \sin(\pi t/r)E_i(t)$ along any geodesic $c(t) = \exp_x(ty)$, $0 \leq t \leq r$, where $E_i(t)$ are parallel vector fields along c,

$$\mathcal{I}_c(V_i, V_i) = 0, \qquad i = 1, \cdots, n-1.$$

By Lemma 12.1.3, $r = \mathbf{c}_y$ and each $V_i(t)$ is a Jacobi field along c. That is,

$$0 = \mathrm{D}_{\dot c}\mathrm{D}_{\dot c}V_i(t) + \mathbf{R}_{\dot c}(V_i(t)) = \Big\{-\Big(\frac{\pi}{r}\Big)^2E_i(t) + \mathbf{R}_{\dot c}(E_i(t))\Big\}\sin^2\Big(\frac{\pi}{r}t\Big).$$

This implies

$$\mathbf{R}_{\dot c}(E_i(t)) = \Big(\frac{\pi}{r}\Big)^2E_i(t), \qquad i = 1, \cdots, n-1.$$

In particular, at $t = 0$,

$$\mathbf{R}_y(\mathbf{b}_i) = \mathbf{b}_i, \qquad i = 1, \cdots, n-1.$$

Since y is arbitrary and $\mathbf{b}_i$'s are arbitrary, we conclude that F has constant curvature $\mathbf{K} = \left(\frac{\pi}{r}\right)^2$. Q.E.D.

Chapter 14

Geometry of Hypersurfaces

The purpose of this chapter is to introduce the notions of Hessian and Laplacian on functions. For the level hypersurfaces of a distance function, we the relationship between the Hessian of the distance function and the normal curvature of its level hypersurfaces, that between the Laplacian of the distance and the mean curvature of its level hypersurfaces.

14.1 Hessian and Laplacian

Let (M, F) be a Finsler space and f a C^2 function on M. The *Hessian* of f is a map $\mathrm{D}^2 f : TM \to \mathbb{R}$ defined by

$$\mathrm{D}^2 f(y) := \frac{d^2}{ds^2}\Big(f \circ c\Big)\Big|_{s=0}, \qquad y \in T_xM,$$

where $c : (-\varepsilon, \varepsilon) \to M$ is the geodesic with $\dot{c}(0) = y \in T_xM$. In local coordinates,

$$\begin{aligned} \mathrm{D}^2 f(y) &= \frac{\partial^2 f}{\partial x^i \partial x^j}(x)\dot{c}^i(0)\dot{c}^j(0) + \frac{\partial f}{\partial x^i}(x)\ddot{c}^i(0) \\ &= \frac{\partial^2 f}{\partial x^i \partial x^j}(x) y^i y^j - 2\frac{\partial f}{\partial x^i}(x) G^i(y). \end{aligned} \qquad (14.1)$$

From the definition, we have

$$\mathrm{D}^2 f(\lambda y) = \lambda^2 \mathrm{D}^2 f(y), \qquad \lambda > 0.$$

In general, $\mathrm{D}^2 f(y)$ is not quadratic in $y \in T_xM$.

The Hessian of a C^∞ distance function ρ on an open subset $\mathcal{U} \subset M$ has special geometric meanings. First, the gradient $\nabla\rho$ is a unit vector field on $\mathcal{U}$. It induces a C^∞ Riemannian metric on $\mathcal{U}$,

$$\hat{F}(y) := \sqrt{\mathbf{g}_{\nabla\rho}(y,y)}, \qquad y \in TM.$$

By Lemma 3.2.1,

$$F(\nabla\rho) = 1 = \hat{F}(\hat{\nabla}\rho).$$

Thus $\nabla\rho = \hat{\nabla}\rho$ is also the gradient of ρ with respect to $\hat{F}$. However, the Hessians $\mathrm{D}^2\rho$ and $\hat{\mathrm{D}}^2\rho$ of ρ with respect to F and $\hat{F}$ are not equal.

Lemma 14.1.1 *Let (M,F) be a Finsler space. For a C^∞ distance function ρ on an open subset $\mathcal{U}$,*

$$\mathrm{D}^2\rho(y) = \hat{\mathrm{D}}^2\rho(y) - \mathbf{T}_{\nabla\rho}(y), \qquad y \in T_x\mathcal{U}, \tag{14.2}$$

where $\hat{\mathrm{D}}^2\rho$ denotes the Hessian of ρ with respect to $\hat{F} = \sqrt{\mathbf{g}_{\nabla\rho}}$.

Proof: Let $\hat{\Gamma}^i_{jk}(x)$ denote the Christoffel symbols of the Levi-Civita connection $\hat{\mathrm{D}}$. By (14.1), the Hessian of $\hat{\mathrm{D}}^2\rho$ is given by

$$\hat{\mathrm{D}}^2\rho(y) = \frac{\partial^2\rho}{\partial x^i\partial x^j}(x)y^iy^j - 2\frac{\partial\rho}{\partial x^i}(x)\hat{G}^i(y),$$

where $2\hat{G}^i(y) := \hat{\Gamma}^i_{jk}(x)y^jy^k$.

$$\hat{\mathrm{D}}^2\rho(y) - \mathrm{D}^2\rho(y) = 2\Big\{G^i(y) - \hat{G}^i(y)\Big\}\frac{\partial\rho}{\partial x^i}(x).$$

By definition (3.15) and (10.1), we have

$$\begin{aligned}
\mathbf{T}_{\nabla\rho}(y) &= \mathbf{g}_{\nabla\rho}\Big(\mathrm{D}_yY - \hat{\mathrm{D}}_yY,\ \nabla\rho\Big)\\
&= d\rho\Big(\mathrm{D}_vV - \hat{\mathrm{D}}_vV\Big)\\
&= 2\Big\{G^i(y) - \hat{G}^i(y)\Big\}\frac{\partial\rho}{\partial x^i}(x)\\
&= \hat{\mathrm{D}}^2\rho(y) - \mathrm{D}^2\rho(y),
\end{aligned}$$

where Y is an arbitrary extension of y. Q.E.D.

Now we introduce the Laplacian on a Finsler m space. Let $(M,F,d\mu)$ be a Finsler m space. For a C^k $(k \geq 1)$ vector field X on $(M,d\mu)$, the divergence $\mathrm{div}(X)$ is a C^{k-1} function on M. On the other hand, for a C^k ($k \geq 2$) function f on (M, F), the gradient ∇f is C^{k-1} on $\mathcal{U} := \{df \neq 0\}$ and C^0 at $M - \mathcal{U}$. Therefore, $\mathrm{div}(\nabla f)$ is a C^{k-2} function on $\mathcal{U}$. Define Δf on $\mathcal{U}$ by

$$\Delta f := \mathrm{div}(\nabla f). \tag{14.3}$$

Let $d\mu = \sigma(x) dx^1 \cdots dx^n$ and $\nabla f = \nabla^i f \frac{\partial}{\partial x^i}$. The Laplacian of f is expressed by

$$\Delta f = \frac{1}{\sigma} \frac{\partial}{\partial x^i}\Big(\sigma \nabla^i f\Big).$$

For any $\phi \in C_o^\infty(\mathcal{U})$,

$$\mathrm{div}(\phi \nabla f) = \phi\, \Delta f + d\phi(\nabla f).$$

Applying the divergence formula (2.6) to the above identity yields

$$\int_M \phi\, \Delta f d\mu = -\int_M d\phi(\nabla f) d\mu, \qquad \phi \in C_o^\infty(\mathcal{U}). \tag{14.4}$$

This leads to the definition of Δf on the whole M in the distribution sense.

$$\int_M (\Delta f)\phi d\mu := -\int_M d\phi(\nabla f) d\mu, \quad \forall \phi \in C_o^\infty(M). \tag{14.5}$$

We call Δ the *Laplacian* on functions. Compare [Sh4] and [BePa].

Assume that $df \neq 0$ on an open subset $\mathcal{U} \subset M$. The gradient of f is given by

$$\nabla f = A^i(df) \frac{\partial}{\partial x^i}, \tag{14.6}$$

where $A^i(\eta)$ are given in a standard local coordinate system (x^i, η_i) in T^*M by

$$A^i(\eta) = \frac{1}{2} \frac{\partial [F^{*2}]}{\partial \eta_i}(\eta) = g^{*ij}(\eta)\eta_j, \qquad \eta \neq 0.$$

The Laplacian is locally expressed by

$$\begin{aligned} \Delta f &= \frac{1}{\sigma(x)} \frac{\partial}{\partial x^i}\Big(\sigma(x) A^i(df)\Big) \\ &= \frac{1}{\sigma(x)} \frac{\partial}{\partial x^i}\Big(\sigma(x) g^{*ij}(df_x) \frac{\partial f}{\partial x^j}\Big). \end{aligned} \tag{14.7}$$

Thus the Laplacian is a fully non-linear elliptic operator.

The variational problem of the canonical energy functional also gives rise to the Laplacian. Let H^1 denote the Hilbert space of all L^2 functions on M such that $df \in L^2$. Denote by H^1_o the space of functions $u \in \mathrm{H}^1$ with $\int_M u d\mu = 0$, if $\partial M = \emptyset$ and with $u_{|\partial M} = 0$ if $\partial M \neq \emptyset$. The *canonical energy functional* $\mathcal{E}$ on H^1_o is defined by

$$\mathcal{E}(u) := \frac{\int_M [F^*(du)]^2 d\mu}{\int_M u^2 d\mu},$$

where F^* denotes the dual Finsler metric, namely, $F^*(\xi) := \sup_{F(y)=1} \xi(y)$. For functions $u, \varphi \in \mathrm{H}^1_o$,

$$\frac{d}{d\varepsilon}\Big[F^{*2}(du + \varepsilon d\varphi)\Big]\Big|_{\varepsilon=0} = \frac{\partial [F^{*2}]}{\partial \eta_i}(du)\frac{\partial \varphi}{\partial x^i} = 2A^i(du)\frac{\partial \varphi}{\partial x^i} = 2d\varphi(\nabla u).$$

Thus, for any $u \in \mathrm{H}^1_o$ with $\int_M u^2 d\mu = 1$,

$$d_u\mathcal{E}(\varphi) = 2\int d\varphi(\nabla u)\, d\mu - 2\lambda \int u\varphi\, d\mu \qquad \forall \varphi \in \mathrm{H}^1_o \tag{14.8}$$

where $\lambda = \mathcal{E}(u)$. From (14.5) and (14.8) it follows that a function $u \in \mathrm{H}^1_o$ satisfies $d_u\mathcal{E} = 0$ with $\lambda = \mathcal{E}(u)$ if and only if

$$\Delta u + \lambda u = 0.$$

In this case, λ and u are called an *eigenvalue* and an *eigenfunction* of $(M, F, d\mu)$, respectively. It is shown in [GeSh] that any eigenfunction f is $C^{1,\alpha}$ for some $0 < \alpha < 1$. Moreover, f is C^∞ on the open subset $\{df \neq 0\}$. Examples show that the eigenfunctions (of the Dirichlet problem) are at most $C^{1,1}$. Thus the best regularity of eigenfunctions one can expect is $C^{1,1}$.

Assume that M is compact without boundary. Recall the definition of the first eigenvalue in (4.8).

$$\lambda_1(M) := \inf_{u \in C^\infty(M)} \frac{\int_M [F^*(du)]^2 d\mu}{\inf_{\lambda \in \mathrm{R}} \int_M |u - \lambda|^2 d\mu}. \tag{14.9}$$

We claim that $\lambda_1 = \lambda_1(M)$ is the smallest eigenvalue of $(M, F, d\mu)$.

For any function $u \in C^\infty(M)$,

$$\int_M |u-\lambda|^2 d\mu = \int_M u^2 d\mu - 2\lambda \int_M u d\mu + \lambda^2 \int_M d\mu.$$

Thus

$$\inf_{\lambda \in \mathrm{R}} \int_M |u-\lambda|^2 d\mu = \int_M \Big[u - \mu(M)^{-1} u\Big]^2 d\mu.$$

Since $C^\infty(M)$ is dense in H^1_o, the above identity implies

$$\inf_{u \in C^\infty(M)} \frac{\int_M [F^*(du)]^2 d\mu}{\inf_{\lambda \in \mathrm{R}} \int_M |u-\lambda|^2 d\mu} = \inf_{u \in \mathrm{H}^1_o(M)} \frac{\int_M [F^*(du)]^2 d\mu}{\int_M u^2 d\mu}.$$

Thus $\lambda_1(M)$ is the minimum of the energy functional E, that is, the smallest eigenvalue of M.

Lemma 14.1.2 *Let $(M, F, d\mu)$ be a Finsler m space and ρ a C^∞ distance function on an open subset. Then*

$$\Delta\rho = \hat{\Delta}\rho - \mathbf{S}(\nabla\rho), \tag{14.10}$$

where $\hat{\Delta}$ denotes the Laplacian of $\hat{g} := \mathbf{g}_{\nabla\rho}$.

Proof: Let $d\mu$ and $dV_{\hat{g}}$ denote the volume forms of F and $\hat{g}$, respectively. In a local coordinate system (x^i), write

$$d\mu = \sigma(x) dx^1 \wedge \cdots \wedge dx^n, \qquad dV_{\hat{g}} = \hat{\sigma}(x) dx^1 \wedge \cdots \wedge dx^n.$$

Let $\nabla\rho = \nabla^i\rho \frac{\partial}{\partial x^i}$. By Lemma 3.2.1, we know that $\nabla\rho$ is also the gradient of ρ with respect to F and $\hat{g}$, i.e., $\nabla\rho = \hat{\nabla}\rho$. For the Riemannian metric $\hat{g}$,

$$\hat{\sigma}(x) = \sqrt{\det\Big(\hat{g}_x(\frac{\partial}{\partial x^i}|_x, \frac{\partial}{\partial x^j}|_x)\Big)}.$$

By (7.25) and (7.29), the S-curvature of F is given by

$$\mathbf{S}(\nabla\rho) = \nabla\rho\Big[\ln\frac{\hat{\sigma}}{\sigma}\Big].$$

Equation (14.3) gives

$$\hat{\Delta}\rho = \frac{1}{\hat{\sigma}} \frac{\partial}{\partial x^i}\Big(\hat{\sigma}\nabla^i\rho\Big) = \frac{\partial \nabla^i\rho}{\partial x^i} + \nabla\rho\Big(\ln\hat{\sigma}\Big).$$

Thus

$$\begin{aligned}\Delta\rho &= \frac{\partial \nabla^i\rho}{\partial x^i} + \nabla\rho\Big(\ln\sigma\Big)\\ &= \frac{\partial \nabla^i\rho}{\partial x^i} + \nabla\rho\Big(\ln\hat{\sigma}\Big) - \nabla\rho\Big(\ln\frac{\hat{\sigma}}{\sigma}\Big)\\ &= \hat{\Delta}\rho - \mathbf{S}(\nabla\rho).\end{aligned}$$

Q.E.D.

The relation between the Hessian and the Laplacian becomes simpler when the Finsler metric is Riemannian. Suppose that F is Riemannian, i.e., $F(y) = \sqrt{g_{ij}(x)y^iy^j}$. Note that g_{ij}, $g^{*ij} = g^{ij}$, Γ^k_{ij}, etc. are functions of x only. By (14.1) and (14.7), we have

$$\mathrm{D}^2 f(y) = \Big\{\frac{\partial^2 f}{\partial x^i \partial x^j} - \frac{\partial f}{\partial x^k}\Gamma^k_{ij}\Big\} y^i y^j, \tag{14.11}$$

$$\Delta f = g^{ij}\Big\{\frac{\partial^2 f}{\partial x^i\partial x^j} + \Big[g_{jk}\frac{\partial g^{kl}}{\partial x^l} + \frac{\partial}{\partial x^j}\Big(\ln\sigma\Big)\Big]\frac{\partial f}{\partial x^i}\Big\}, \tag{14.12}$$

where $\sigma := \sqrt{\det(g_{ij})}$. Note

$$\frac{\partial}{\partial x^j}\Big(\ln\sigma\Big) = \frac{1}{2}g^{kl}\frac{\partial g_{kl}}{\partial x^j}.$$

Thus

$$g_{jk}\frac{\partial g^{kl}}{\partial x^l} + \frac{\partial}{\partial x^j}\Big(\ln\sigma\Big) = -g^{kl}\frac{\partial g_{jk}}{\partial x^l} + \frac{1}{2}g^{kl}\frac{\partial g_{kl}}{\partial x^j} = -g^{kl}g_{jt}\Gamma^t_{kl}. \tag{14.13}$$

Plugging (14.13) into (14.12) yields

$$\Delta f = g^{ij}\Big\{\frac{\partial^2 f}{\partial x^i\partial x^j} - \frac{\partial f}{\partial x^k}\Gamma^k_{ij}\Big\}. \tag{14.14}$$

Let $\{\mathbf{b}_i\}_{i=1}^n$ be an orthonormal basis for T_xM. (14.14) can be expressed by

$$\Delta f = \sum_{i=1}^n \mathrm{D}^2 f(\mathbf{b}_i). \tag{14.15}$$

In the above sense, the Laplacian Δf is the mean value of the Hessian $\mathrm{D}^2 f$.

14.2 Normal Curvature

The purpose of this section is to introduce the notion of normal curvature for submanifolds in a Finsler space. We will use it to study the level hypersurfaces of a distance function on a Finsler space.

Let (M, F) be a Finsler space and $\varphi : N \to M$ be a submanifold. Let $\bar{F} := \varphi^* F$ denote the induced Finsler metric on N. For simplicity, we denote $d\varphi(T_{\bar{x}}N)$ by $T_x N$, where $x = \varphi(\bar{x})$. For a tangent vector $y \in T_x N$, there is a unique geodesic $\bar{c}$ in $(N, \bar{F})$ such that the curve $c(t) := \varphi \circ \bar{c}(t)$ satisfies $\dot{c}(0) = y$. Define

$$\mathbf{A}(y) := -\mathrm{D}_{\dot{c}}\dot{c}(0).$$

From the definition, $\mathbf{A}(y)$ has the following homogeneous property

$$\mathbf{A}(\lambda y) = \lambda^2 \mathbf{A}(y), \qquad \lambda > 0.$$

$\mathbf{A}(y)$ is called the *normal curvature* of N in the direction of $y \in T_x N$.

Let $(\bar{x}^a)$ be a local coordinate system in N and (x^i) a local coordinate system in M. Let $y = \dot{c}(0) = d\varphi(\bar{y})$, where $\bar{y} = \dot{\bar{c}}(0)$. We have

$$\begin{aligned}
\mathrm{D}_{\dot{c}}\dot{c}(0) &= \Big\{ \frac{d^2 c^i}{dt^2}(0) + 2G^i(y) \Big\} \frac{\partial}{\partial x^i}|_x \\
&= \Big\{ \frac{\partial^2 \varphi^i}{\partial \bar{x}^a \partial \bar{x}^b}(\bar{x}) \bar{y}^a \bar{y}^b + \frac{\partial x^i}{\partial \bar{x}^a}(\bar{x}) \frac{d^2 \bar{c}^a}{dt^2} + 2G^i(y) \Big\} \frac{\partial}{\partial x^i}|_x \\
&= \Big\{ \frac{\partial^2 \varphi^i}{\partial \bar{x}^a \partial \bar{x}^b}(\bar{x}) \bar{y}^a \bar{y}^b - 2 \frac{\partial x^i}{\partial \bar{x}^a}(\bar{x}) \bar{G}^a(\bar{y}) + 2G^i(y) \Big\} \frac{\partial}{\partial x^i}|_x .
\end{aligned}$$

Then the normal curvature $\mathbf{A}(y)$ in the direction $y = d\varphi(\bar{y})$ is given by

$$\mathbf{A}(y) = -\Big\{ \frac{\partial^2 \varphi^i}{\partial \bar{x}^a \partial \bar{x}^b}(\bar{x}) \bar{y}^a \bar{y}^b - 2 \frac{\partial \varphi^i}{\partial \bar{x}^a} \bar{G}^a(\bar{y}) + 2G^i(y) \Big\} \frac{\partial}{\partial x^i}|_x .$$

Assume that N is a hypersurface in M. Let $\mathbf{n}$ denote a normal vector of N at $x \in N$. Define

$$\Lambda_{\mathbf{n}}(y) := \mathbf{g_n}\Big(\mathbf{n}, \mathbf{A}(y)\Big), \qquad y \in T_x N.$$

$\Lambda_{\mathbf{n}}(y)$ is called the *normal curvature* of N in the direction y with respect to $\mathbf{n}$. There are exactly two normal vectors $\mathbf{n}, \mathbf{n}' \in T_x M$. In general, $\mathbf{n}' \neq -\mathbf{n}$ and $\Lambda_{\mathbf{n}} \neq -\Lambda_{\mathbf{n}'}$.

Let $\mathbf{n} = \mathbf{n}^i \frac{\partial}{\partial x^i}|_x \in T_xM$. $\Lambda_{\mathbf{n}}(y)$ is expressed by

$$\Lambda_{\mathbf{n}}(y) := -\mathbf{n}^j g_{ij}(\mathbf{n})\Big\{ \frac{\partial^2 \varphi^i}{\partial \bar{x}^a \partial \bar{x}^b}(\bar{x})\bar{y}^a\bar{y}^b - 2\frac{\partial \varphi^i}{\partial \bar{x}^a}\bar{G}^a(\bar{y}) + 2G^i(y)\Big\}. \tag{14.16}$$

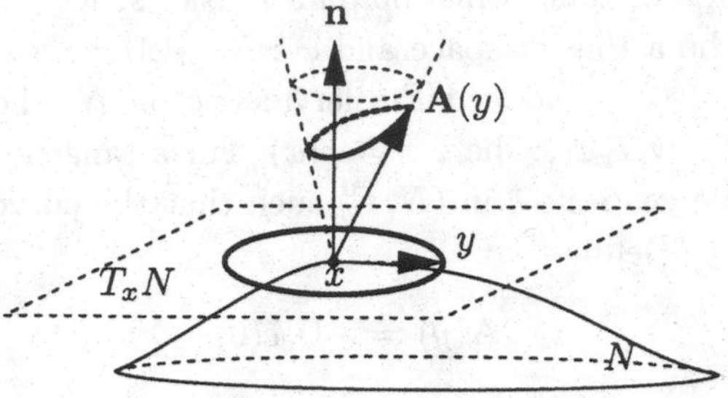

Note that $\mathbf{A}(y)$ is not orthogonal to N with respect to $\mathbf{g}_{\mathbf{n}}$. Thus $\Lambda_{\mathbf{n}}(y)$ does not capture all the data of $\mathbf{A}(y)$.

Example 14.2.1 *Consider the indicatrix* $\mathrm{S} = F^{-1}(1)$ *in a Minkowski space* (V, F). *Let* $\varphi : \mathrm{S} \to \mathrm{V}$ *denote the natural embedding. For* $y \in \mathrm{S}$, *identify* $T_y\mathrm{S}$ *with a hyperplane*

$$\mathrm{W}_y := \Big\{ v \in \mathrm{V},\ \mathbf{g}_y(y, v) = 0 \Big\} \subset \mathrm{V}.$$

Define $\dot{g}$ *and* $\dot{\mathbf{C}}$ *by*

$$\dot{g}(u, v) := \mathbf{g}_y(u, v), \qquad \dot{g}(\dot{\mathbf{C}}(u, v), w) := \mathbf{C}_y(u, v, w), \tag{14.17}$$

where $u, v, w \in T_y\mathrm{S} = \mathrm{W}_y$. *Fix an arbitrary basis* $\{\mathbf{b}_i\}_{i=1}^n$ *for* V. *The vector-valued function* $\varphi = \varphi^i \mathbf{b}_i$ *satisfies the following Varga equation [Va]*

$$\dot{\mathrm{D}}^2\varphi + d\varphi(\dot{\mathbf{C}}) + \dot{g}\,\varphi = 0.$$

In local coordinate system $(\bar{x}^a)$ *in* S,

$$\frac{\partial^2 \varphi^i}{\partial \bar{x}^a \partial \bar{x}^b} + M^c_{ab}\frac{\partial \varphi^i}{\partial \bar{x}^c} + \dot{g}_{ab}\varphi^i = 0,$$

where M^c_{ab} *are some local functions on* S. *It is easy to see that* $\mathbf{n} = \varphi$, *since* $F^2(\varphi) = 1$. *Namely,*

$$\mathbf{g}_{\mathbf{n}}\Big(\mathbf{n}, \frac{\partial \varphi}{\partial \bar{x}^a}\Big) = \mathbf{n}^j g_{ij}(\mathbf{n})\frac{\partial \varphi^i}{\partial \bar{x}^a} = 0.$$

Observe that

$$\begin{aligned}
\Lambda_{\mathbf{n}}(y) &= -\mathbf{n}^j g_{ij}(\mathbf{n})\Big\{\frac{\partial^2\varphi^i}{\partial\bar{x}^a\partial\bar{x}^b}(\bar{x})\bar{y}^a\bar{y}^b - 2\frac{\partial\varphi^i}{\partial\bar{x}^a}\bar{G}^a(\bar{y})\Big\}\\
&= -\mathbf{n}^j g_{ij}(\mathbf{n})\Big\{-M^c_{ab}\frac{\partial\varphi^i}{\partial\bar{x}^c} - \dot{g}_{ab}\varphi^i - 2\frac{\partial\varphi^i}{\partial\bar{x}^a}\bar{G}^a(\bar{y})\Big\}\\
&= -\mathbf{n}^j g_{ij}(\mathbf{n})\varphi^i\,\dot{g}(y,y)\\
&= \dot{g}(y,y).
\end{aligned}$$

We obtain

$$\Lambda_{\mathbf{n}}(y) = \dot{g}(y,y), \qquad \forall y \in T_x\mathrm{S}. \tag{14.18}$$

♮

Let N be a level hypersurface of a C^∞ distance function ρ on an open subset $\mathcal{U} \subset M$. By (3.15),

$$d\rho(v) = \mathbf{g}_{\nabla\rho}(\nabla\rho, v), \qquad v \in TM. \tag{14.19}$$

This implies that

$$\mathbf{g}_{\nabla\rho}(\nabla\rho, v) = 0, \qquad v \in TN.$$

Thus $\mathbf{n} := \nabla\rho|_N$ is a normal vector to N.

Proposition 14.2.1 *Let ρ be a C^∞ distance function on an open subset $\mathcal{U} \subset (M,F)$ and $N := \rho^{-1}(s) \subset \mathcal{U}$. With respect to the normal vector $\mathbf{n} := \nabla\rho_x$ at $x \in N$, the normal curvature $\Lambda_{\mathbf{n}}$ satisfies*

$$\mathrm{D}^2\rho(y) = \Lambda_{\mathbf{n}}(y), \qquad \forall y \in T_xN. \tag{14.20}$$

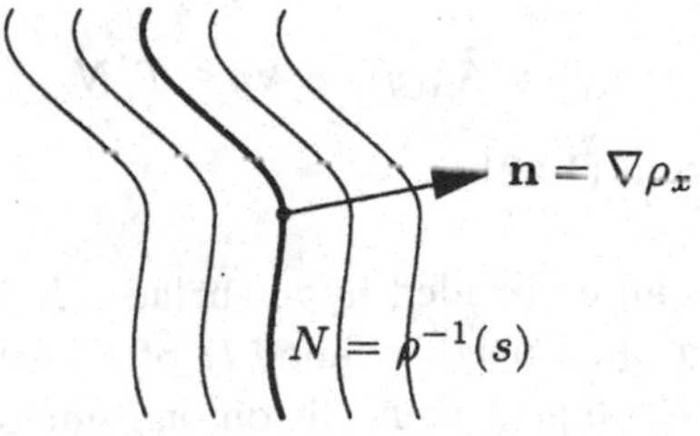

Proof. Let $\varphi : N \to \mathcal{U}$ be the natural embedding and $\bar{F} = \varphi^* F$ the induced Finsler metric. For a vector $y = d\varphi(\bar{y}) \in T_x N$, where $\bar{y} \in T_{\bar{x}} N$, let $\bar{c}(t)$ be the geodesic in $(N, \bar{F})$ with $\frac{d\bar{c}}{dt}(0) = \bar{y}$. Then the curve $c(t) := \varphi \circ \bar{c}(t)$ satisfies $\dot{c}(0) = y$. Since $\rho(c(t)) = constant$, differentiating it twice with respect to t yields

$$\frac{\partial^2 \rho}{\partial x^i \partial x^j}(x) y^i y^j = -\frac{\partial \rho}{\partial x^i}(x) \frac{\partial^2 \varphi^i}{\partial \bar{x}^a \partial \bar{x}^b}(\bar{x}) \bar{y}^a \bar{y}^b + \frac{\partial \rho}{\partial x^i}(x) \frac{\partial \varphi^i}{\partial \bar{x}^a}(\bar{x}) \bar{G}^a(\bar{y}), \quad (14.21)$$

where we have used the geodesic equation (5.5) of $\bar{c}$. On the other hand, by the definition of $\mathrm{D}^2\rho$,

$$\mathrm{D}^2\rho(y) = \frac{\partial^2 \rho}{\partial x^i \partial x^j}(x) y^i y^j - 2\frac{\partial \rho}{\partial x^i}(x) G^i(y). \quad (14.22)$$

Plugging (14.21) into (14.22) yields

$$\mathrm{D}^2\rho(y) = -\frac{\partial \rho}{\partial x^i}(x) \Big\{ \frac{\partial^2 \varphi^i}{\partial \bar{x}^a \partial \bar{x}^b}(\bar{x}) \bar{y}^a \bar{y}^b - 2\frac{\partial \varphi^i}{\partial \bar{x}^a} \bar{G}^a(\bar{y}) + 2G^i(y) \Big\}. \quad (14.23)$$

In virtue of (14.16) and (14.23), we conclude that

$$\mathrm{D}^2\rho(y) = d\rho\Big(\mathbf{A}(y)\Big) = \mathbf{g}_{\nabla\rho}\Big(\nabla\rho, \mathbf{A}(y)\Big) = \Lambda_{\mathbf{n}}(y), \qquad y \in T_x N.$$

This proves (14.20). Q.E.D.

Corollary 14.2.2 *Let ρ be a C^∞ distance function on an open subset $\mathcal{U} \subset (M, F)$ and $\hat{g} := \mathbf{g}_{\nabla\rho}$. Let $\Lambda_{\mathbf{n}}$ and $\hat{\Lambda}_{\mathbf{n}}$ denote the normal curvature of $N := \rho^{-1}(s)$ in $(\mathcal{U}, F)$ and $(\mathcal{U}, \hat{g})$ with respect to $\mathbf{n} := \nabla\rho_x = \hat{\nabla}\rho_x$ at $x \in N$, respectively. Then*

$$\Lambda_{\mathbf{n}}(y) = \hat{\Lambda}_{\mathbf{n}}(y) - \mathbf{T}_{\mathbf{n}}(y), \qquad \forall y \in T_x N \quad (14.24)$$

Proof. By Lemma 3.2.1, we know that ρ is also a distance function of $\hat{g}$. Thus (14.20) holds for $\hat{g}$, i.e.,

$$\hat{\mathrm{D}}^2\rho(y) = \hat{\Lambda}_{\mathbf{n}}(y), \qquad \forall y \in T_x N. \quad (14.25)$$

Then (14.24) follows from (14.2). Q.E.D.

Let $N \subset (M, F)$ be an embedded hypersurface. N is said to be *convex* at $x \in N$ if there is an open neighborhood $\mathcal{U}$ of x such that all geodesics in $\mathcal{U}$, issuing from x and tangent to N, lie on one side of N. In addition, if these geodesics do not intersect N except for x, then N is said to be *strictly convex* at x. Note that the indicatrix of a Minkowski space is strictly convex. More general, we have the following

Theorem 14.2.3 *Let $N \subset (M, F)$ be an embedded hypersurface and $\Lambda_{\mathbf{n}}$ the normal curvature of N at $x \in N$ with respect to a normal vector $\mathbf{n}$. If N is convex at $x \in N$, then either $\Lambda_{\mathbf{n}} \geq 0$ or $\Lambda_{\mathbf{n}} \leq 0$. If $\Lambda_{\mathbf{n}} > 0$ or $\Lambda_{\mathbf{n}} < 0$, then N is strictly convex at x.*

Proof: Suppose that N is convex at x. Let $\mathcal{U}$ be an open neighborhood of x, in which all geodesics issuing from x and tangent to N at x, lie on one side of N. Choose a smaller $\mathcal{U}$ if necessary, we may assume that there is a C^∞ distance function ρ on $\mathcal{U}$ such that $N \cap \mathcal{U} = \rho^{-1}(0)$. Further we may assume that all these geodesics lie in $\mathcal{U}^+ := \{x : \rho(x) \geq 0\}$.

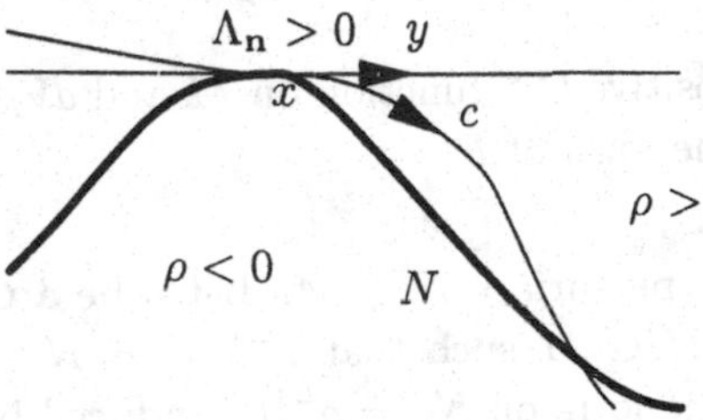

Let $y \in T_xN$ and $c(t)$ be the geodesic in $\mathcal{U}$ with $\dot{c}(0) = y$. By assumption, $\rho \circ c(t) \geq 0$ for all small t and $\rho \circ c(0) = \rho(x) = 0$. Thus for $\mathbf{n} = \nabla \rho_x$

$$\Lambda_{\mathbf{n}}(y) = \mathrm{D}^2\rho(y) = \frac{d^2}{dt^2}\Big[\rho \circ c(t)\Big]\Big|_{t=0} \geq 0.$$

This proves the first statement.

Now we assume that $\Lambda_{\mathbf{n}} \geq \delta > 0$ for $\mathbf{n} = \nabla \rho_x$. For any normal geodesic c in M with $\dot{c}(0) = y \in T_xN$,

$$\frac{d^2}{dt^2}\Big[\rho \circ c(t)\Big]|_{t=0} = \mathrm{D}^2\rho(y) = \Lambda_{\mathbf{n}}(y) \geq \delta > 0.$$

Thus, there is a small number $\varepsilon > 0$ such that

$$\rho \circ c(t) \geq \frac{1}{4}\delta t^2, \qquad \forall t \in (-\varepsilon, \varepsilon).$$

Clearly, ε can be taken independent of the unit vectors $y \in T_xN$. Therefore all geodesics in a small neighborhood $\mathcal{U}$ with $\dot{c}(0) = y \in T_xN$ lie on one side of N, which intersect N only at x. This proves the second statement. Q.E.D.

14.3 Mean Curvature

In this section, we are going to introduce the notion of mean curvature for hypersurfaces in a Finsler m space. Since every Finsler metric induces the Busemann-Hausdorff volume form, the mean curvature is of course defined for hypersurfaces in a Finsler space. Let $(M, F, d\mu)$ be an n-dimensional Finsler m space. Express $d\mu$ by

$$d\mu = \phi_\mu dV_F, \tag{14.26}$$

where ϕ_μ is a positive C^∞ function on M and dV_F denotes the Busemann-Hausdorff volume form of F.

Consider a hypersurface $N \subset M$. Let ρ be a C^∞ distance function on an open subset $\mathcal{U} \subset M$ such that $\rho^{-1}(s) = N \cap \mathcal{U}$ for some s. Let $d\nu_t$ denote the volume form on $N_t := \rho^{-1}(t)$ induced by $d\mu$ and $d\nu := d\nu_s$. Let $c(t)$ be an integral curve of $\nabla\rho$ with $c(0) \in N_s$. We have $\rho(c(t)) = t$, hence $c(s+\varepsilon) \in N_{s+\varepsilon}$ for small $\varepsilon > 0$. By definition, the flow ϕ_ε of $\nabla\rho$ satisfies

$$\phi_\varepsilon(c(s)) := c(s+\varepsilon).$$

Thus,

$$\phi_\varepsilon : N \cap \mathcal{U} = N_s :\to N_{s+\varepsilon}.$$

The pull-back $(n-1)$-form $(\phi_\varepsilon)^* d\nu_{s+\varepsilon}$ is a multiple of $d\nu_s$. Thus there is a function $\Theta(x, \varepsilon)$ on N such that

$$(\phi_\varepsilon)^* d\nu_{s+\varepsilon}|_x = \Theta(x, \varepsilon) d\nu_s|_x, \qquad \forall x \in N, \tag{14.27}$$

Note

$$\Theta(x, 0) = 1, \qquad \forall x \in N. \tag{14.28}$$

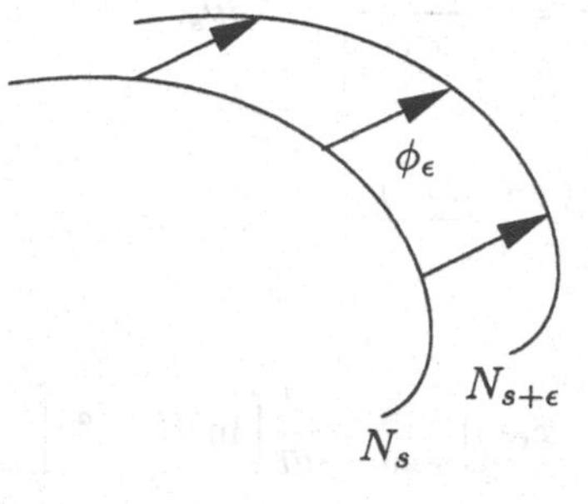

Set

$$\Pi_{\mathbf{n}} := \frac{\partial}{\partial \varepsilon}\Big(\ln \Theta(x,\varepsilon)\Big)_{|\varepsilon=0}. \tag{14.29}$$

$\Pi_{\mathbf{n}}$ is called the *mean curvature* of N at x with respect to $\mathbf{n} := \nabla\rho_x$.

Proposition 14.3.1 *Let ρ be a C^∞ distance function on an open subset $\mathcal{U} \subset (M,F)$. Let $\Pi_{\mathbf{n}}$ be the mean curvature of $N := \rho^{-1}(s)$ at x with respect to $\mathbf{n} := \nabla\rho_x$. Then*

$$\Delta\rho(x) = \Pi_{\mathbf{n}}, \qquad \forall x \in N. \tag{14.30}$$

Proof. Take a special local coordinate system in M,

$$\varphi = (t, x^a) : \mathcal{U} \subset M \to \mathrm{R}^n$$

such that $\rho \circ \varphi^{-1}(t,x^a) = t$ and $\nabla\rho = \frac{\partial}{\partial t}$. Write

$$\nabla\rho = Y^1\frac{\partial}{\partial t} + Y^a\frac{\partial}{\partial x^a}$$

with $Y^1 = 1$ and $Y^a = 0$. Put

$$d\mu = \sigma(t,x^a)dt \wedge dx^2 \wedge \cdots \wedge dx^n.$$

The induced volume form $d\nu_t$ on $N_t := \rho^{-1}(t)$ is given by

$$d\nu_t = \sigma(t,x^a)dx^2 \wedge \cdots \wedge dx^n.$$

Observe that $\phi_\varepsilon : (s,x^a) \to (s+\varepsilon, x^a)$ so that

$$\begin{aligned}(\phi_\varepsilon)^* d\nu_{s+\varepsilon} &= \sigma(s+\varepsilon,x^a)dx^2 \cdots \wedge dx^n \\ &= \frac{\sigma(s+\varepsilon,x^a)}{\sigma(s,x^a)}\sigma(s,x^a)dx^2 \cdots \wedge dx^n\end{aligned}$$

$$= \frac{\sigma(s+\varepsilon, x^a)}{\sigma(s, x^a)} d\nu_s.$$

Thus

$$\Theta(x, \varepsilon) = \frac{\sigma(s+\varepsilon, x^a)}{\sigma(s, x^a)}, \qquad \forall x = \varphi^{-1}(s, x^a) \in N.$$

This gives

$$\Pi_{\mathbf{n}} = \frac{d}{d\varepsilon}\Big[\ln \Theta(x, \varepsilon)\Big]_{|\varepsilon=0} = \frac{d}{dt}\Big[\ln \sigma(t, x^a)\Big]_{|t=s}. \tag{14.31}$$

On the other hand, by (14.3), we have

$$\Delta\rho(x) = \frac{\partial}{\partial t}\Big[\ln \sigma(t, x^a)\Big]_{|t=s}. \tag{14.32}$$

Now (14.30) follows from (14.31) and (14.32). Q.E.D.

Corollary 14.3.2 *Let ρ be a C^∞ distance function on an open subset $\mathcal{U} \subset (M, F)$ and $\hat{g} := \mathbf{g}_{\nabla\rho}$. Let $\Pi_{\mathbf{n}}$ and $\hat{\Pi}_{\mathbf{n}}$ denote the mean curvature of $N := \rho^{-1}(s)$ in $(\mathcal{U}, F)$ and $(\mathcal{U}, \hat{g})$ with respect to $\mathbf{n} = \nabla\rho|_N$, respectively. Then*

$$\Pi_{\mathbf{n}} = \hat{\Pi}_{\mathbf{n}} - \mathbf{S}(\mathbf{n}). \tag{14.33}$$

Proof: Let $\hat{\Delta}$ denote the Laplacian of $\hat{g}$. By Lemma 3.2.1, we know that ρ is also a distance function of $\hat{g}$. Then Proposition 14.3.1 holds for $\hat{g}$, that is,

$$\hat{\Delta}\rho = \hat{\Pi}_{\mathbf{n}}. \tag{14.34}$$

By (14.10), (14.30) and (14.34), one obtains (14.33). Q.E.D.

We have defined the normal curvature and the mean curvatures. We are going to show that for hypersurfaces in a Berwald space with the Busemann-Hausdorff measure, the mean curvature is the mean value of the normal curvature.

Proposition 14.3.3 *Let N be a hypersurface in an n-dimensional Berwald space (M, F). Then for a normal vector $\mathbf{n}$ at $x \in N$,*

$$\Pi_{\mathbf{n}} = \sum_{i=1}^{n-1} \Lambda_{\mathbf{n}}(\mathbf{b}_i), \tag{14.35}$$

where $\{\mathbf{b}_i\}_{i=1}^{n-1}$ *is an orthonormal basis for* T_xN *with respect to* $\mathbf{g}_\mathbf{n}$.

Proof. There is a C^∞ distance function ρ in a neighborhood $\mathcal{U}$ of x, such that $\rho^{-1}(0) = \mathcal{U} \cap N$ and $\mathbf{n} := \nabla\rho_x$ is the normal vector at $x \in N$. Let $\hat{\mathrm{D}}^2\rho$ and $\hat{\Delta}\rho$ denote the Hessian and Laplacian of ρ with respect to $\hat{g} := \mathbf{g}_{\nabla\rho}$, respectively. By Propositions 14.2.1 and 14.3.1, we have

$$\begin{aligned} \hat{\Lambda}_\mathbf{n} &= \hat{\mathrm{D}}^2\rho, \\ \hat{\Pi}_\mathbf{n} &= \hat{\Delta}\rho. \end{aligned}$$

Let c be the geodesic in M passing through $x \in N$ with $\dot{c}(0) = \mathbf{n}$. For small $t > 0$,

$$\rho \circ c(t) = t.$$

Thus

$$\hat{\mathrm{D}}^2\rho(\mathbf{n}) = \frac{d^2}{dt^2}\Big[\rho \circ c(t)\Big]|_{t=0} = 0.$$

Take an orthonormal basis $\{\mathbf{b}_i\}_{i=1}^n$ with $\mathbf{b}_n = \mathbf{n}$ for $(T_xM, \mathbf{g}_\mathbf{n})$. It follows from (14.15) that

$$\hat{\Pi}_\mathbf{n} = \hat{\Delta}\rho(x) = \sum_{i=1}^{n} \hat{\mathrm{D}}^2\rho(\mathbf{b}_i) = \sum_{i=1}^{n-1} \hat{\Lambda}_\mathbf{n}(\mathbf{b}_i). \tag{14.36}$$

Since F is a Berwald metric, $\mathbf{T} = 0$. By Proposition 7.3.1, $\mathbf{S} = 0$. Then (14.35) follows from (14.24), (14.33) and (14.36). Q.E.D.

14.4 Shape Operator

In order to study the relationship between the normal curvature and the Riemann curvature, we introduce the notion of shape operator and derive the Riccati equation for the shape operators along a geodesic which is orthogonal to the level surfaces of a distance function.

Let (M, F) be a Finsler space and N an embedded hypersurface in M. Let ρ be a C^∞ distance function in a neighborhood $\mathcal{U}$ of N such that $\rho^{-1}(s) = \mathcal{U} \cap N$. Let $\hat{\mathrm{D}}$ denote the Levi-Civita connection of $\hat{g} := \mathbf{g}_{\nabla\rho}$. For the normal vector $\mathbf{n} := \nabla\rho|_N$, $\hat{g}(\mathbf{n}, \mathbf{n}) = 1$. Thus

$$0 = w\Big[\hat{g}(\mathbf{n}, \mathbf{n})\Big] = 2\hat{g}\Big(\hat{\mathrm{D}}_w\mathbf{n}, \mathbf{n}\Big), \qquad \forall w \in T_xN.$$

This implies that

$$\hat{\mathrm{D}}_w \mathbf{n} \in T_x N, \qquad \forall w \in T_x N.$$

Define the *shape operator* $\mathrm{S} : T_x N \to T_x N$ by

$$\mathrm{S}(w) := \hat{\mathrm{D}}_w \mathbf{n}, \qquad w \in T_x N. \tag{14.37}$$

We claim that the shape operator is self-adjoint with respect to $\mathbf{g}_\mathbf{n}$, i.e.,

$$\mathbf{g}_\mathbf{n}(\mathrm{S}(u), v) = \mathbf{g}_\mathbf{n}(u, \mathrm{S}(v)), \qquad \forall u, v \in T_x N. \tag{14.38}$$

To prove (14.38), take two vector fields U and V in a neighborhood of x in M such that $U_x = u, V_x = v$. U and V can be chosen so that they are tangent to N. Thus

$$\hat{\mathrm{D}}_U V|_x - \hat{\mathrm{D}}_V U|_x = [U, V]|_x = 0.$$

Then (14.38) follows from the following observation,

$$\begin{aligned}
\mathbf{g}_\mathbf{n}\Big(\mathrm{S}(u), v\Big) &= \hat{g}\Big(\hat{\mathrm{D}}_u \mathbf{n}, v\Big) \\
&= -\hat{g}\Big(\mathbf{n}, \hat{\mathrm{D}}_U V\Big)|_x \\
&= -\hat{g}\Big(\mathbf{n}, \hat{\mathrm{D}}_V U\Big)|_x \\
&= \hat{g}\Big(u, \hat{\mathrm{D}}_v \mathbf{n}\Big) \\
&= \mathbf{g}_\mathbf{n}\Big(u, \mathrm{S}(v)\Big).
\end{aligned}$$

The normal curvature of N in $(\mathcal{U}, \hat{g})$ is determined by the shape operator. More precisely,

Lemma 14.4.1 *Let $\hat{\Lambda}$ denote the normal curvature of N in $(\mathcal{U}, \hat{g})$ with respect to* $\mathbf{n}$. *Then*

$$\hat{\Lambda}_\mathbf{n}(w) = \mathbf{g}_\mathbf{n}\Big(\mathrm{S}(w), w\Big), \qquad \forall w \in TN. \tag{14.39}$$

Proof: Let $\xi : (-\varepsilon, \varepsilon) \to N$ be a geodesic curve in $(\mathcal{U}, \hat{g})$ with $\dot{\xi}(0) = w$. Observe that

$$\begin{aligned}
\hat{\Lambda}_\mathbf{n}(w) &= \hat{\mathrm{D}}^2 \rho(w) \\
&= \frac{d^2}{ds^2}\Big[\rho\Big(\xi(s)\Big)\Big]|_{s=0}
\end{aligned}$$

$$
\begin{aligned}
&= \frac{d}{ds}\Big[d\rho\Big(\dot{\xi}(s)\Big)\Big]|_{s=0} \\
&= \frac{d}{ds}\Big[\hat{g}\Big(\nabla\rho, \dot{\xi}(s)\Big)\Big]|_{s=0} \\
&= \hat{g}\Big(\hat{\mathrm{D}}_w \mathbf{n}, w\Big) \\
&= \mathbf{g_n}\Big(\mathrm{S}(w), w\Big).
\end{aligned}
$$

This proves the lemma. Q.E.D.

Let $N_t := \rho^{-1}(t)$ be a family of hypersurfaces in $\mathcal{U}$. Fix an integral curve $c(t)$ of $\nabla\rho$ with $\rho \circ c(t) = t$, $\forall t \in [a, b]$. For small t, $\dot{c}(t)$ is orthogonal to N_t at $c(t)$ with respect to $\mathbf{g}_{\dot{c}(t)}$.

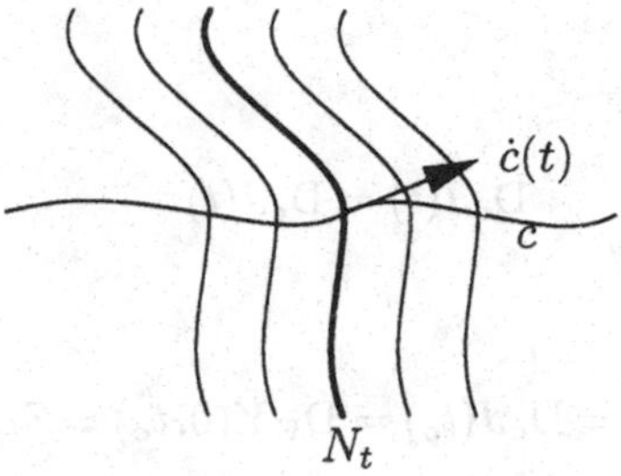

Let

$$\mathrm{S}_t : T_{c(t)}N_t \to T_{c(t)}N_t$$

denote the shape operator of N_t at $c(t)$. Define

$$\dot{\mathrm{S}}_t : T_{c(t)}N_t \to T_{c(t)}N_t$$

by

$$\dot{\mathrm{S}}_t(U) := \mathrm{D}_{\dot{c}(t)}\Big[\mathrm{S}_t(U)\Big] - \mathrm{S}_t\Big(\mathrm{D}_{\dot{c}(t)}U\Big), \tag{14.40}$$

where $U = U(t)$ is an arbitrary vector field along c. When U is parallel,

$$\dot{\mathrm{S}}_t(U) := \mathrm{D}_{\dot{c}(t)}\Big[\mathrm{S}_t(U)\Big].$$

We have the following *Riccati equation*,

Lemma 14.4.2

$$\dot{\mathrm{S}}_t + \mathrm{S}_t^2 + \mathbf{R}_{\dot{c}(t)} = 0. \tag{14.41}$$

Proof. Fix a number t_o. Take a C^∞ curve $\xi(s)$, $-\varepsilon < s < \varepsilon$ in N_{t_o} with $\dot{\xi}(0) = v \in T_{c(t_o)}N_{t_o}$. Take a variation of c, $H : (-\varepsilon, \varepsilon) \times [a, b] \to M$, such that each curve $c_s(t) = H(s, t)$ is a geodesic with $c_s(t_o) = \xi(s) \in N_{t_o}$. Let

$$V(s,t) := \frac{\partial H}{\partial s}(s,t), \qquad Y(s,t) := \frac{\partial H}{\partial t}(s,t).$$

Note that $Y(s,t) = \nabla\rho|_{H(s,t)}$ is normal to N_t and $J(t) := V(0,t)$ is a Jacobi field along c with $J(t_o) = v$. Let D and $\hat{\mathrm{D}}$ denote the connection of F and $\hat{F} = \sqrt{\mathbf{g}_{\nabla\rho}}$, respectively. By (6.14),

$$\mathrm{D}_{\nabla\rho} = \hat{\mathrm{D}}_{\nabla\rho}.$$

Restricting it to c yields

$$\mathrm{D}_{\dot{c}}J(t) = \hat{\mathrm{D}}_{\dot{c}}J(t).$$

Observe that

$$\mathrm{D}_{\dot{c}}J(t_o) = \hat{\mathrm{D}}_{\dot{c}}J(t_o) = \hat{\mathrm{D}}_V Y(0, t_o) = \mathrm{S}_{t_o}(v).$$

Thus

$$\begin{aligned}
\mathbf{R}_{\dot{c}(t_o)}(v) &= -\mathrm{D}_{\dot{c}}\mathrm{D}_{\dot{c}}J(t_o) \\
&= -\mathrm{D}_{\dot{c}}\Big[\mathrm{S}_t(J(t))\Big]_{|t=t_o} \\
&= -\dot{\mathrm{S}}_{t_o}(v) - \mathrm{S}_{t_o}\Big(\mathrm{D}_{\dot{c}}J(t_o)\Big) \\
&= -\dot{\mathrm{S}}_{t_o}(v) - \mathrm{S}_{t_o}^2(v).
\end{aligned}$$

This gives (14.41). Q.E.D.

Consider a family of hypersurfaces N_t in a Finsler space (M, F) and a geodesic $c(t)$ such that

(1) $c(t) \in N_t$;
(2) $\dot{c}(t)$ is a normal vector to N_t.

Let $J(t)$ be a Jacobi field along c such that $J(t)$ and $D_{\dot{c}}J(t)$ are $\mathbf{g}_{\dot{c}(t)}$-orthogonal to $\dot{c}(t)$. By the above arguments, we see that the shape operator is given by

$$S_t(J(t)) = D_{\dot{c}}J(t). \tag{14.42}$$

It follows from (14.39) and (14.42) that

$$\hat{\Lambda}_t(J(t)) = \mathbf{g}_{\dot{c}(t)}\Big(\hat{D}_{\dot{c}}J(t), J(t)\Big). \tag{14.43}$$

Now we study the shape operators of the regular metric t-sphere $\tilde{S}(p,t)$ in the cut-domain $\mathcal{D}_p$,

$$\tilde{S}(p,t) = \mathcal{D}_p \cap S(p,t).$$

Let $\rho(x) := d(p,x)$. ρ is a C^∞ distance function on $\mathcal{D}_p$. We can view $\tilde{S}(p,t)$ as a level hypersurface of ρ,

$$\tilde{S}(p,t) = \rho^{-1}(t).$$

The gradient $\nabla\rho$ restricted to $\tilde{S}(p,t)$ is a normal vector field along $\tilde{S}(p,t)$. Fix a unit vector $y \in S_pM$. Let $c(t) := \exp_p(ty)$, $0 \le t < \mathbf{i}_y$. For a vector $v \in T_pM$, let $J(t)$ be the Jacobi field along c satisfying

$$J(0) = 0, \qquad D_{\dot{c}}J(0) = v.$$

Assume that $\mathbf{g}_y(y,v) = 0$. By the Gauss Lemma (Lemma 11.2.1), we know that $J(t)$ and $D_{\dot{c}}J(t)$ are $\mathbf{g}_{\dot{c}}$-orthogonal to $\dot{c}(t)$ for all $0 \le t < \varepsilon$,

$$\mathbf{g}_{\dot{c}(t)}\Big(\dot{c}(t), J(t)\Big) = 0 = \mathbf{g}_{\dot{c}(t)}\Big(\dot{c}(t), D_{\dot{c}}J(t)\Big).$$

Let S_t be the shape operator of $S(p,t)$ at $c(t)$. By (14.42),

$$S_t(J(t)) = D_{\dot{c}}J(t). \tag{14.44}$$

Let $V = V(t)$ be a parallel vector field along c with $V(0) = v$. Then $V(t)$ is $\mathbf{g}_{\dot{c}(t)}$-orthogonal to $\dot{c}(t)$ for all $t \in [0,\varepsilon)$,

$$\mathbf{g}_{\dot{c}(t)}\Big(\dot{c}(t), V(t)\Big) = 0.$$

Lemma 14.4.3 *Let $V(t)$ be a parallel vector field along c with $V(0) = v$. Assume that v is $\mathbf{g}_y$-orthogonal to y. The shape operator S_t of $\tilde{S}(p,t)$ at $c(t)$ has the following Taylor expansion*

$$\mathrm{S}_t(V(t)) = \frac{1}{t}V(t) - \frac{1}{3}\mathrm{R}(t)t + o(t), \tag{14.45}$$

where $\mathrm{R}(t)$ is a parallel vector field along c with $\mathrm{R}(0) = \mathbf{R}_y(u)$.

Proof: Let $\{E_i(t)\}_{i=1}^n$ be a $\mathbf{g}_{\dot{c}}$-orthonormal frame along c with $E_n(t) = \dot{c}(t)$. Let J be a Jacobi field along c with $J(0) = 0$ and $\mathrm{D}_{\dot{c}}J(0) = v$. Put

$$J(t) = J^i(t)E_i(t)$$

and

$$\mathbf{R}_{\dot{c}(t)}(E_j(t)) = \mathrm{R}^i_j(t)E_i(t).$$

J satisfies

$$\ddot{J}^i(t) + \mathrm{R}^i_j(t)J^j(t) = 0,$$

with $J^i(0) = 0$ and $\dot{J}^i(0) = v^i$. This implies

$$J^i(t) = v^i t - \frac{1}{6}\mathrm{R}^i_j(0)v^j t^3 + o(t^3). \tag{14.46}$$

Let

$$\mathrm{S}^i_j(t)E_i(t) := \mathrm{S}_t(E_j(t)).$$

It follows from (14.44) that for $t > 0$,

$$\mathrm{S}^i_j(t)\Big[v^j t - \frac{1}{6}\mathrm{R}^j_k(0)v^k t^3 + o(t^3)\Big] = v^i - \frac{1}{2}\mathrm{R}^i_j(0)v^j t^2 + o(t^2).$$

Thus

$$\mathrm{S}^i_j(t) = \frac{1}{t}\Big\{\delta^i_j - \frac{1}{3}\mathrm{R}^i_j(0)t^2 + o(t^2)\Big\}.$$

This implies (14.45). Q.E.D.

Now we take a look at the normal curvature and the mean curvature of the regular metric t-sphere $\tilde{S}(p,t)$ in the cut-domain $\mathcal{D}_p$. Fix a unit vector $y \in \mathrm{S}_pM$ and let $c(t) := \exp_p(ty)$, $0 \le t < \mathbf{i}_y$. Denote by $\hat{\Lambda}_t$ and $\hat{\Pi}_t$ the normal curvature and the mean curvature of $\tilde{S}(p,t)$ at $c(t)$ in $(\mathcal{D}_p, \hat{g})$ with respect to the normal vector $\dot{c}(t)$, where $\hat{g} := \mathbf{g}_{\nabla\rho}$ is the Riemannian metric induced by the distance function $\rho(x) = d(p,x)$ from p. Let $v \in T_pM$ with $\mathbf{g}_y(y,v) = 0$ and $V(t)$ denote the parallel vector field along c with $V(0) = v$. We know that $V(t)$ is always tangent to $\tilde{S}(p,t)$ at $c(t)$ for any $0 < t < \mathbf{i}_y$. From (14.39) and (14.45), we obtain

$$\hat{\Lambda}_t(V) = \frac{1}{t}\mathbf{g}_y(v,v) - \frac{1}{3}\mathbf{g}_y\Big(\mathbf{R}_y(v), v\Big)t + o(t), \qquad (14.47)$$

$$\hat{\Pi}_t = \frac{n-1}{t} - \frac{1}{3}\mathbf{Ric}(y)t + o(t). \qquad (14.48)$$

Define $\dot{\mathbf{T}}_y : T_xM \to T_xM$ by

$$\mathbf{T}_{\dot{c}(t)}(V(t)) = \mathbf{T}_y(v) + \dot{\mathbf{T}}_y(v)t + o(t).$$

By (14.24), we obtain the following

Proposition 14.4.4 *Let $V = V(t)$ be a parallel vector field along a geodesic $c(t) = \exp_x(ty)$, $0 \le t < \varepsilon$. The normal curvature Λ_t of $S(p,t)$ at $c(t)$ has the following Taylor expansion,*

$$\Lambda_t(V(t)) = \frac{1}{t}\mathbf{g}_y(v,v) - \mathbf{T}_y(v) - \frac{1}{3}\Big[\mathbf{R}_y(v) + 3\dot{\mathbf{T}}_y(v)\Big]t + o(t). \qquad (14.49)$$

Now we consider a Finsler m space $(M, F, d\mu)$. For a vector $y \in T_xM$, define $\dot{\mathbf{S}}(y)$ by

$$\mathbf{S}(\dot{c}(t)) = \mathbf{S}(y) + \dot{\mathbf{S}}(y)t + o(t),$$

where $c(t)$ denotes the geodesic with $\dot{c}(0) = y$. By (14.33), we obtain the following

Proposition 14.4.5 *Let $(M, F, d\mu)$ be a Finsler m space. The mean curvature Π_t of $S(p,t)$ at $c(t)$ have the following forms*

$$\Pi_t = \frac{n-1}{t} - \mathbf{S}(y) - \frac{1}{3}\Big[\mathbf{Ric}(y) + 3\dot{\mathbf{S}}(y)\Big]t + o(t). \qquad (14.50)$$

Chapter 15

Geometry of Metric Spheres

Metric spheres in a Finsler space are the level hypersurface of a distance function from a point. In Section 14.4, we studied the geometry of small metric spheres. In this chapter, we will study metric spheres at large scale. We will show that the normal (resp. mean) curvature is under control by the flag (resp. Ricci) curvature and the T-curvature (resp. S-curvature).

15.1 Estimates on the Normal Curvature

Let (M, F) be a positively complete Finsler space and $p \in M$. The regular metric t-sphere $\tilde{S}(p,t)$ is defined by

$$\tilde{S}(p,t) = S(p,t) \cap \mathcal{D}_p,$$

where $\mathcal{D}_p$ denotes the cut-domain at p. The distance function $\rho(x) := d(p,x)$ is C^∞ on $\mathcal{D}_p \setminus \{p\}$ such that $S(p,t) = \rho^{-1}(t)$ and $\tilde{S}(p,t) = \rho^{-1}(t) \cap \mathcal{D}_p$. The gradient $\nabla\rho$ is a C^∞ unit geodesic field on $\mathcal{D}_p \setminus \{p\}$, hence it induces a Riemannian metric $\hat{g} := \mathbf{g}_{\nabla\rho}$ on $\mathcal{D}_p \setminus \{p\}$. Let $\hat{\nabla}\rho$ denotes the gradient of ρ with respect to $\hat{g}$. According to Lemma 3.2.1, $\nabla\rho = \hat{\nabla}\rho$ is also a unit geodesic field of $\hat{F} := \sqrt{\hat{g}}$. Further, $\mathbf{n} = \nabla\rho|_{\tilde{S}(p,t)}$ is a normal vector to $\tilde{S}(p,t)$ in $(\mathcal{D}_p, F)$ and $(\mathcal{D}_p, \hat{F})$ respectively.

By definition, the flag curvature satisfies the bound $\mathbf{K} \le \lambda$ if for any tangent plane $P \subset T_xM$ and any $y \in P \setminus \{0\}$,

$$\mathbf{K}(P, y) \le \lambda.$$

This is equivalent to saying that

$$\mathbf{g}_y\Big(\mathbf{R}_y(v), v\Big) \le \lambda \Big\{\mathbf{g}_y(y,y)\mathbf{g}_y(v,v) - \mathbf{g}_y(y,v)\mathbf{g}_y(y,v)\Big\},$$

where $y, v \in TM \setminus \{0\}$. Similarly, we define the bound $\mathbf{K} \ge \lambda$.

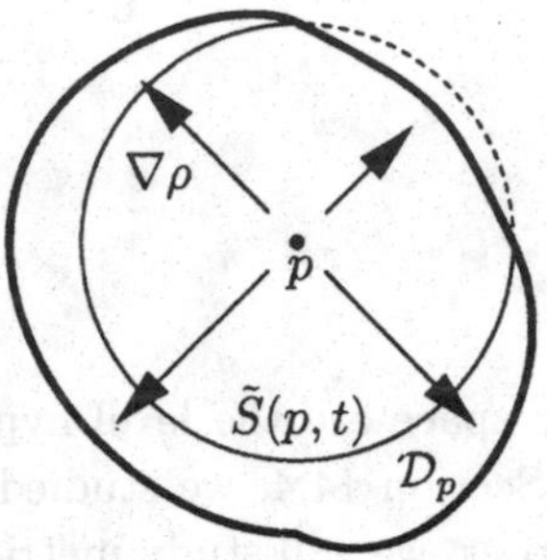

Theorem 15.1.1 *Let (M, F) be a positively complete Finsler space and $p \in M$. Let $\hat{\Lambda}_t$ denote the normal curvature of $\tilde{S}(p,t)$ in $(\mathcal{D}_p, \hat{g})$ with respect to the outward-pointing normal vector.*

(a) Suppose that $\mathbf{K} \le \lambda$. Then

$$\hat{\Lambda}_t \ge \frac{\mathbf{s}'_\lambda(t)}{\mathbf{s}_\lambda(t)}\, \hat{g}.$$

(b) Suppose that $\mathbf{K} \ge \lambda$. Then

$$\hat{\Lambda}_t \le \frac{\mathbf{s}'_\lambda(t)}{\mathbf{s}_\lambda(t)}\, \hat{g}.$$

Proof: Fix a unit vector $y \in \mathrm{S}_pM$ and let

$$c(t) := \exp_p(ty), \qquad 0 \le t < \mathbf{i}_y.$$

Take an arbitrary vector $v \in T_pM$ with $\mathbf{g}_y(y, v) = 0$. Let $J(t)$ be the Jacobi field along c with

$$J(0) = 0, \qquad \mathrm{D}_{\dot{c}}J(0) = v \in T_pM.$$

By the Gauss Lemma 11.2.1, $J(t)$ and $\mathrm{D}_{\dot{c}}J(t)$ are tangent to $\tilde{S}(p,t)$ at $c(t)$ for any t with $0 < t < \mathbf{i}_y$.

We first prove (a) under the assumption that $\mathbf{K} \leq \lambda$. By (14.43),

$$\hat{\Lambda}_t(J(t)) = \mathbf{g}_{\dot{c}(t)}\Big(\mathrm{D}_{\dot{c}}J(t), J(t)\Big).$$

Let

$$\phi(t) := \frac{\hat{\Lambda}_t(J(t))}{\hat{g}_{c(t)}(J(t), J(t))} = \frac{\mathbf{g}_{\dot{c}(t)}\Big(\mathrm{D}_{\dot{c}}J(t), J(t)\Big)}{\mathbf{g}_{\dot{c}(t)}(J(t), J(t))}.$$

Using the Jacobi equation (11.2), we obtain

$$\phi' = -\frac{\mathbf{g}_{\dot{c}}(\mathbf{R}_{\dot{c}}(J), J)}{\mathbf{g}_{\dot{c}}(J, J)} + \frac{\mathbf{g}_{\dot{c}}(\mathrm{D}_{\dot{c}}J, \mathrm{D}_{\dot{c}}J)\mathbf{g}_{\dot{c}}(J, J) - \mathbf{g}_{\dot{c}}(\mathrm{D}_{\dot{c}}J, J)^2}{\mathbf{g}_{\dot{c}}(J, J)^2} - \phi^2. \quad (15.1)$$

By the Schwartz inequality,

$$\mathbf{g}_{\dot{c}}(\mathrm{D}_{\dot{c}}J, \mathrm{D}_{\dot{c}}J)\mathbf{g}_{\dot{c}}(J, J) - \mathbf{g}_{\dot{c}}(\mathrm{D}_{\dot{c}}J, J)^2 \geq 0,$$

we obtain

$$\phi' \geq -\lambda - \phi^2. \quad (15.2)$$

To estimate ϕ, we introduce a new function

$$\phi_\lambda(t) := \frac{\mathbf{s}'_\lambda(t)}{\mathbf{s}_\lambda(t)}. \quad (15.3)$$

ϕ_λ satisfies the following ODE,

$$\phi'_\lambda + \phi^2_\lambda = -\lambda. \quad (15.4)$$

Let

$$f(t) := e^{\int(\phi(t)+\phi_\lambda(t))dt}\Big[\phi(t) - \phi_\lambda(t)\Big].$$

It follows from (15.2) and (15.4) that

$$f'(t) = e^{\int(\phi(t)+\phi_\lambda(t))dt}\Big[\phi'(t) + \phi^2(t) - \phi'_\lambda(t) - \phi^2_\lambda(t)\Big] \geq 0.$$

Thus f is non-decreasing. By Proposition 11.2.3, we have

$$\phi(t) = \frac{1}{t} - \frac{1}{3}\frac{\mathbf{g}_y(\mathbf{R}_y(v), v)}{\mathbf{g}_y(v, v)}t + o(t). \quad (15.5)$$

Thus

$$f(t) \geq \lim_{t\to 0+} f(t) = 0.$$

This implies that $\phi(t) \geq \phi_\lambda(t)$. Namely,

$$\frac{\hat{\Lambda}_t(J(t))}{\hat{g}_{c(t)}(J(t), J(t))} = \frac{\mathbf{g}_{\dot{c}}(\mathrm{D}_{\dot{c}}J, J)}{\mathbf{g}_{\dot{c}}(J, J)} \geq \frac{\mathbf{s}'_\lambda(t)}{\mathbf{s}_\lambda(t)}, \qquad 0 < t < \mathbf{i}_y. \tag{15.6}$$

Now we prove (b). Let $E(t)$ be a parallel unit vector field along c with respect to $\mathbf{g}_{\dot{c}(t)}$. Let

$$\phi(t) := \hat{\Lambda}_t(E(t)).$$

By (14.39), we have

$$\phi(t) = \mathbf{g}_{\dot{c}(t)}\Big(\mathrm{S}_t(E(t)), E(t)\Big).$$

Let S_t denote the shape operator of $\tilde{S}(p, t)$ at $c(t)$, where $0 < t < \mathbf{i}_y$. The Riccati equation (14.41) gives

$$\frac{d}{dt}\Big[\mathbf{g}_{\dot{c}}(\mathrm{S}_t(E), E)\Big] = -\mathbf{g}_{\dot{c}}\Big(\mathbf{R}_{\dot{c}}(E), E\Big) - \mathbf{g}_{\dot{c}}\Big(\mathrm{S}_t(E), \mathrm{S}_t(E)\Big). \tag{15.7}$$

By the Schwartz inequality

$$\mathbf{g}_{\dot{c}}\Big(\mathrm{S}_t(E), E\Big)^2 \leq \mathbf{g}_{\dot{c}}\Big(\mathrm{S}_t(E), \mathrm{S}_t(E)\Big),$$

we obtain

$$\phi' \leq -\lambda - \phi^2.$$

Let

$$f(t) := e^{\int(\phi + \phi_\lambda)dt}\Big[\phi(t) - \phi_\lambda(t)\Big],$$

where ϕ_λ is defined in (15.3). We have

$$f'(t) = e^{\int(\phi+\phi_\lambda)dt}\Big[\phi'(t) + \phi^2(t) - \phi'_\lambda(t) - \phi^2_\lambda(t)\Big] \leq 0. \tag{15.8}$$

Since $\phi(t)$ has the Taylor expansion (15.5), from (15.8), we obtain

$$f(t) \leq \lim_{t \to 0^+} f(t) = 0.$$

Thus $\phi(t) \leq \phi_\lambda(t)$. Namely,

$$\hat{\Lambda}_t(E(t)) \geq \frac{\mathbf{s}'_\lambda(t)}{\mathbf{s}_\lambda(t)}, \qquad 0 < t < \mathbf{i}_y.$$

Q.E.D.

Remark 15.1.2 *Let $\hat{\mathrm{D}}^2\rho$ denote the Hessian of ρ with respect to the Riemannian metric $\hat{g} = \mathbf{g}_{\nabla\rho}$ on the cut-domain $\mathcal{D}_p$. Let $\hat{\Lambda}_{\nabla\rho}$ denote the normal curvature of $\tilde{S}(p,t)$ in $(\mathcal{D}_p, \hat{g})$ with respect to the outward-pointing normal vector $\nabla\rho$. According to (14.25), we have*

$$\hat{\mathrm{D}}^2\rho(v) = \hat{\mathrm{D}}^2\rho(v^\perp) = \hat{\Lambda}_{\nabla\rho}(v^\perp), \qquad \forall v \in T_xM, \tag{15.9}$$

where $v^\perp := v - \hat{g}(\nabla\rho_x, v)\nabla\rho_x$. The estimates on $\hat{\Lambda}_{\nabla\rho}$ in Theorem 15.1.1 also give estimates on $\hat{\mathrm{D}}^2\rho$. For instance, if $\mathbf{K} \le \lambda$, then

$$\hat{\mathrm{D}}^2\rho(v) \ge \frac{\mathbf{s}'_\lambda(\rho)}{\mathbf{s}_\lambda(\rho)}\, \hat{g}(v^\perp, v^\perp), \qquad v \in T_xM, \tag{15.10}$$

Recall that the T-curvature satisfies

$$\mathbf{T}_{\lambda y} = \lambda \mathbf{T}_y, \qquad \mathbf{T}_y(y) = 0, \qquad \lambda > 0.$$

We say that the T-curvature satisfies the bound $\mathbf{T} \ge -\delta$ if

$$\mathbf{T}_y(u) \ge -\delta\Big\{\mathbf{g}_y(u,u) - \Big[\mathbf{g}_y\Big(u, \frac{y}{F(y)}\Big)\Big]^2\Big\}F(y),$$

where $y, u \in TM \setminus \{0\}$. Similarly, we define the bound $\mathbf{T} \le \delta$.

Equation (14.24) and Theorem 15.1.1 imply

Theorem 15.1.3 *Let (M, F) be a positively complete Finsler space and $p \in M$. Let $\hat{\Lambda}_t$ denote the normal curvature of $\tilde{S}(p,t)$ in $(\mathcal{D}_p, F)$ with respect to the outward-pointing normal vector $\mathbf{n}$. Then*

(i) if $\mathbf{K} \le \lambda$ and $\mathbf{T} \le \delta$,

$$\Lambda_t \ge \Big\{\frac{\mathbf{s}'_\lambda(t)}{\mathbf{s}_\lambda(t)} - \delta\Big\}\mathbf{g}_{\mathbf{n}}.. \tag{15.11}$$

(ii) if $\mathbf{K} \ge \lambda$ and $\mathbf{T} \ge -\delta$,

$$\Lambda_t \le \Big\{\frac{\mathbf{s}'_\lambda(t)}{\mathbf{s}_\lambda(t)} + \delta\Big\}\mathbf{g}_{\mathbf{n}}. \tag{15.12}$$

Remark 15.1.4 *Let* $\mathrm{D}^2\rho$ *denote the Hessian of* ρ *with respect to* F. *By (14.2), we have*

$$\mathrm{D}^2\rho = \hat{\mathrm{D}}^2\rho - \mathbf{T}_{\nabla\rho}. \tag{15.13}$$

In virtue of (15.9) and (15.13), we see that the estimates on $\hat{\Lambda}_t$ *in Theorem 15.1.1 also give estimates on* $\mathrm{D}^2\rho$. *For instance, if* $\mathbf{K} \leq \lambda$ *and* $\mathbf{T} \leq \delta$, *then*

$$\mathrm{D}^2\rho(v) \geq \Big\{\frac{\mathbf{s}_\lambda'(\rho)}{\mathbf{s}_\lambda(\rho)} - \delta\Big\} g_{\nabla\rho}(v^\perp, v^\perp), \qquad \forall v \in T_xM \tag{15.14}$$

where $v^\perp := v - \mathbf{g}_{\nabla\rho_x}(\nabla\rho_x, v)\nabla\rho_x$.

15.2 Convexity of Metric Balls

Let (M, F) be a complete reversible Finsler space. It is proved by J. H. C. Whitehead [Wh] that at any point $p \in M$, there is a small $r > 0$ such that $B(p,r)$ is strictly convex, that is, for any minimizing geodesic $c(t), a \leq t \leq b$ with $c(a), c(b) \in B(p,r)$, the whole curve c is contained in $B(p,r)$.

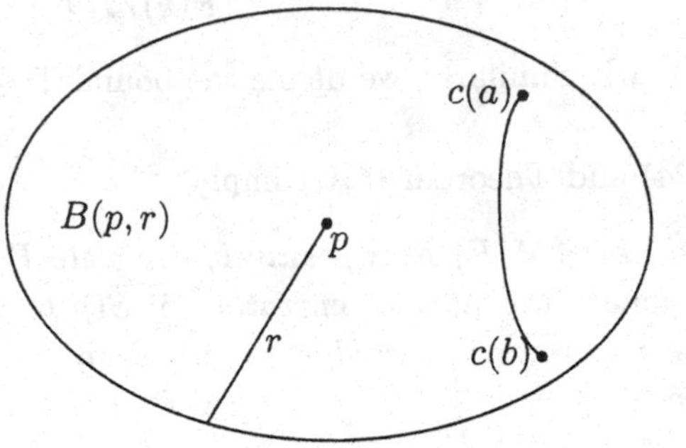

We can estimate the size of the convex balls in M under certain curvature bounds. Let $\lambda, \delta \in \mathrm{R}$. Define $r_o = r_o(\lambda, \delta)$ to be the first zero of

$$\mathbf{s}_\lambda'(t) - \delta\mathbf{s}_\lambda(t) = 0.$$

Let

$$r_1 = r_1(p, \lambda, \delta) := \min\Big(r_o,\ \mathbf{i}_p\Big).$$

Let $c(t), 0 \leq t \leq a$, be an arbitrary minimizing unit speed geodesic with $c(0), c(a) \in B(p, r_1/2)$. Note that

$$a \leq d(p, c(0)) + d(p, c(a)) < r_1.$$

We claim that c is contained in $B(p, r_1)$. Let t be an arbitrary number with $0 \leq t \leq a$. If $t \leq a/2$,

$$d(p, c(t)) \leq d(p, c(0)) + d(c(0), c(t)) < r_1/2 + t \leq r_1.$$

If $r > a/2$, then

$$d(p, c(t)) \leq d(p, c(a)) + d(c(a), c(t)) < r_1/2 + (a - t) \leq r_1.$$

We conclude that $c(t) \in B(p, r_1)$ for any $0 \leq t \leq a$. We will show that c is actually contained in $B(p, r_1/2)$.

Theorem 15.2.1 *Let (M, F) be a complete reversible Finsler space. Suppose that the flag curvature and the tangent curvature satisfy the bounds*

$$\mathbf{K} \leq \lambda, \qquad \mathbf{T} \leq \delta.$$

Then the metric ball $B(p, r)$ is strictly convex for any $r < r_1/2$.

Proof. Let $c : [a, b] \to M$ be a minimizing geodesic with $c(a), c(b) \in B(p, r)$. We may assume that c does not pass through p. By assumption, $c \subset B(p, r_1)$. Let $\rho(x) = d(p, x)$. By (15.14) the function $f(t) := \rho \circ c(t)$ satisfies

$$f''(t) = \mathrm{D}^2\rho\Big(\dot{c}(t)\Big) \geq \Big\{\frac{\mathbf{s}'_\lambda(f(t))}{\mathbf{s}_\lambda(f(t))} - \delta\Big\} g_{\nabla\rho}\Big(\dot{c}(t)^\perp, \dot{c}(t)^\perp\Big) \geq 0,$$

where

$$\dot{c}(t)^\perp := \dot{c}(t) - \mathbf{g}_{\nabla\rho}(\nabla\rho, \dot{c}(t))\nabla\rho.$$

We conclude that

$$f(t) \leq \max(f(a), f(b)), \qquad \forall t \in (a, b).$$

This proves the theorem. Q.E.D.

Assume that $\lambda \leq 0$. We have

$$\lim_{t\to\infty} \frac{\mathbf{s}'_\lambda(t)}{\mathbf{s}_\lambda(t)} = \sqrt{-\lambda}.$$

If $\sqrt{-\lambda} - \delta \geq 0$, then $s'_\lambda(t) - \delta s_\lambda(t) \neq 0$ for all $t > 0$. By this observation, we obtain the following

Corollary 15.2.2 *Let (M, F) be a complete reversible Finsler space. Let (M, F) be a positively complete simply connected Finsler space. Suppose that M is simply connected and the flag curvature and the S-curvature satisfy the bounds*

$$\mathbf{K} \leq \lambda, \qquad \mathbf{T} \leq \delta,$$

where $\lambda \leq 0$, $\delta \geq 0$ with $\sqrt{-\lambda} - \delta \geq 0$. any metric ball $B(p, r)$ is strictly convex.

15.3 Estimates on the Mean Curvature

In the previous section, we have obtained some estimates on the normal curvature of metric spheres under certain bounds on the flag curvature. The mean curvature of a metric sphere can viewed as the mean of the normal curvature. In this section, we will show that the mean curvature is under control by the Ricci curvature and the S-curvature.

Let $(M, F, d\mu)$ be a positively complete Finsler m space and $p \in M$. The distance function $\rho(x) := d(p, x)$ induces a Riemannian metric $\hat{g} := \mathbf{g}_{\nabla\rho}$ on the cut-domain $\mathcal{D}_p \subset M$.

Theorem 15.3.1 *Let (M, F) be an n-dimensional positively complete Finsler space. Let $\hat{\Pi}_t$ denote the mean curvature of $\tilde{S}(p, t)$ in $(\mathcal{D}_p, \hat{g})$ with respect to the outward-pointing normal vector.*

(i) *Suppose that $\mathbf{K} \leq \lambda$. Then*

$$\hat{\Pi}_t \geq (n-1)\frac{s'_\lambda(t)}{s_\lambda(t)}. \tag{15.15}$$

(ii) *Suppose that $\mathbf{Ric} \geq (n-1)\lambda$. Then*

$$\hat{\Pi}_t \leq (n-1)\frac{s'_\lambda(t)}{s_\lambda(t)}. \tag{15.16}$$

Proof: (i) follows from Theorem 15.1.1 (i). Now we prove (ii). Let S_t be the shape operator of the metric sphere $\tilde{S}(p, t)$. Fix a unit tangent vector $y \in S_pM$ and let $c(t) := \exp_p(ty)$, $t \geq 0$, and $\{E_i(t)\}_{i=1}^n$ be a $\mathbf{g}_{\dot{c}(t)}$-orthonormal frame along c with $E_n(t) = \dot{c}(t)$. It follows from (15.7) that

$$\frac{d}{dt}\Big[\sum_{i=1}^{n-1} \mathbf{g}_{\dot{c}}\Big(\mathrm{S}_t(E_i), E_i\Big)\Big] = -\mathbf{Ric}(\dot{c}) - \sum_{i=1}^{n-1} \mathbf{g}_{\dot{c}}\Big(\mathrm{S}_t(E_i), \mathrm{S}_t(E_i)\Big). \quad (15.17)$$

By Proposition 14.3.3, the mean curvature $\hat{\Pi}_t$ is the mean value of the normal curvature $\hat{\Lambda}_t$,

$$\hat{\Pi}_t = \sum_{i=1}^{n-1} \hat{\Lambda}_t\Big(E_i(t)\Big).$$

Let

$$\phi(t) := \frac{1}{n-1}\hat{\Pi}_t. \quad (15.18)$$

By (14.39), we have

$$\phi(t) = \frac{1}{n-1}\sum_{i=1}^{n-1} \mathbf{g}_{\dot{c}}\Big(\mathrm{S}_t(E_i), E_i\Big).$$

Observe that

$$\Big\{\sum_{i=1}^{n-1} \mathbf{g}_{\dot{c}}\Big(\mathrm{S}_t(E_i), E_i\Big)\Big\}^2 \leq (n-1)\sum_{i=1}^{n-1} \mathbf{g}_{\dot{c}}\Big(\mathrm{S}_t(E_i), \mathrm{S}_t(E_i)\Big)^2$$

Namely,

$$\phi^2(t) \leq \frac{1}{n-1}\sum_{i=1}^{n-1} \mathbf{g}_{\dot{c}}\Big(\mathrm{S}_t(E_i), \mathrm{S}_t(E_i)\Big). \quad (15.19)$$

By (15.17) and (15.19), we obtain

$$\phi' \leq -\lambda - \phi^2.$$

Take a Jacobi field $J(t)$ along c such that $J(0) = 0$ and $\mathrm{D}_{\dot{c}}J(0) = v$, where v satisfies $\mathbf{g}_y(y, v) = 0$. By the Gauss Lemma (Lemma 11.2.1), we know that both $J(t)$ and $\mathrm{D}_{\dot{c}}J(t)$ are $\mathbf{g}_{\dot{c}(t)}$-orthogonal to $\dot{c}(t)$. By (14.43),

$$\hat{\Lambda}_t(J(t)) = \mathbf{g}_{\dot{c}(t)}\Big(\hat{\mathrm{D}}_{\dot{c}}J(t), J(t)\Big).$$

Using the Taylor expansions in Proposition 11.2.3, we obtain

$$\phi(t) = \frac{1}{t} - \frac{1}{3(n-1)}\mathbf{Ric}(y)t + o(t). \tag{15.20}$$

By the standard argument as in the proof of Theorem 15.1.1(ii), one obtains

$$\phi(t) \leq \frac{\mathbf{s}'_\lambda(t)}{\mathbf{s}_\lambda(t)}.$$

This gives (ii). Q.E.D.

Recall that the S-curvature is positively homogeneous of degree one,

$$\mathbf{S}(\lambda y) = \lambda \mathbf{S}(y), \qquad \lambda > 0.$$

We say that the S-curvature satisfies the bound $\mathbf{S} \geq -(n-1)\delta$ if

$$\mathbf{S}(y) \geq -(n-1)\delta F(y), \qquad \forall y \in TM \setminus \{0\}.$$

Similarly, we define the lower bound $\mathbf{S} \leq (n-1)\delta$.

By (14.33) and Theorem 15.3.1, we immediately obtain the following

Theorem 15.3.2 *Let $(M, F, d\mu)$ be an n-dimensional positively complete Finsler m space. Let Π_t denote the mean curvature of $\tilde{S}(p,t)$ in $(\mathcal{D}_p, F)$ with respect to the outward-pointing normal vector.*

(i) *Suppose that*

$$\mathbf{K} \leq \lambda, \qquad \mathbf{S} \leq (n-1)\delta.$$

Then

$$\Pi_t \geq (n-1)\frac{\mathbf{s}'_\lambda(t)}{\mathbf{s}_\lambda(t)} - (n-1)\delta. \tag{15.21}$$

(ii) *Suppose that*

$$\mathbf{Ric} \geq (n-1)\lambda, \qquad \mathbf{S} \geq -(n-1)\delta.$$

Then

$$\Pi_t \leq (n-1)\frac{\mathbf{s}'_\lambda(t)}{\mathbf{s}_\lambda(t)} + (n-1)\delta. \tag{15.22}$$

In virtue of (14.30) and (15.22), we obtain the following

Theorem 15.3.3 *Let $(M, F, d\mu)$ be an n-dimensional Finsler m space. Suppose that the Ricci curvature and the S-curvature satisfy the bounds,*

$$\mathbf{Ric} \geq (n-1)\lambda, \qquad \mathbf{S} \geq -(n-1)\delta.$$

Then the distance function $\rho(x) := d(p, x)$ satisfies

$$\Delta\rho \leq (n-1)\frac{\mathbf{s}'_\lambda(\rho)}{\mathbf{s}_\lambda(\rho)} + (n-1)\delta.$$

In virtue of (14.30) and (15.21), we obtain the following

Theorem 15.3.4 *Let $(M, F, d\mu)$ be an n-dimensional Finsler m space. Suppose that the flag curvature and the S-curvature satisfy the bounds*

$$\mathbf{K} \leq \lambda, \qquad \mathbf{S} \leq (n-1)\delta.$$

Then

$$\Delta\rho \geq (n-1)\frac{\mathbf{s}'_\lambda(\rho)}{\mathbf{s}_\lambda(\rho)} - (n-1)\delta. \tag{15.23}$$

Below is an application of Theorem 15.3.4.

Theorem 15.3.5 *Let $(M, F, d\mu)$ be an n-dimensional positively complete simply connected Finsler m space. Suppose that for constants $\lambda \leq 0$ and $\delta \geq 0$ with $\sqrt{-\lambda} - \delta > 0$, the flag curvature and the S-curvature satisfy the bounds*

$$\mathbf{K} \leq \lambda, \qquad \mathbf{S} \leq (n-1)\delta \tag{15.24}$$

Then for any bounded regular domain $\Omega \subset M$,

$$\mu(\Omega) \leq \frac{1}{(n-1)(\sqrt{-\lambda} - \delta)}\nu(\partial\Omega).$$

Proof. Let $\rho(x) = d(p, x)$, where $p \notin \Omega$. Then ρ is C^∞ on Ω.

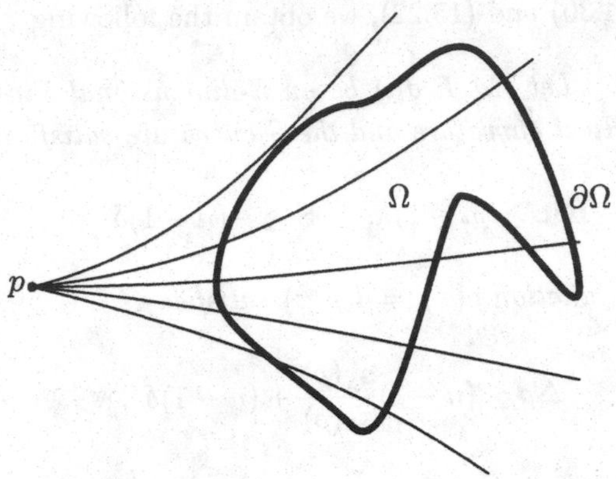

By definition,

$$\Delta \rho = \operatorname{div}(\nabla \rho).$$

By Theorem 2.4.2, we obtain

$$\int_{\Omega} \Delta \rho \, d\mu = \int_{\partial\Omega} \mathbf{g}_{\mathbf{n}}(\mathbf{n}, \nabla \rho) \, d\nu \leq \nu(\partial\Omega),$$

where $\mathbf{n}$ is the outer normal along $\partial\Omega$. By (15.23), we have

$$\Delta \rho \geq (n-1)(\sqrt{-\lambda} - \delta) > 0.$$

Therefore

$$(n-1)(\sqrt{-\lambda} - \delta)\mu(\Omega) \leq \int_{\Omega} \Delta \rho \, d\mu \leq \nu(\partial\Omega).$$

This proves the theorem. Q.E.D.

Remark 15.3.6 *By Theorem 15.3.5, one can obtain an upper bound on the first eigenvalue of a Finsler space satisfying (15.24).*

15.4 Metric Spheres in a Convex Domain

In this section, we shall estimate the mean curvature of a metric sphere from below under certain upper bounds on the Ricci curvature and the S-curvature. The metric spheres under our consideration lie inside a convex domain at a point.

For a unit vector $y \in \mathrm{S}_pM$, the *convex value* $\hat{\mathbf{i}}_y$ of y is defined to the largest $0 < r < \mathbf{i}_y$ such that the shape operator S_t of th regular metric sphere $\tilde{S}(p,t)$ has nonnegative eigenvalues at $\exp_p(ty) \in \tilde{S}(p,t)$ for all $0 < t < r$. Let

$$\hat{\mathcal{D}}_p := \Big\{ \exp_p(ty),\ 0 \le t < \hat{\mathbf{i}}_y \Big\}.$$

$\hat{\mathcal{D}}_p$ is called the *convex domain* of p. From the definition, we know that $\hat{\mathcal{D}}_p \subset \mathcal{D}_p$. For $t > 0$, let

$$\hat{S}(p,t) := S(p,t) \cap \hat{\mathcal{D}}_p.$$

By definition, $\hat{S}(p,t) \subset \tilde{S}(p,t)$.

Let $y \in \mathrm{S}_pM$ and $c(t) := \exp_p(ty)$, $0 \le t < \hat{\mathbf{i}}_y$. Take an arbitrary $\mathbf{g}_{\dot{c}(t)}$-orthonormal basis $\{E_i(t)\}_{i=1}^{n-1}$ for $T_x\hat{S}(p,t)$. By definition, the symmetric matrix (h_{ij}) has nonnegative eigenvalues

$$h_{ij} := \mathbf{g}_{\dot{c}}\Big(\mathrm{S}_t(E_i), E_j\Big).$$

Thus

$$\sum_{i,k=1}^{n} h_{ik}h_{ki} \le \Big(\sum_{i=1} h_{ii}\Big)^2.$$

This implies that

$$\sum_{i=1}^{n-1} \mathbf{g}_{\dot{c}}\Big(\mathrm{S}_t(E_i), \mathrm{S}_t(E_i)\Big) \le \Big[\sum_{i=1}^{n-1} \mathbf{g}_{\dot{c}}\Big(\mathrm{S}_t(E_i), E_i\Big)\Big]^2. \tag{15.25}$$

Assume that

$$\mathbf{Ric} \le (n-1)\lambda.$$

Let $\hat{\Pi}_t$ denote the mean curvature of $\hat{S}(p,t)$ at $c(t)$ in $(\hat{\mathcal{D}}_p, \hat{g})$ and

$$\phi(t) := \frac{1}{n-1}\hat{\Pi}_t.$$

It follows from (15.17) and (15.25) that

$$\phi' \geq -\lambda - (n-1)\phi^2.$$

Let

$$\psi_\lambda(t) := \frac{1}{n-1}\frac{\mathbf{s}'_{(n-1)\lambda}(t)}{\mathbf{s}_{(n-1)\lambda}(t)}.$$

Then

$$\psi'_\lambda = -\lambda - (n-1)\psi_\lambda^2.$$

It follows from (15.20) that

$$\phi(t) - \psi_\lambda(t) \approx \frac{n-2}{n-1}t^{-1} + O(t).$$

We may assume that $n > 2$ (the case when $n = 2$ is covered by Theorem 15.3.1(i)). Then for sufficiently small $t_o > 0$,

$$\phi(t_o) - \psi_\lambda(t_o) > 0.$$

Consider the following function

$$f(t) := e^{\int(\phi+\psi_\lambda)dt}\Big[\phi(t) - \psi_\lambda(t)\Big].$$

We obtain

$$f'(t) = e^{\int(\phi+\psi_\lambda)dt}\Big[\phi'(t) + \phi^2(t) - \psi'_\lambda(t) - \psi_\lambda^2(t)\Big] \geq 0.$$

Thus

$$f(t) \geq f(t_o), \qquad t_o < t < \hat{\mathbf{i}}_y.$$

This implies

$$\phi(t) - \psi_\lambda(t) \geq \Big[\phi(t_o) - \psi_\lambda(t_o)\Big]e^{\int_{t_o}^{t}(\phi+\psi_\lambda)ds} > 0.$$

Since $t_o > 0$ can be arbitrary small, we have proved the following

Theorem 15.4.1 *Let (M, F) be an n-dimensional positively complete Finsler space. Let $\hat{\Pi}_t$ denote the mean curvature of $\hat{S}(p,t)$. Suppose that the Ricci curvature satisfies the bound*

$$\mathbf{Ric} \leq (n-1)\lambda.$$

Then

$$\hat{\Pi}_t \geq \frac{\mathbf{s}'_{(n-1)\lambda}(t)}{\mathbf{s}_{(n-1)\lambda}(t)}.$$

The above theorem is a slight modification of Ding's estimate on the Laplacian of the distance function on a Riemannian space [Di] (see also [Xi]).

Remark 15.4.2 Suppose that

$$\mathbf{Ric} \leq (n-1)\lambda, \qquad \mathbf{S} \leq (n-1)\delta.$$

It follows from Theorem 15.4.1 that the Laplacian of $\rho = d(p, \cdot)$ on the convex domain $\hat{\mathcal{D}}_p$ and the mean curvature Π_ρ of $\hat{S}(p, \rho)$ satisfy

$$\Delta\rho = \Pi_\rho \geq \frac{\mathbf{s}'_{(n-1)\lambda}(\rho)}{\mathbf{s}_{(n-1)\lambda}(\rho)} - (n-1)\delta. \tag{15.26}$$

Remark 15.4.3 *Let $\hat{\Delta}\rho$ denote the Laplacian of ρ with respect to $\hat{g}$. According to (14.34), we know that the estimates on $\hat{\Pi}_{\mathbf{n}}$ in Theorem 15.3.1 also give estimates on $\hat{\Delta}\rho$.*

Chapter 16

Volume Comparison Theorems

In the previous chapter, we estimate the mean curvature of metric spheres from above. By a similar argument, we can estimate the mean curvature of the level hypersurfaces of a distance function. In this chapter, we will show that there is a close relationship between the growth of the mean curvature and the growth of the volume of metric spheres. Then we establish a volume comparison theorem for the volume of metric balls under a lower Ricci curvature bound and a lower S-curvature bound. There are applications of this volume comparison theorem. Throughout this chapter, we always set

$$\omega_n := \mathrm{Vol}(\mathbb{B}^n), \qquad \sigma_{n-1} := \mathrm{Vol}(\mathbb{S}^{n-1}).$$

16.1 Volume of Metric Balls

In this section, we will estimate the volume of metric balls in a Finsler m space. Let $(M, F, d\mu)$ be an n-dimensional Finsler m space. We first take a look at the Busemann-Hausdorff volume form dV_F and its induced volume form dA_F on hypersurfaces.

For a vector $y \in \mathrm{S}_pM$, let $\{\mathbf{b}_i\}_{i=1}^n$ be a basis for T_pM such that

$$\mathbf{b}_1 = y, \qquad \mathbf{g}_y(y, \mathbf{b}_i) = 0, \quad i = 2, \cdots, n.$$

Extend $\{\mathbf{b}_i\}_{i=1}^n$ to a global frame on T_pM in a natural way. We see that $\mathbf{b}_i$'s are tangent to S_pM for $2 \le i \le n$. Let $\{\theta^i\}_{i=1}^n$ denote the basis for T_p^*M dual to $\{\mathbf{b}_i\}_{i=1}^n$. Express dV_F at p by

$$dV_p = \sigma_F(p)\theta^1 \wedge \cdots \wedge \theta^n,$$

where

$$\sigma_F(p) = \frac{\mathrm{Vol}(\mathbb{B}^n)}{\mathrm{Vol}\Big\{(y^i) \in \mathrm{R}^n \;\Big|\; F(y^i \mathbf{b}_i) < 1\Big\}}.$$

Extending $\{\theta^i\}_{i=1}^n$ to a global coframe on T_pM in a natural way, we obtain a volume form dV_p on T_pM. The induced volume form dA_p by dV_p on S_pM is given by

$$dA_p = \sigma_F(p)\theta^2 \wedge \cdots \wedge \theta^n.$$

We have

$$\int_{\mathrm{S}_pM} dA_p = \sigma_{n-1}.$$

The volume form $d\mu$ can be expressed as

$$d\mu = \phi \, dV_F.$$

At point $p \in M$, let $d\nu_p$ denote the induced volume form by $d\mu_p$ on S_pM. We have

$$d\nu_p = \phi(p)dA_p.$$

Thus

$$\int_{\mathrm{S}_pM} d\nu_p = \phi(p) \int_{\mathrm{S}_pM} dA_p = \phi(p)\sigma_{n-1}. \tag{16.1}$$

For a point $p \in M$ and $t \geq 0$, let

$$\Sigma_p^t := \Big\{ y \in \mathrm{S}_pM \;\Big|\; \mathbf{i}_y > t \Big\} \subset \mathrm{S}_pM.$$

Σ_p^t is an open subset in S_pM. Note that $\Sigma_p^t = \mathrm{S}_pM$ for $0 \leq t < \mathbf{i}_x$ and

$$\Sigma_p^{t'} \subset \Sigma_p^{t''}, \qquad t'' < t'.$$

Let $\mathcal{D}_p$ denote the cut-domain at p and $S(p,t)$ the metric sphere of radius t around p. The regular part

$$\tilde{S}(p,t) := S(p,t) \cap \mathcal{D}_p$$

is a C^∞ hypersurface. We call it the *regular metric sphere* of radius t around p. Define $\varphi_t : \Sigma_p^t \to \tilde{S}(p,t)$ by

$$\varphi_t(y) := \exp_p(ty), \qquad y \in \Sigma_p^t,\ t \geq 0.$$

φ_t is a diffeomorphism.

Let dA_t denote the induced volume form by dV_F on $\tilde{S}(P,t)$. Then $d\nu_t = \phi\, dA_t$ is the induced volume form on $\tilde{S}(p,t)$ by $d\mu = \phi\, dV_F$. Define $\eta_t : \mathrm{S}_pM \to [0,\infty)$ by

$$(\varphi_t)^*\Big(d\nu_t|_{\varphi_t(y)}\Big) = \eta_t(y)\,\Big(d\nu_p|_y\Big), \qquad y \in \Sigma_p^t \tag{16.2}$$

and

$$\eta_t(y) = 0, \qquad y \in \mathrm{S}_pM \setminus \Sigma_p^t.$$

We have

$$\eta_t(y) = t^{n-1}\Big[1 + o(t)\Big]. \tag{16.3}$$

We prove (16.3) as follows. Fix a vector $y \in \mathrm{S}_pM$ and let

$$c(t) := \varphi_t(y), \qquad 0 \leq t \leq \mathbf{i}_y.$$

Take an arbitrary basis $\{\mathbf{b}_i\}_{i=1}^n$ for T_pM as above and define

$$E_i(t) := (\exp_p)_{*|ty}(\mathbf{b}_i), \qquad t \geq 0,\ i = 1, \cdots, n.$$

$E_i(t)$'s are C^∞ vector fields along c. Let $\{\theta^i(t)\}_{i=1}^n$ denote the basis for $T_{c(t)}M$ dual to $\{E_i(t)\}_{i=1}^n$ and

$$d\mu|_{c(t)} = \sigma(t)\theta^1(t) \wedge \cdots \wedge \theta^n(t),$$

where

$$\sigma(t) := \frac{\omega_n \phi(c(t))}{\mathrm{Vol}\Big\{(y^i) \in \mathrm{R}^n \Big|\ F\Big(y^i E_i(t)\Big) < 1\Big\}}.$$

We have

$$d\nu_t = \sigma(t)\theta^2(t) \wedge \cdots \wedge \theta^n(t).$$

Since $\{E_i(t)\}_{i=1}^n$ are C^∞ at $t = 0$, we conclude that $\sigma(t)$ is C^∞ on $[0, \mathbf{i}_y)$. Observe that

$$(\varphi_t)^*\theta^i(t)(\mathbf{b}_j) = \theta^i(t)\Big((\varphi_t)_*(\mathbf{b}_j)\Big) = t\,\theta^i(t)\Big(E_j(t)\Big) = t\,\delta^i_j = t\,\theta^i(\mathbf{b}_j).$$

Thus

$$(\varphi_t)^*\theta^i(t) = t\,\theta^i.$$

We obtain

$$\begin{aligned}(\varphi_t)^* d\nu_t &= \sigma(t)(\varphi_t)^*\Big[\theta^2(t)\wedge\cdots\wedge\theta^n(t)\Big]\\ &= t^{n-1}\sigma(t)\,\theta^2\wedge\cdots\wedge\theta^n\\ &= t^{n-1}\frac{\sigma(t)}{\sigma(0)}\sigma(0)\,\theta^2\wedge\cdots\wedge\theta^n\\ &= t^{n-1}\frac{\sigma(t)}{\sigma(0)}\,d\nu_p.\end{aligned}$$

Thus

$$\eta_t(y) = \frac{\sigma(t)}{\sigma(0)} = \frac{\mathrm{Vol}\Big\{(y^i)\in \mathrm{R}^n \;\Big|\; F\Big(y^i\mathbf{b}_i\Big) < 1\Big\}}{\mathrm{Vol}\Big\{(y^i)\in \mathrm{R}^n \;\Big|\; F\Big(y^iE_i(t)\Big) < 1\Big\}}t^{n-1}.$$

Thus (16.3) holds.

For a small number $\varepsilon > 0$, define

$$\phi_\varepsilon(x) := \varphi_{t+\varepsilon}\circ\varphi_t^{-1}(x), \qquad x\in\tilde S(p,t). \tag{16.4}$$

For a point $x\in\tilde S(p,t)$, there is an open neighborhood $\mathcal U$ of x in $\tilde S(p,t)$ and a small number $\delta > 0$ such that ϕ_ε is defined on $\mathcal U$ for all $0\le\varepsilon\le\delta$.

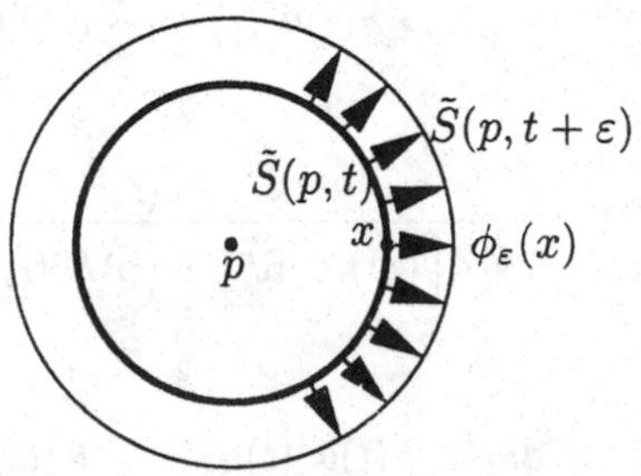

As in (14.27), define $\Theta(x,\varepsilon)$ by

$$(\phi_\varepsilon)^*\Big(d\nu_{t+\varepsilon}|_{\phi_\varepsilon(x)}\Big) = \Theta(x,\varepsilon)d\nu_t|_x, \qquad x\in\tilde S(p,t).$$

By (16.2) and (16.4) we have

$$\Theta(x, \varepsilon) = \frac{\eta_{t+\varepsilon}(y)}{\eta_t(y)}, \qquad x = \varphi_t(y). \tag{16.5}$$

Let Π_t denote the mean curvature of $\tilde{S}(p,t)$ at x with respect to the outward-pointing normal vector. In virtue of (14.29) and (16.5), we have

$$\Pi_t = \frac{d}{d\varepsilon}\Big[\ln \Theta(x, \varepsilon)\Big]\Big|_{\varepsilon=0} = \frac{d}{dt}\Big(\ln \eta_t(y)\Big). \tag{16.6}$$

Let

$$\chi(t) := \Big[e^{\delta t}\mathbf{s}_\lambda(t)\Big]^{n-1}.$$

By Theorem 15.3.2 and (16.6), we obtain

$$\Pi_t = \frac{d}{dt}\Big(\ln \eta_t(y)\Big) \le \frac{d}{dt}\Big(\ln \chi(t)\Big). \tag{16.7}$$

This implies

$$\frac{d}{dt}\Big(\frac{\eta_t(y)}{\chi(t)}\Big) \le 0. \tag{16.8}$$

Thus the quotient $h(t) := \eta_t(y)/\chi(t)$ is non-increasing.

Now we study the induced volume $\nu\Big(\tilde{S}(p,t)\Big)$ of the regular metric sphere $\tilde{S}(p,t)$. By (16.2), we have

$$\nu\Big(\tilde{S}(p,t)\Big) = \int_{\mathrm{S}_pM} \eta_t(y) d\nu_p. \tag{16.9}$$

Applying the co-area formula to the distance function $\rho(x) := d(p,x)$, we obtain

$$\mu\Big(D(p,t)\Big) - \mu\Big(D(p,t) \cap \mathcal{D}_p\Big) - \int_0^t \nu(\tilde{S}(p,s))ds.$$

Thus

$$\mu\Big(B(p,t)\Big) = \int_0^t \Big[\int_{\mathrm{S}_pM} \eta_s(y) d\nu_p\Big] ds. \tag{16.10}$$

It follows from (16.3) that

$$\nu\Big(S(p,t)\Big) = \int_{S_pM} \eta_t(y)d\nu_p = \phi(p)\sigma_{n-1}t^{n-1}\Big[1+o(t)\Big]. \tag{16.11}$$

Now we are ready to prove the following

Theorem 16.1.1 ([Sh2]) *Let $(M, F, d\mu)$ be an n-dimensional positively complete Finsler m space. Assume that*

$$\mathbf{Ric} \geq (n-1)\lambda, \qquad \mathbf{S} \geq -(n-1)\delta. \tag{16.12}$$

Then the quotients

$$\frac{\nu\Big(\tilde{S}(p,t)\Big)}{\Big[e^{\delta t}\mathbf{s}_\lambda(t)\Big]^{n-1}} \quad \text{and} \quad \frac{\mu\Big(B(p,t)\Big)}{\int_0^t \Big[e^{\delta s}\mathbf{s}_\lambda(s)ds\Big]^{n-1}} \tag{16.13}$$

are non-increasing. In particular, we have

$$\mu\Big(B(p,t)\Big) \leq \phi_\mu(p)\omega_n \int_0^t \Big[e^{\delta s}\mathbf{s}_\lambda(s)\Big]^{n-1} ds.$$

Proof. From (16.8) we see that $\eta_t(y)/\chi(t)$ is non-increasing.

$$\eta_{t_2}(y)\chi(t_1) \leq \eta_{t_1}(y)\chi(t_2), \qquad 0 < t_1 < t_2. \tag{16.14}$$

Let $A(t) := \nu(\tilde{S}(p,t))$. Integrating (16.14) over S_pM with respect to $d\nu_p$ yields

$$A(t_2)\chi(t_1) \leq A(t_1)\chi(t_2), \qquad 0 < t_1 < t_2. \tag{16.15}$$

Inequality (16.15) implies that the first quotient in (16.13) is non-increasing.

Rewrite (16.15) as follows

$$A(t'')\chi(t') \leq A(t')\chi(t''), \qquad 0 \leq t' < t''. \tag{16.16}$$

Integrating (16.16) with respect to t'' over $[t_1, t_2]$, then integrating the resulting inequality with respect to t' over $[0, t_1]$, we obtain the following inequality

$$\int_0^{t_1} \chi(t)dt \int_{t_1}^{t_2} A(t)dt \leq \int_0^{t_1} A(t)dt \int_{t_1}^{t_2} \chi(t)dt. \tag{16.17}$$

Inequality (16.17) implies that the second quotient in (16.13) is non-increasing. This completes the proof. Q.E.D.

The above comparison theorem on the Busemann-Hausdorff volume is due to Bishop [BiCr] and Gromov [GLP][Gr4] in the Riemannian case.

Integrating (16.15) with respect to t_1 over $[0, t_2]$ yields

$$\frac{\nu\Big(\tilde{S}(p,t)\Big)}{\mu\Big(B(p,t)\Big)} \leq \frac{\Big[e^{\delta u}\mathbf{s}_\lambda(u)\Big]^{n-1}}{\int_0^t \Big[e^{\delta u}\mathbf{s}_\lambda(u)\Big]^{n-1} du}. \tag{16.18}$$

In other words,

$$\frac{d}{dt}\Big[\ln \mu\Big(B(p,t)\Big)\Big] \leq \frac{d}{dt}\Big[\ln \int_0^t \chi(u)du\Big].$$

By (14.50) and (16.6), we obtain a Taylor expansion for $\eta_t(y)$ which is better than (16.3)

$$\eta_t(y) = t^{n-1}\Big[1 - \mathbf{S}(y)t - \frac{1}{6}\Big(\mathbf{Ric}(y) + 3\dot{\mathbf{S}}(y) - 3\mathbf{S}^2(y)\Big)t^2 + o(t^2)\Big]. \tag{16.19}$$

Since $\mathcal{D}_p$ contains the metric ball $B(p, \mathbf{i}_x)$, $\tilde{S}(p,t) = S(p,t)$, $0 < t < \mathbf{i}_x$. By (16.9) and (16.10), we obtain a Taylor expansion for $\nu\Big(S(p,t)\Big)$ which is better than (16.11),

$$\begin{aligned}
\nu\Big(S(p,t)\Big) &= \phi_\mu(p)\sigma_{n-1}t^{n-1}\Big\{1 - \mathbf{s}(p)t - \frac{\mathbf{r}(p) + 3\mathbf{h}(p)}{6n}t^2 + o(t^2)\Big\},\\
\mu\Big(B(p,t)\Big) &= \phi_\mu(p)\omega_n t^n\Big\{1 - \frac{n\mathbf{s}(p)}{n+1}t - \frac{\mathbf{r}(p) + 3\mathbf{h}(p)}{6(n+2)}t^2 + o(t^2)\Big\},
\end{aligned}$$

where

$$\begin{aligned}
\mathbf{r}(p) :&= \frac{n}{\sigma_{n-1}}\int_{\mathbf{S}_pM} \mathbf{Ric}(y)d\nu_p\\
\mathbf{s}(p) :&= \frac{1}{\sigma_{n-1}}\int_{\mathbf{S}_pM} \mathbf{S}(y)d\nu_p,\\
\mathbf{h}(p) :&= \frac{n}{\sigma_{n-1}}\int_{\mathbf{S}_pM} \Big[\dot{\mathbf{S}}(y) - \mathbf{S}^2(y)\Big]d\nu_p.
\end{aligned}$$

Note that if F is reversible, i.e., $F(-y) = F(y)$, then $\mathbf{S}(-y) = -\mathbf{S}(y)$. Thus $\mathbf{s}(p) = 0$.

16.2 Volume of Tubular Neighborhoods

In this section, we will estimate the volume of tubular neighborhoods of a hypersurface under lower bounds on the Ricci curvature and the S-curvature.

Let $(M, F, d\mu)$ be an n-dimensional Finsler m space. We still assume that

$$\mathbf{Ric} \geq (n-1)\lambda, \qquad \mathbf{S} \geq -(n-1)\delta.$$

Consider a hypersurface $N \subset M$ and a normal field $\mathbf{n}$ along N. For the sake of simplicity, we assume that N is closed. Take a small neighborhood $\mathcal{U}$ of N. N divides $\mathcal{U}$ into two parts. Let $\mathcal{U}_+$ denote the part toward which $\mathbf{n}$ points and $\mathcal{U}_-$ the other part. Choosing a smaller $\mathcal{U}$ if necessary, we may assume that the following function

$$\rho(x) := \begin{cases} d(N, x) & x \in \mathcal{U}_+ \\ -d(x, N) & x \in \mathcal{U}_+ \end{cases}$$

is a C^∞ distance function on $\mathcal{U}$ of N with $\rho^{-1}(0) = N$ and $\mathbf{n} = \nabla\rho|_N$. ρ induces a Riemannian metric $\hat{g} := \mathbf{g}_{\nabla\rho}$ on $\mathcal{U}$. Let Π_t denote the mean curvature of $N_t := \rho^{-1}(t)$ in $(\mathcal{U}, F)$ with respect to $\nabla\rho|_{N_t}$ and $\hat{\Pi}_{\mathbf{n}}$ denote the mean curvature of N in $(\mathcal{U}, \hat{g})$ with respect to $\mathbf{n}$. Assume that

$$\hat{\Pi}_{\mathbf{n}} \leq (n-1)\tau. \tag{16.20}$$

We will show that the mean curvature Π_t of N_t in $(\mathcal{U}, F)$ satisfies

$$\Pi_t \leq (n-1)\frac{d}{dt}\Big\{ \ln \Big[e^{\delta t}\Big(\tau \mathbf{s}_\lambda(t) + \mathbf{s}'_\lambda(t) \Big)\Big]\Big\}. \tag{16.21}$$

We prove (16.21) as follows. Define $\varphi_t : N \to M$ by

$$\varphi_t(z) := \exp_z(t\mathbf{n}_z), \qquad z \in N.$$

For small $t \geq 0$, $\varphi_t(N) \subset N_t$.

Fix a point $z \in N$. Let $c(t) = \varphi_t(z)$, $t \geq 0$. Let $\hat{\Pi}_t$ and $\hat{\Lambda}_t$ denote the mean curvature and the normal curvature of N_t at $c(t)$ in $(\mathcal{U}, \hat{g})$ with respect to the normal vector $\nabla\rho = \hat{\nabla}\rho$, respectively. Let

$$\phi(t) := \frac{1}{n-1}\hat{\Pi}_t.$$

Since $\hat{g}$ is a Riemannian metric, the mean curvature $\hat{\Pi}_t$ is the mean value of the normal curvature $\hat{\Lambda}_t$. See Proposition 14.3.3. Let S_t denote the shape operator of $N_t = \rho^{-1}(t)$ at $c(t) \in N_t$. By (14.39), we have

$$\phi(t) = \frac{1}{n-1}\sum_{i=1}^{n-1} \mathbf{g}_{\dot{c}}\Big(\mathrm{S}_t(E_i), E_i\Big),$$

where $\{E_i(t)\}$ is a frame along c.

As we see in the proof of Theorem 15.3.1, the Riccati equation (14.41) implies

$$\phi' \leq -\lambda - \phi^2. \tag{16.22}$$

Let

$$\phi_o(t) := \frac{d}{dt}\Big[\ln\Big(\tau \mathbf{s}_\lambda(t) + \mathbf{s}'_\lambda(t)\Big)\Big].$$

Note that

$$\phi'_o = -\lambda - \phi_o^2. \tag{16.23}$$

Let

$$f(t) := e^{\int_0^t(\phi(s)+\phi_o(s))ds}\Big[\phi(t) - \phi_o(t)\Big].$$

By (16.22) and (16.23), we have

$$f'(t) = e^{\int_0^t(\phi(s)+\phi_o(s))ds}\Big[\phi' + \phi^2 - \phi'_o - \phi_o^2\Big] \leq 0.$$

Note that

$$f(0) = \frac{1}{n-1}\hat{\Pi}_{\mathrm{n}} - \tau \leq 0.$$

Thus

$$f(t) \leq f(0) \leq 0, \qquad t \geq 0.$$

That is,

$$\hat{\Pi}_t \leq (n-1)\phi_o(t). \tag{16.24}$$

By (14.33) and (16.24), we have

$$\begin{aligned}\Pi_t &= \hat{\Pi}_t - \mathbf{S}(\dot{c}(t)) \leq (n-1)\phi_o + (n-1)\delta \\ &= (n-1)\frac{d}{dt}\Big\{ \ln \Big[e^{\delta t}\Big(\tau \mathbf{s}_\lambda(t) + \mathbf{s}'_\lambda(t)\Big)\Big]\Big\}. \end{aligned} \tag{16.25}$$

This proves (16.21). Q.E.D.

Now we study the growth of the volume of the tubular neighborhoods of N,

$$U_t(N) := \Big\{ x \in \mathcal{U}^+ \;\Big|\; d(N, x) < t \Big\}.$$

For a point $z \in N$, let d_z denote the largest number $r > 0$ such that

$$\rho\Big(\varphi_t(z)\Big) = d\Big(N, \varphi_t(z)\Big) = t, \qquad 0 < t < r.$$

Define $\eta_t(z)$ by

$$(\varphi_t)^*\Big(d\nu_t|_{\varphi_t(z)}\Big) = \eta_t(z)\Big(d\nu|_z\Big), \tag{16.26}$$

where $d\nu_t$ denotes the induced volume form on N_t by $d\mu$. Set $\eta_t(z) = 0$ for $t \geq d_z$.

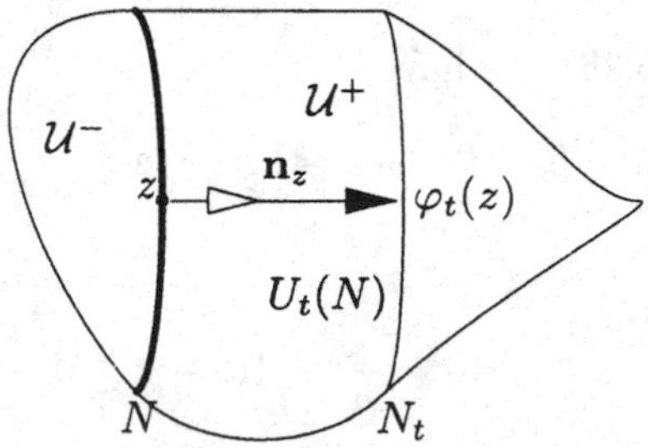

We have

$$\nu_t(N_t) = \int_N \eta_t(z) d\nu,$$

and

$$\mu\Big(U_t(N)\Big) = \int_0^t \nu_s(N_s)ds = \int_0^t \Big[\int_N \eta_s(z)d\nu\Big]ds.$$

For a point $x := \varphi_t(z) \in N_t$ and small $\varepsilon > 0$, define $\Theta(x,\varepsilon)$ as in (14.27) by

$$(\varphi_\varepsilon)^*\Big(d\nu_{t+\varepsilon}|_{\varphi_{t+\varepsilon}(z)}\Big) = \Theta(x,\varepsilon)d\nu_t|_x.$$

By a similar argument as for (16.5), we obtain

$$\Theta(x,\varepsilon) = \frac{\eta_{t+\varepsilon}(z)}{\eta_t(z)}, \qquad x = \varphi_t(z).$$

By (14.29), we obtain

$$\Pi_t = \frac{d}{d\varepsilon}\Big[\ln\Theta(x,\varepsilon)\Big]\Big|_{\varepsilon=0} = \frac{d}{dt}\Big[\ln\eta_t(z)\Big]. \tag{16.27}$$

It follows from (16.25) and (16.27) that

$$\frac{d}{dt}\Big[\ln\eta_t(z)\Big] = \Pi_t \le (n-1)\frac{d}{dt}\Big\{\ln\Big[e^{\delta t}\Big(\tau \mathbf{s}_\lambda(t) + \mathbf{s}'_\lambda(t)\Big)\Big]\Big\}.$$

Thus the following ratio

$$\frac{\eta_t(y)}{\Big[e^{\delta t}\Big(\tau \mathbf{s}_\lambda(t) + \mathbf{s}'_\lambda(t)\Big)\Big]^{n-1}}$$

is non-increasing.

By the same argument as above for the volume of metric balls, we obtain the following

Theorem 16.2.1 *Let (M,F) be an n-dimensional positively complete Finsler space. Let $N \subset M$ be a closed hypersurface and $\mathbf{n}$ a normal vector field on N. Suppose*

$$\mathbf{Ric} \ge (n-1)\lambda, \qquad \mathbf{S} \ge -(n-1)\delta, \qquad \hat{\Pi}_{\mathbf{n}} \le (n-1)\tau.$$

Then the following ratio

$$\frac{\mu\Big(U_t(N)\Big)}{\nu(N)\int_0^t \Big[e^{\delta u}\Big(\tau\, \mathbf{s}_\lambda(u) + \mathbf{s}'_\lambda(u)\Big)\Big]^{n-1} du} \tag{16.28}$$

is non-increasing on $[0, \infty)$. *In particular, when* N *is the boundary of* M,

$$\mu(M) \leq \nu(N) \int_0^d \left[e^{\delta u} \left(\tau \, \mathbf{s}_\lambda(u) + \mathbf{s}'_\lambda(u) \right) \right]^{n-1} du, \tag{16.29}$$

where $d = \sup_{x \in M} d(N, x)$.

The above result in the Riemannian case is due to Heintze-Karcher [HeKa].

16.3 Gromov Simplicial Norms

In order to find some relations between the geometry and the topology of Riemannian manifolds, M. Gromov introduced many new topological quantities. Among them are the so-called Gromov simplicial volume [Gr3]. In this section, we will give a brief discussion on the Gromov simplicial volume and its relationship with the Ricci curvature and the S-curvature.

Let M be a topological space. Denote by $\mathrm{C}_k(M)$ the k-th complex of real singular chains $c = \sum_i r_i \sigma_i$, where $\sigma_i : \Delta^k \to M$ are k-dimensional simplices and r_i are real numbers which are all, but finite, zero. Let $\mathrm{H}_k(M)$ denote the real singular homology of M.

The natural L^1-norm $\|\cdot\|_1$ on $\mathrm{C}_k(M)$ is defined by

$$\|c\| = \sum_i |r_i|, \qquad \text{for } \; c = \sum_i r_i \sigma_i \in \mathrm{C}_k(M).$$

This L^1-norm induces a pseudo-norm $\|\cdot\|$ in $\mathrm{H}_k(M)$. We call $\|\cdot\|$ the *Gromov simplicial norm*. For a class $z \in \mathrm{H}_k(M)$, $\|z\|$ is given by

$$\|z\| := \inf_{z=[c]} \|c\|_1.$$

Assume that M is an n-dimensional closed oriented manifold. For the fundamental class $[M] \in \mathrm{H}_n(M)$, we set

$$\|M\| := \|[M]\|.$$

The constant $\|M\|$ is called the *Gromov simplicial volume*.

Assume that (M, F) is an oriented reversible Finsler space. For a C^∞ immersed submanifold $\sigma : N \to M$, define

$$\mathrm{Vol}_F(\sigma) := \mathrm{Vol}_{\sigma^* F}(N).$$

This notion can be extended to all singular chains as follows. For a singular simplex $\sigma : \Delta^k \to M$, define $\mathrm{Vol}_F(\sigma)$ as the infimal of $\mathrm{Vol}_G(\Delta^k)$, where G is a reversible Finsler metric on Δ^k such that $\sigma : (\Delta^k, G) \to (M, F)$ is distance-decreasing. Set $\mathrm{Vol}_F(\sigma) = \infty$ if no such G exists. For a singular chain $c = \sum_i r_i \sigma_i \in \mathrm{C}_k(M)$, define

$$\mathrm{Vol}_F(c) := \sum_i |r_i| \mathrm{Vol}_F(\sigma_i).$$

For a homology class $z \in \mathrm{H}_k(M)$, we set

$$\mathrm{Vol}_F(z) := \inf_{z=[c]} \mathrm{Vol}_F(c).$$

Then for the fundamental class $[M] \in \mathrm{H}_n(M)$,

$$\mathrm{Vol}_F(M) = \mathrm{Vol}_F([M]).$$

Gromov proved the following

Lemma 16.3.1 ([Gr3]) *Let (M, F) be an n-dimensional oriented closed reversible Finsler space. For any class $z \in \mathrm{H}_k(M)$,*

$$\|z\| \leq k! \min_{r>0} \sup_{\tilde{x} \in \tilde{M}} \left[\frac{A_{\tilde{F}}(\tilde{S}(\tilde{x}, r))}{\mathrm{Vol}_{\tilde{F}}(B(\tilde{x}, r))} \right]^k \mathrm{Vol}_F(z),$$

where $B(\tilde{x}, r)$ and $\tilde{S}(\tilde{x}, r)$ denote the metric ball and the regular metric sphere of radius r around $\tilde{x}$ in the universal cover $\pi : \tilde{M} = M$ respectively and $\tilde{F} := \pi^* F$. In particular,

$$\|M\| \leq n! \min_{r>0} \sup_{\tilde{x} \in \tilde{M}} \left[\frac{A_{\tilde{F}}(\tilde{S}(\tilde{x}, r))}{\mathrm{Vol}_{\tilde{F}}(B(\tilde{x}, r))} \right]^n \mathrm{Vol}_F(M).$$

By the volume comparison theorem (Theorem 16.1.1), we can easily prove the following

Theorem 16.3.2 *Let (M, F) be an n-dimensional oriented closed reversible Finsler space. Suppose that the Ricci curvature and the S-curvature*

satisfy the bounds:

$$\mathbf{Ric} \geq -(n-1), \qquad \mathbf{S} \geq -(n-1). \tag{16.30}$$

Then for any homology class $z \in \mathrm{H}_k(M)$,

$$\|z\| \leq k!(n-1)^k 2^k \mathrm{Vol}_F(z).$$

In particular,

$$\|M\| \leq n!(n-1)^n 2^n \mathrm{Vol}_F(M).$$

Proof. Let $\tilde{F} = \pi^* F$ denote the lifted Finsler metric on the universal cover $\pi : \tilde{M} \to M$. The Ricci curvature **Ric** and the S-curvature of $\tilde{F}$ still satisfy the same bounds as in (16.30). By (16.18), we have

$$\frac{A_{\tilde{F}}(\tilde{S}(\tilde{x}, r))}{\mathrm{Vol}_{\tilde{F}}(B(\tilde{x}, r))} \leq \frac{\Big[e^r \sinh(r)\Big]^{n-1}}{\int_0^r \Big[e^t \sinh(t)\Big]^{n-1} dt}.$$

Note that

$$\lim_{r \to \infty} \frac{\Big[e^r \sinh(r)\Big]^{n-1}}{\int_0^r \Big[e^t \sinh(t)\Big]^{n-1} dt} = 2(n-1).$$

It follows from Lemma 16.3.1 that for any $z \in \mathrm{H}_k(M)$,

$$\|z\| \leq k!(n-1)^k 2^k \mathrm{Vol}_F(z).$$

Q.E.D.

Let (M, F) be an n-dimensional closed oriented reversible Finsler space. An open subset $U \subset M$ is said to be *amenable* if for any connected component $i : W \subset U \to M$, the image

$$i_*\Big(\pi_1(W, x)\Big) \subset \pi_1(M, x), \qquad x \in W$$

is an amenable subgroup of $\pi_1(M)$. Hence if $\pi_1(M)$ is amenable, then every open subsets in M are amenable. M. Gromov proved that if the manifold is covered by amenable open subsets $\{U_i\}$ such that every point is contained at most n open subsets U_i, then the simplicial volume vanishes.

Assume that the Ricci curvature and the S-curvature of F satisfy the bounds (16.30). Following the line in [Gr3], one can construct an $(n-1)$-dimensional polyhedron P^{n-1} and a map $f : M \to P^{n-1}$ such that $f^{-1}(\mathcal{U})$ of any star neighborhood $\mathcal{U}$ in P^{n-1} is amenable, when the volume of F is sufficiently small. Since P^{n-1} is covered by its star neighborhoods of vertices and each point in P^{n-1} is contained in at most n star neighborhoods, we conclude that M can be covered by amenable open subsets $\{U_i\}$ (the pre-images of star neighborhoods) such that every point x is contained in at most $n = \dim M$ open subsets U_i. This implies $\|M\| = 0$. More precisely,

Theorem 16.3.3 *Let (M, F) be an n-dimensional closed oriented reversible Finsler space satisfying*

$$\mathbf{Ric} \geq -(n-1), \qquad \mathbf{S} \geq -(n-1). \tag{16.31}$$

There is a number $\varepsilon(n) > 0$ such that if for some $0 < \nu \leq 1$,

$$\mathrm{Vol}_F(B(x, \rho_o)) \leq \varepsilon(n)\, \nu^n, \qquad \forall x \in M \tag{16.32}$$

then there exist an $(n-1)$-dimensional polyhedron P^{n-1} and a continuous map $f : M \to P^{n-1}$ such that for any star neighborhood $\mathcal{U} \subset P^{n-1}$, the pre-image $f^{-1}(\mathcal{U})$ is almost nilpotent. Further

$$\mathrm{Diam}(f^{-1}(\mathcal{U})) \leq C(n)\nu.$$

Thus $\|M\| = 0$.

According to Theorem 16.3.3, if M is an n-dimensional oriented closed manifold with $\|M\| \neq 0$ and F is a reversible Finsler metric on M satisfying the curvature bounds (16.31), then there is a positive number $\varepsilon(n)$ depending on n such that

$$\mathrm{Vol}_F(M) > \varepsilon(n).$$

16.4 Estimates on the Expansion Distance

In many cases, metric measure spaces concentrate to a point if the dimension goes to infinity. A natural question arises: under what conditions the dimension of metric measure spaces in a concentrating sequence must go to infinity? We shall give an answer to this problem for Finsler m spaces.

Let $(M, F, d\mu)$ be an n-dimensional compact Finsler m space. Let dV_F denote the Busemann-Hausdorff volume form of F and $\mathbf{S}$ denote the S-curvature of (F, dV_F).

Put

$$d\mu = \phi_\mu dV_F.$$

The S-curvature $\mathbf{S}_\mu$ of $(F, d\mu)$ is related to $\mathbf{S}$ by

$$\mathbf{S}_\mu = \mathbf{S} - \frac{d\phi_\mu}{\phi_\mu}. \tag{16.33}$$

We have

Theorem 16.4.1 ([GuSh]) *Given numbers $\lambda < 0, \delta > 0, d > 0$ and $\tau > 0$. Let $X = (M, F, d\mu)$ be an n-dimensional compact reversible Finsler m space. Suppose the Finsler metric F satisfies*

$$\mathbf{Ric} \geq (n-1)\lambda, \quad \mathbf{S} \geq -(n-1)\delta, \quad \mathrm{Diam}(M) \leq d \tag{16.34}$$

and the volume form $d\mu = \phi_\mu dV_F$ satisfies

$$\left|\frac{d\phi_\mu}{\phi_\mu}\right| \leq (n-1)\tau, \tag{16.35}$$

where ω_n denotes the volume of the standard unit ball in $\mathbb{R}^n$. Then for any $0 < \varepsilon < 1/(1 + e^{nkd})$,

$$\mathrm{ExDist}(M; \varepsilon) \geq e^{-2\kappa d}\Big[(1-\varepsilon)^{\frac{1}{n}} - e^{\kappa d}\varepsilon^{\frac{1}{n}}\Big]\Big[\frac{\mathrm{Vol}_F(M)}{\omega_n}\Big]^{\frac{1}{n}} > 0, \tag{16.36}$$

where $\kappa := \sqrt{-\lambda} + \delta + \tau$.

Proof. From the definition of expansion distance, we know that if for some $\rho > 0$, there exists a compact subset $A \subset M$ with $\mu(A) \geq \varepsilon V$, satisfying

$$\mu\Big(U_\rho(A)\Big) < (1-\varepsilon)V,$$

then

$$\mathrm{ExDist}(M; \varepsilon) \geq \rho.$$

Let

$$d := \mathrm{Diam}(M), \quad V := \mu(M), \quad v := \left[\frac{\mathrm{Vol}_F(M)}{\omega_n}\right]^{\frac{1}{n}}.$$

By assumption and (16.33), the S-curvature $\mathbf{S}_\mu$ of $(F, d\mu)$ is bounded

$$\mathbf{S}_\mu \geq \mathbf{S} - \frac{d\phi_\mu}{\phi_\mu} \geq -(n-1)(\delta + \tau).$$

Let $p \in M$ such that $\phi_\mu(p) = \inf \phi_\mu$. Then by assumption,

$$V = \mu(M) \geq \phi_\mu(p)\mathrm{Vol}_F(M) = \omega_n \phi_\mu(p) v^n. \tag{16.37}$$

Take a small ball $A = B(p, r)$ such that

$$\mu(A) = V\varepsilon.$$

Then

$$U_\rho(A) = B(p, r + \rho) = B(p, d_\rho), \qquad \rho > 0,$$

where $d_\rho := \min(d, r + \rho)$. By Theorem 16.1.1, we have

$$\mu(A) = V\varepsilon \leq \phi_\mu(p)\mathcal{V}(r), \tag{16.38}$$

and

$$\mu\Big(U_\rho(A)\Big) \leq \frac{\mathcal{V}(d_\rho)}{\mathcal{V}(r)} V\varepsilon, \tag{16.39}$$

where

$$\mathcal{V}(r) := \sigma_{n-1} \int_0^r \left[e^{(\delta+\tau)t} \mathbf{s}_\lambda(t)\right]^{n-1} dt.$$

From (16.38), we obtain

$$V\varepsilon = \mu(A) \leq \phi_\mu(p)\mathcal{V}(r) \leq \phi_\mu(p)\omega_n e^{n\kappa d} r^n, \tag{16.40}$$

where $\kappa := \sqrt{-\lambda} + \delta + \tau$. From (16.37) and (16.40), we obtain a lower bound on r:

$$r \geq \left(\frac{V\varepsilon}{\phi_\mu(p)\omega_n e^{n\kappa d}}\right)^{\frac{1}{n}} \geq v e^{-kd} \varepsilon^{1/n}. \tag{16.41}$$

Inequality (16.41) implies that for any $\rho > 0$,

$$\frac{\mathcal{V}(d_\rho)}{\mathcal{V}(r)} < e^{n\kappa d_\rho}\Big(\frac{d_\rho}{r}\Big)^n \le e^{n\kappa d}\Big(1+\frac{\rho}{r}\Big)^n \le e^{n\kappa d}\Big[1+v^{-1}e^{kd}\varepsilon^{-1/n}\rho\Big]^n. \tag{16.42}$$

Let ρ_o be the number satisfying

$$e^{n\kappa d}\Big[1+v^{-1}e^{kd}\varepsilon^{-1/n}\rho_o\Big]^n V\varepsilon = V(1-\varepsilon). \tag{16.43}$$

ρ_o is given by

$$\rho_o = e^{-2\kappa d}\Big[(1-\varepsilon)^{\frac{1}{n}} - e^{kd}\varepsilon^{\frac{1}{n}}\Big]\Big[\frac{\mathrm{Vol}_F(M)}{\omega_n}\Big]^{\frac{1}{n}}.$$

Assume that $0 < \varepsilon < 1/(1+e^{nkd})$. Then $\rho_o > 0$. It follows from (16.39) and (16.42) that for any $0 < \rho < \rho_o$,

$$\mu\Big(U_\rho(A)\Big) \le \frac{\mathcal{V}(d_\rho)}{\mathcal{V}(r)}V\varepsilon \le \frac{\mathcal{V}(d_{\rho_o})}{\mathcal{V}(r)}V\varepsilon < V(1-\varepsilon).$$

This implies that

$$\mathrm{ExDist}(M;\varepsilon) > \rho.$$

Letting $\rho \to \rho_o$, we obtain

$$\mathrm{ExDist}(M;\varepsilon) \ge \rho_o.$$

Q.E.D.

Corollary 16.4.2 *Given numbers $\lambda < 0, \delta > 0, d > 0$ and $\tau > 0$, let $X_i = (M_i, F_i, d\mu_i)$ be a sequence of reversible Finsler m spaces satisfying (16.34) and (16.35) for λ, δ, d and τ. Suppose that for any $0 < \varepsilon < \frac{1}{2}$,*

$$\lim_{i\to\infty} \mathrm{ExDist}(X_i;\varepsilon) = 0,$$

then either $n_i = \dim M_i \to \infty$ or there is a subsequence i_k such that $n_{i_k} \equiv n$ and

$$\lim_{k\to\infty} \mathrm{Vol}_{F_{i_k}}(M_{i_k}) = 0.$$

Remark 16.4.3 *According to [Gr3], for a compact oriented manifold M, the Gromov simplicial volume $\|M\|$ does not vanish if M admits a Riemannian metric of negative curvature. But $\|M\| = 0$ if $\pi_1(M)$ is almost nilpotent. According to Theorem 16.3.3, for any n-dimensional closed oriented reversible Finsler space (M, F) with $\|M\| \neq 0$, if*

$$\mathbf{Ric} \geq -(n-1), \qquad \mathbf{S} \geq -(n-1),$$

then

$$\mathrm{Vol}_F(M) \geq \varepsilon(n).$$

Thus if all M_i in Corollary 16.4.2 satisfy $\|M_i\| \neq 0$, then the second case does not occur, i.e., the dimensions of M_i must approach infinity.

Chapter 17

Morse Theory of Loop Spaces

The Morse theory plays an important role in differential topology. In order to study the relationship between the topology and the Riemann curvature, we consider the canonical energy functional on the loop space at a point. Since the loop space at a point is infinite-dimensional, the Morse theory can not be directly applied to this case. This problem can be solved by approximations.

17.1 A Review on the Morse Theory

In this section we shall give a quick review on the Morse theory. Technical details will be omitted. One is referred to [Mi] for details.

Let (M, F) be an n-dimensional Finsler manifold and $f : M \to \mathrm{R}$ a C^2 function. A point $x \in M$ is called a *critical point* of f if $df_x = 0$. A number $a \in \mathrm{R}$ is called a *critical value* of f if $f^{-1}(a)$ contains a critical point of f. It is known that the set of critical values has zero measure in R (the Sard Theorem).

In a local coordinate system, the Hessian of f at a point x is given by

$$\mathrm{D}^2 f(y) = \frac{\partial^2 f}{\partial x^i \partial x^j}(x) y^i y^j - 2G^i(y)\frac{\partial f}{\partial x^k}(x), \qquad y = y^i \frac{\partial}{\partial x^i}|_x.$$

At a critical point x, $\frac{\partial f}{\partial x^i}(x) = 0$, hence

$$\mathrm{D}^2 f(y) = \frac{\partial^2 f}{\partial x^i \partial x^j}(x) y^i y^j, \qquad y = y^i \frac{\partial}{\partial x^i}|_x.$$

We see that $\mathrm{D}^2 f$ is a quadratic form on $T_x M$. A critical point x is said to be *non-degenerate* if D^2f is non-degenerate on T_xM. The *index* of f at a critical point x is defined to be the *maximal dimension* of subspaces in T_xM, on which D^2f is negative definite.

At a non-degenerate critical point x_o of f, there exists a local coordinate system (x^i) such that

$$f(x) = f(x_o) - x_1^2 - \cdots - x_k^2 + x_{k+1}^2 + \cdots + x_n^2,$$

where $k = \text{ind}_{x_o}(f)$. Thus, non-degenerate critical points are isolated. A C^∞ function f on M is called a *Morse function* if it has only non-degenerate critical points.

Proposition 17.1.1 *Let f be a Morse function on a manifold M.*

(i) *If $f^{-1}[a,b]$ is compact, and a, b are regular values, then $f^{-1}[a,b]$ has the homotopy type of a finite CW complex obtained from $f^{-1}(a)$ by attaching one cell of dimension λ for each critical point of index λ.*
(ii) *If $f^{-1}[a,c]$ is compact for all $c < b$ and a is a regular value, then $f^{-1}[a,b)$ has the homotopy type of a (possibly infinite) CW complex obtained from $f^{-1}(a)$ by attaching one cell of dimension λ for each critical point of index λ.*

Proposition 17.1.1 is very important. However, functions arising from geometry, might have degenerate critical points. To overcome this problem, we approximate them by Morse functions. Given a C^∞ function f on an open subset $\mathcal{U} \subset \mathbb{R}^n$, let $x \in \mathcal{U}$ be a critical point of f. There is an open neighborhood U_x of x and $\varepsilon > 0$ such that if h is a Morse function on U satisfying

$$\|D^2f - D^2h\|_{C^0} < \varepsilon$$

then for any critical point $z \in U$ of h

$$\text{ind}_z(h) \geq \text{ind}_x(f).$$

Proposition 17.1.2 *Let f be a C^∞ function on a manifold M. Suppose that $f^{-1}[a,b]$ is compact for some regular values $a, b \in \mathrm{R}$. Then f can be approximated by a C^∞ Morse function h on M such that $h^{-1}[a,b] = f^{-1}[a,b]$ and $h = f$ near the boundary $\partial(f^{-1}[a,b])$. Further, if $\text{ind}_x(f) \geq \lambda$ for all critical points x of f on $f^{-1}[a,b]$, then $\text{ind}_x(h) \geq \lambda$ for all critical points x of h on $h^{-1}[a,b]$.*

We now quote some important facts from the homotopy theory. Let Y be a Hausdorff topological space and $\{X_i\}_{i=0}^{\infty}$ be an increasing closed subsets in Y such that $Y = \bigcup_{i=0}^{\infty} X_i$. Suppose that each X_i has the homotopy type of a CW complex with cells of dimension $\geq \lambda$. Then Y also has the homotopy type of a CW complex with cells of dimension $\geq \lambda$.

Let $X \subset Y$ be Hausdorff topological spaces. Suppose that Y has the homotopy type of a CW complex obtained from X by attaching cells of dimension $\geq \lambda$. Then, for any $k \leq \lambda - 1$, every map $\phi : \mathbb{D}^k \to Y$ with $\phi(\partial \mathbb{D}^k) \subset X$ can be deformed to a map $\phi_1 : \mathbb{D}^k \to X$. Thus the kth relative homotopy groups

$$\pi_k(Y, X) = 0, \qquad k \leq \lambda - 1.$$

In a special case when $\lambda = 2$, every curve $c : [0,1] \to Y$ with $c(0), c(1) \in X$ can be deformed to a curve $c^* : [0,1] \to X$. With this in mind, one can prove the following path-deformation theorem.

Proposition 17.1.3 *Let $f : M \to \mathrm{R}$ be a C^∞ function (possibly with degenerate critical points). Suppose that $f^{-1}(-\infty, \delta]$ is compact for all $\delta < b$. Then for any C^∞ curve $c : [0,1] \to f^{-1}(-\infty, b)$ and any $\varepsilon > 0$, there is a homotopy c_s of c with*

$$c_0 = c, \qquad c_s(0) = c(0), \qquad c_s(1) = c(1), \qquad 0 \leq s \leq 1,$$

such that c_1 is a C^∞ curve and

$$f \circ c_1 \leq \max(d, m) + \varepsilon,$$

where $d := \max\Big(f \circ c(0), f \circ c(1)\Big)$ and m denotes the largest critical value of f with index one or zero.

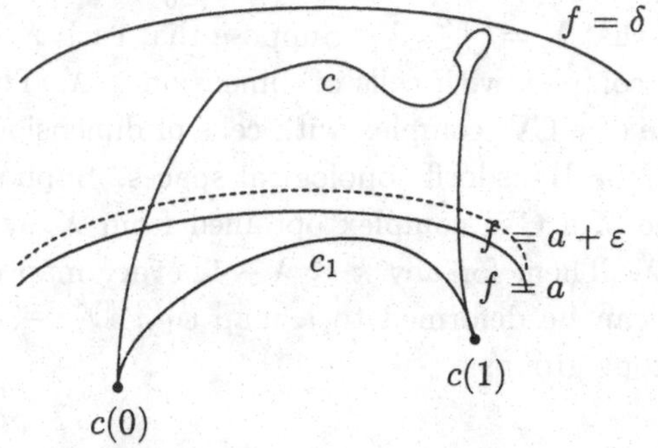

Proof: Let $a = \max(m, d)$. Assume that $a + \varepsilon < b$. By the Sard Lemma, one can find two regular values a^*, b^* of f such that $a < a^* < a+\varepsilon < b^* < b$ and $c \subset f^{-1}(-\infty, b^*)$. It follows from Proposition 17.1.2 that f can be approximated by a C^∞ function $g : M \to \mathrm{R}$ such that

$$g^{-1}[a^*, b^*] = f^{-1}[a^*, b^*]$$

and g has only non-degenerate critical points of index ≥ 2 in $g^{-1}[a^*, b^*]$. Therefore, $f^{-1}[a^*, b^*]$ has the homotopy type of a CW complex obtained from $f^{-1}(a^*)$ by attaching cells each of dimension ≥ 2. Thus, there is a homotopy c_s, $0 \leq s \leq 1$, of c with $c_0 = c$, fixing two endpoints, such that $c_1 \subset f^{-1}(a^*)$, i.e.,

$$f \circ c_1 \leq a^* < a + \varepsilon.$$

c_1 might not be a C^∞ curve, but one can always deform c_1 to a C^∞ curve c_1^* with

$$f \circ c_1^* < a + \varepsilon.$$

Q.E.D.

17.2 Indexes of Geodesic Loops

Before we study the energy functional on a loop space, we discuss some basic properties of the index form along a geodesic.

Let (M, F) be a Finsler space and $c : [a, b] \to M$ a geodesic. Recall that the index form $\mathcal{I}_c$ along c is defined by

$$\mathcal{I}_c(U, V) := \int_a^b \Big\{ \mathbf{g}_{\dot{c}}\Big(\mathrm{D}_{\dot{c}}U, \mathrm{D}_{\dot{c}}V\Big) - \mathbf{g}_{\dot{c}}\Big(\mathbf{R}_{\dot{c}}(U), V\Big)\Big\} dt, \qquad (17.1)$$

where U, V are piecewise C^∞ vector fields along c.

Let $\mathcal{W}_c$ denote the space of all piecewise C^∞ vector fields V along c with $V(a) = V(b) = 0$. $\mathcal{I}_c$ is a symmetric quadratic form on $\mathcal{W}_c$. Define $\mathrm{ind}(\mathcal{I}_c)$ to be the maximal dimension of subspaces of $\mathcal{W}_c$ on which $\mathcal{I}_c$ is negative definite. Set

$$\mathrm{ind}(c) := \mathrm{ind}(\mathcal{I}_c). \qquad (17.2)$$

We call $\mathrm{ind}(c)$ the *index* of c. The index of a geodesic is independent of positively oriented reparametrizations. That is, for any geodesic $c(t)$, $a \le t \le b$, and a new parametrization, $t(s) = c + ds$, where $d > 0$, the resulting geodesic $\tilde{c}(s) := c \circ t(s)$

$$\mathrm{ind}(c) = \mathrm{ind}(\tilde{c}).$$

The geodesic c is said to be *degenerate* if $\mathcal{I}_c$ is degenerate, i.e., there is a non-zero vector field $J \in \mathcal{W}_c$ such that

$$\mathcal{I}_c(J, V) = 0, \qquad V \in \mathcal{W}_c.$$

We see that such a vector field J must be a C^∞ Jacobi field along c with $J(a) = 0 = J(b)$. Thus c is degenerate if and only if $c(b)$ is conjugate to $c(a)$ along c.

Fix any partition $a = t_0 < \cdots < t_k = b$ such that each portion of c,

$$c_i(t) := c(t), \qquad t_i \le t \le t_{i+1}.$$

contains no conjugate points to $c(t_i)$ along c_i. By Lemma 12.1.2, for any piecewise C^∞ vector field $V \ne 0$ along c_i with $V(t_i) = V(t_{i+1}) = 0$,

$$\mathcal{I}_{c_i}(V, V) > 0.$$

Let $\mathcal{J}_c$ denote the subspace of piecewise C^∞ Jacobi fields $J \in \mathcal{W}_c$ along c such that J is C^∞ on each $[t_i, t_{i+1}]$. Each $J|_{[t_i, t_{i+1}]}$ is uniquely determined by $J(t_i)$ and $\mathrm{D}_{\dot{c}}J(t_i)$. Thus

$$\dim \mathcal{J}_c < \infty.$$

Let $\tilde{\mathcal{I}}_c$ denote the restriction of $\mathcal{I}_c$ on $\mathcal{J}_c$. Define the index $\mathrm{ind}(\tilde{\mathcal{I}}_c)$ of the quadratic form $\tilde{\mathcal{I}}_c$ in a usual way. We have

Lemma 17.2.1 *For any C^∞ geodesic $c(t)$, $a \le t \le b$ and any partition $a = t_0 < \cdots < t_k = b$ as above.*

$$\mathrm{ind}(c) = \mathrm{ind}(\mathcal{I}_c) = \mathrm{ind}(\tilde{\mathcal{I}}_c).$$

Moreover, c is degenerate if and only if $\mathcal{I}_c$ is degenerate on $\mathcal{J}_c$.

The proof of Lemma 17.2.1 will be completed in the following two lemmas. We denote by $\mathcal{V}_c$ the set of piecewise C^∞ vector fields V along c satisfying $V(t_i) = 0$, $0 \le i \le k$. For any $V \in \mathcal{W}_c$, let $J \in \mathcal{J}_c$ such that $J(t_i) = V(t_i)$. Then $V - J \in \mathcal{V}_c$. Since $c(t_{i+1})$ is not conjugate to $c(t_i)$ along $c_i = c|_{[t_i,t_{i+1}]}$, $\mathcal{V}_c \cap \mathcal{J}_c = \{0\}$. Then we obtain the following direct decomposition

$$\mathcal{W}_c = \mathcal{V}_c \oplus \mathcal{J}_c.$$

Lemma 17.2.2 *For any $V \in \mathcal{V}_c$ and $J \in \mathcal{J}_c$,*

$$\mathcal{I}_c(V, J) = 0.$$

Proof. Observe that

$$\begin{aligned} \mathcal{I}_c(V,J) &= \int_a^b \Big\{ \mathbf{g}_{\dot c}\Big(\mathrm{D}_{\dot c}V, \mathrm{D}_{\dot c}J\Big) - \mathbf{g}_{\dot c}\Big(\mathbf{R}_{\dot c}(V), J\Big)\Big\} dt \\ &= \int_a^b \frac{d}{dt}\Big[\mathbf{g}_{\dot c}\Big(V, \mathrm{D}_{\dot c}J\Big)\Big] dt \\ &= \sum_{i=0}^{k-1} \mathbf{g}_{\dot c}\Big(V, \mathrm{D}_{\dot c}J\Big)\Big|_{t_i^-}^{t_{i+1}^+} = 0. \end{aligned}$$

Q.E.D.

Now we prove the second lemma.

Lemma 17.2.3 *$\mathcal{I}_c|_{\mathcal{V}_c}$ is positive definite.*

Proof. By the choice of the partition of $[0, 1]$, any smooth Jacobi field J along $c_i := c|_{[t_i,t_{i+1}]}$ with $J(t_i) = 0, J(t_{i+1}) = 0$ must be trivial. According to Lemma 12.1.2, for any piecewise C^∞ vector field $V \ne 0$ along c_i with $V(t_i) = 0$ and $V(t_{i+1}) = 0$,

$$\mathcal{I}_{c_i}(V, V) > 0$$

Thus for any $0 \neq V \in \mathcal{V}_c$,

$$\mathcal{I}_c(V,V) = \sum_{i=0}^{k-1} \mathcal{I}_{c_i}(V,V) > 0.$$

Q.E.D.

By Lemmas 17.2.2 and 17.2.3, we see that the index of $\mathcal{I}_c$ on $\mathcal{W}_c$ is determined by that of $\tilde{\mathcal{I}}_c = \mathcal{I}_c|_{\mathcal{J}_c}$ and $\mathcal{I}_c$ is degenerate on $\mathcal{W}_c$ if and only if $\tilde{I}_c$ is degenerate on $\mathcal{J}_c$. This proves Lemma 17.2.1.

17.3 Energy Functional on a Loop Space

Let (M, F) be complete. For a point $p \in M$, we denote by $\Omega(p)$ the set of piecewise C^∞ curves $c : [0,1] \to M$ with $c(0) = p = c(1)$. There is a canonical functional E defined on $\Omega(p)$ by

$$E(c) = \int_0^1 [F(\dot{c})]^2 dt.$$

E is is called the canonical *energy functional* on $\Omega(p)$. The topology of $\Omega(p)$ is the open-compact topology so that E is continuous.

Fix a loop $c \in \Omega(p)$ and a partition $0 = t_0 < \cdots < t_k = 1$. Consider a curve c_s, $-\varepsilon < s < \varepsilon$ passing through c in Ω, i.e.,

$$c_s \in \Omega(p), \qquad c_0 = c, \quad |s| < \varepsilon. \tag{17.3}$$

We assume that c_s is C^∞ in the sense that $H(s,t) := c_s(t)$ is a piecewise C^∞ map on $(-\varepsilon, \varepsilon) \times [0,1]$ such that H is C^∞ on each $(-\varepsilon,\varepsilon) \times [t_i, t_{i+1}]$. Let

$$V := \frac{\partial H}{\partial s}(s,t), \qquad T := \frac{\partial H}{\partial t}(s,t). \tag{17.4}$$

Note that $V(t) := V(0,t)$ is a piecewise C^∞ vector field along c with $V(0) = V(1) = 0$ and $T(0,t) = \dot{c}(t)$. Thus $V(t) \in \mathcal{W}_c$. We can view $\mathcal{W}_c$ as the tangent space of $\Omega(p)$ at c. Define the *differential* $d_cE : \mathcal{W}_c \to \mathrm{R}$ by

$$d_cE(V) := \frac{d}{ds}\Big[E(c_s)\Big]\Big|_{s=0}, \qquad V = \frac{\partial H}{\partial s}|_{s=0}. \tag{17.5}$$

We have

$$\begin{aligned}
\frac{d}{ds}\Big[E(c_s)\Big]\Big|_{s=0} &= \int_0^1 V\Big[g_T(T,T)\Big]dt\Big|_{s=0} \\
&= 2\int_0^1 \Big\{T\Big[g_T(V,T)\Big] - g_T(V,\mathrm{D}_T T)\Big\}dt\Big|_{s=0} \\
&= -2\int_0^1 \mathbf{g}_{\dot c}\Big(V, \mathrm{D}_{\dot c}\dot c\Big)dt \\
&\quad +2\sum_{i=1}^{k-1}\Big\{\mathbf{g}_{\dot c(t_i^-)}\Big(V(t_i), \dot c(t_i^-)\Big) - \mathbf{g}_{\dot c(t_i^+)}\Big(V(t_i), \dot c(t_i^+)\Big)\Big\}.
\end{aligned}$$

From the above identity, we see that if c is a C^∞ geodesic, then

$$d_c E(V) = 0, \qquad V \in \mathcal{W}_c.$$

By definition, a loop $c \in \Omega(p)$ is called a *critical loop* of E if $d_c E = 0$.

Lemma 17.3.1 *A curve $c \in \Omega(p)$ is a critical loop of E if and only if c is a C^∞ geodesic.*

Proof: Suppose that $c \in \Omega(p)$ is a critical loop of E. First, take a piecewise defined function φ on $[0, 1]$ such that

(a) φ is C^∞ on each $[t_i, t_{i+1}]$, $i = 0, \cdots, k-1$;
(b) $\varphi(t) > 0$, $t_i < t < t_{i+1}$, $i = 0, \cdots, k-1$;
(c) $\varphi(t_i) = 0$, $i = 0, \cdots, k$.

For a vector field $V(t) := \varphi(t)\mathrm{D}_{\dot c}\dot c(t) \in \mathcal{W}_c$,

$$d_c E(V) = -2\int_0^1 \varphi\, \mathbf{g}_{\dot c}\Big(\mathrm{D}_{\dot c}\dot c, \mathrm{D}_{\dot c}\dot c\Big)dt = 0.$$

Thus $\mathrm{D}_{\dot c}\dot c = 0$, i.e., c is a geodesic on each $[t_i, t_{i+1}]$.

Fix i_o with $0 < i_o < k$. For an arbitrary vector $v \in T_{c(t_{i_o})}M$, take a piecewise C^∞ vector field $V(t) \in \mathcal{W}_c$ such that $V(t_{i_o}) = v$ and $V(t_i) = 0$ for all i with $i \neq i_o$. Then

$$d_c E(V) = 2\Big\{\mathbf{g}_{\dot c(t_{i_o}^-)}\Big(v, \dot c(t_{i_o}^-)\Big) - \mathbf{g}_{\dot c(t_{i_o}^+)}\Big(v, \dot c(t_{i_o}^+)\Big)\Big\} = 0.$$

Since v is arbitrary, by Lemma 1.2.4, we conclude that $\dot c(t_{i_o}^+) = \dot c(t_{i_o}^-)$. Therefore c is a C^∞ geodesic on $[0, 1]$ from O.D.E. theory. Q.E.D.

Let c be a C^∞ geodesic loop at p. Formally, we define the *Hessian* $d_c^2 E$ of E at c by

$$d_c^2 E(V) = \frac{d^2}{ds^2}\Big[E(c_s)\Big]\big|_{s=0}, \qquad V \in \mathcal{W}_c,$$

where c_s is a curve in $\Omega(p)$ whose variation field along c is V. By the second variation formula, we have

$$\frac{d^2}{ds^2}\Big[E(c_s)\Big]\Big|_{s=0} = 2\int_0^1 \Big\{\mathbf{g}_{\dot c}\Big(\mathrm{D}_{\dot c}V, \mathrm{D}_{\dot c}V\Big) - \mathbf{g}_{\dot c}\Big(\mathbf{R}_{\dot c}(V), V\Big)\Big\}dt = 2\mathcal{I}_c(V,V).$$

From the above identity, we see that the index $\mathrm{ind}_c E$ of E is equal to that of $\mathcal{I}_c$, respectively.

$$\mathrm{ind}_c E = \mathrm{ind}(\mathcal{I}_c) = \mathrm{ind}(c). \tag{17.6}$$

17.4 Approximation of Loop Spaces

To apply the Morse theory to the energy functional E on a loop space $\Omega(p)$, we approximate $\Omega(p)$ by finite-dimensional manifolds. For a number $\delta > 0$, let

$$\Omega^\delta(p) := E^{-1}(-\infty, \delta] = \Big\{c \in \Omega(p), \ \ E(c) < \delta\Big\}.$$

Since $B(p, \sqrt{\delta})$ is precompact, there is a positive number $i_o > 0$ such that the injectivity radius satisfies

$$\mathbf{i}_x \geq i_o, \qquad \forall x \in B(p, \sqrt{\delta}).$$

Fix a partition $0 = t_0 < \cdots < t_k = 1$ such that

$$t_{i+1} - t_i < \frac{i_o^2}{\delta}, \qquad 0 \leq i \leq k-1. \tag{17.7}$$

We denote by $\Omega^\delta(t_0, \cdots, t_k)$ the set of loops $c \in \Omega^\delta(p)$ such that $c|_{[t_i, t_{i+1}]}$ is a minimizing geodesic. For a loop $c \in \Omega(p)$,

$$E^\delta(c) \ = \ \sum_{i=1}^{k-1} \int_{t_i}^{t_{i+1}} [F(\dot c)]^2 dt$$

$$= \sum_{i=1}^{k-1} \frac{1}{t_{i+1}-t_i}\Big[\int_{t_i}^{t_{i+1}} F(\dot{c})dt\Big]^2$$

$$= \sum_{i=1}^{k-1} \frac{d\Big(c(t_i), c(t_{i+1})\Big)^2}{t_{i+1}-t_i}.$$

Denote by E^δ the restriction of E to $\Omega^\delta(t_0, \cdots, t_k)$. By assumption, for any $c \in \Omega^\delta(t_0, \cdots, t_k)$, $E(c) < \delta$, i.e.,

$$E^\delta(c) = \sum_{i=0}^{k-1} \frac{d\Big(c(t_i), c(t_{i+1})\Big)^2}{t_{i+1}-t_i} < \delta. \tag{17.8}$$

Let N denote the set of $(x_1, \cdots, x_{k-1}) \in M \times \cdots \times M$ such that

$$\sum_{i=0}^{k-1} \frac{d(x_i, x_{i+1})^2}{t_{i+1}-t_i} < \delta, \tag{17.9}$$

where $x_0 = p, x_k = q$. Hence, for any $(x_1, \cdots, x_{k-1}) \in N$,

$$d(x_i, x_{i+1}) < \sqrt{(t_{i+1}-t_i)\delta} < i_o, \qquad i = 0, \cdots, k-1. \tag{17.10}$$

For each $0 \le i \le k-1$, there is a unique minimizing geodesic c_i from x_i to x_{i+1}. They form a piecewise C^∞ geodesic $c = \bigcup_{i=0}^{k-1} c_i \in \Omega^\delta(t_0, \cdots, t_k)$. Define a map $\Phi : \Omega^\delta(t_0, \cdots, t_k) \to N$ by

$$\Phi(c) := \Big(c(t_1), \cdots, c(t_{k-1})\Big).$$

By the above argument, we can see that Φ is one-to-one and onto. Since N is a C^∞ manifold without boundary, we can endow a manifold structure on $\Omega^\delta(t_o, \cdots, t_k)$ by Φ so that Φ is a diffeomorphism.

Lemma 17.4.1 *There is a deformation retraction of $\Omega^\delta(p)$ onto $\Omega^\delta(t_0, \cdots, t_k)$ such that $\Omega^{\delta'}(p)$ is mapped onto $\Omega^\delta(t_0, \cdots, t_k) \cap \Omega^{\delta'}(p)$ for any $\delta' < \delta$.*

Proof. Let $c \in \Omega^\delta(p)$. Observe that for any $t \in [t_i, t_{i+1}]$,

$$\begin{aligned} d\Big(c(t_i), c(t)\Big) &\le L(c|_{[t_i,t]}) \le \int_{t_i}^{t_{i+1}} F(\dot{c})dt \\ &\le \sqrt{\int_{t_i}^{t_{i+1}} F(\dot{c})^2 dt}\sqrt{t_{i+1}-t_i} \\ &\le \sqrt{E(c)}\sqrt{t_{i+1}-t_i} < \sqrt{(t_{i+1}-t_i)\delta} < i_o. \end{aligned}$$

Thus for any $0 \le s \le 1$, there is a unique minimizing geodesic $\sigma_s^i(t)$, $t_i \le t \le t_i(s)$, issuing from $c(t_i)$ to $c(t_i(s))$, where $t_i(s) := t_i + s(t_{i+1} - t_i)$. Define a map $\Psi_s : \Omega^\delta(p) \to \Omega^\delta(p)$ by

$$\Psi_s(c)(t) = \begin{cases} \sigma_s^i(t), & \text{if } t_i \le t \le t_i(s) \\ c(t), & \text{if } t_i(s) \le t \le t_{i+1}. \end{cases}$$

Observe that for $0 < s \le 1$

$$\begin{aligned} \int_{t_i}^{t_i(s)} \left[F\left(\frac{d\sigma_s^i}{dt}\right)\right]^2 dt &= \frac{d\Big(c(t_i), c(t_i(s))\Big)^2}{(t_{i+1} - t_i)s} \\ &\le \frac{\left[\int_{t_i}^{t_i(s)} F(\dot{c})dt\right]^2}{(t_{i+1} - t_i)s} \\ &\le \int_{t_i}^{t_i(s)} F(\dot{c})^2 dt \end{aligned}$$

This implies

$$E(\Psi_s(c)) \le E(c). \tag{17.11}$$

Notice that $\Psi_0 = \mathit{identity}$ and $\Psi_1 : \Omega^\delta(p) \to \Omega^\delta(t_0, \cdots, t_k)$ such that $\Psi_1 = \mathit{identity}$ on $\Omega^\delta(t_0, \cdots, t_k)$. Thus Ψ_s is a deformation retraction of $\Omega^\delta(p)$ onto $\Omega^\delta(t_0, \cdots, t_k)$ with the desired properties. Q.E.D.

For a loop $c \in \Omega^\delta(t_0, \cdots, t_k)$, take an arbitrary C^∞ curve c_s in $\Omega^\delta(t_0, \cdots, t_k)$ with $c_0 = c$. The map $H(s,t) := c_s(t)$ is a piecewise C^∞ geodesic variation and the variation field

$$J(t) := \frac{\partial H}{\partial s}(0, t)$$

is a piecewise C^∞ Jacobi field along c with $J(0) = J(1) = 0$ such that J is C^∞ on each $[t_i, t_{i+1}]$. Thus we can identify the *tangent space* $T_c\Omega^\delta(t_0, \cdots, t_k)$ with $\mathcal{J}_c$.

Consider E^δ as a C^∞ function on the finite-dimensional manifold $\Omega^\delta(t_0, \cdots, t_k)$. Let $c \in \Omega^\delta(p,q)$ be a C^∞ geodesic loop at p. The differential $d_cE^\delta : \mathcal{J}_c \to \mathrm{R}$ is given by

$$d_cE^\delta(J) := \frac{d}{ds}\Big[E^\delta(c_s)\Big]\Big|_{s=0}, \qquad J \in \mathcal{J}_c,$$

where $c_s \in \Omega^\delta(t_0, \cdots, t_k)$ with $c_0 = c$ and its variation field along c is J. We have

$$d_c E^\delta(J) = 2\sum_{i=0}^{k-1}\left\{\mathbf{g}_{\dot{c}(t_i^-)}\left(V_i, \dot{c}(t_i^-)\right) - \mathbf{g}_{\dot{c}(t_i^+)}\left(V_i, \dot{c}(t_i^+)\right)\right\} = d_c E(J). \tag{17.12}$$

By Lemma 1.2.4 and (17.12), we see that

$$d_c E^\delta(J) = 0, \qquad J \in \mathcal{J}_c$$

if and only if $\dot{c}(t_i^-) = \dot{c}(t_i^+)$. Thus c is a critical point of E^δ in $\Omega^\delta(t_0, \cdots, t_k)$ if and only if c is a C^∞ geodesic.

The Hessian $d_c^2 E^\delta$ on $\mathcal{J}_c$ is given by

$$d_c^2 E^\delta(J) = 2\tilde{\mathcal{I}}_c(J, J) = 2\mathcal{I}_c(J, J) = d_c^2 E(J), \qquad J \in \mathcal{J}_c \tag{17.13}$$

where $\tilde{\mathcal{I}}_c$ denotes the restriction of $\mathcal{I}_c$ onto $\mathcal{J}_c$. Define $\mathrm{ind}_c E^\delta$ in a usual way. By Lemma 17.2.1 and (17.13), we see that

$$\mathrm{ind}_c E^\delta = \mathrm{ind}(\tilde{\mathcal{I}}_c) = \mathrm{ind}(c) = \mathrm{ind}(\mathcal{I}_c) = \mathrm{ind}_c E. \tag{17.14}$$

Theorem 17.4.2 *Suppose that all geodesics in $\Omega(p)$ are non-degenerate. Then $\Omega(p)$ has the homotopy type of a CW complex with one cell of dimension λ for each geodesic c of index λ.*

Proof. Let $\delta_k \to \infty$. Construct $\Omega^{\delta_k}(t_0, \cdots, t_k)$ as above. Let E^{δ_k} denote the restriction of E to $\Omega^{\delta_k}(t_0, \cdots, t_k)$. By Lemma 17.4.1, there is a deformation from $\Omega^{\delta_k}(p)$ onto $\Omega^{\delta_k}(t_0, \cdots, t_k)$. Hence, these two spaces are homotopy equivalent to each other. Since $c \in \Omega^{\delta_k}(p)$ is a critical point of E if and only if c is a critical loop of E^{δ_k}, by (17.4) (17.6) and (17.14),

$$\mathrm{ind}_c E^{\delta_k} = \mathrm{ind}(c) \tag{17.15}$$

For any critical loop $c \in \Omega^{\delta_k}$, since c is non-degenerate by assumption, the index form $\mathcal{I}_c$ is non-degenerate on $\mathcal{W}_c$. According to Lemma 17.2.1, $\tilde{I}_c$ is non-degenerate on $\mathcal{J}_c$, hence $d_c^2 E^\delta$ is non-degenerate on $T_c\Omega^{\delta_k}(t_0, \cdots, t_k)$. We conclude that E^{δ_k} is a Morse function on $\Omega^{\delta_k}(t_0, \cdots, t_k)$. Notice that

$$(E^{\delta_k})^{-1}(-\infty, \delta] = \Omega^{\delta_k}(t_0, \cdots, t_k) \cap \Omega^\delta(p)$$

is always compact for any $0 < \delta < \delta_k$. Applying Proposition 17.1.1 to E^{δ_k}, we conclude that

$$\Omega^{\delta_k}(p) \sim \Omega^{\delta_k}(t_0, \cdots, t_k)$$

has the homotopy type of a CW complex with one cell of dimension λ for each geodesic $c \in \Omega^{\delta_k}(p)$ of index λ. Since $\Omega^{\delta_k}(p)$ increases in $\Omega(p)$ and

$$\Omega(p) = \bigcup_k \Omega^{\delta_k}(p),$$

we can conclude that $\Omega(p)$ has the homotopy type of a CW complex with one cell of dimension λ for each geodesic $c \in \Omega(p)$ of index λ. Q.E.D.

Theorem 17.4.2 is beautiful, but in reality, geodesic loops in $\Omega(p)$ might be degenerate. By approximation and Proposition 17.1.2, we obtain the following

Theorem 17.4.3 *Suppose that all indices of geodesics $c \in \Omega(p)$ satisfy*

$$\text{ind}(c) \geq \lambda.$$

Then $\Omega(p)$ has the homotopy type of a CW complex with cells each of dimension $\geq \lambda$.

Proof: The proof is similar to that of Theorem 17.4.2. Take $\delta_k \to \infty$. By (17.15), we know that for any geodesic $c \in \Omega^{\delta_k}(t_0, \cdots, t_k)$

$$\text{ind}_c E^{\delta_k} = \text{ind}_c E = \text{ind}(c) \geq \lambda.$$

By Proposition 17.1.2, we conclude that $\Omega^{\delta_k}(t_0, \cdots, t_k)$ has the homotopy type of a CW complex with cells each of dimension $\geq \lambda$. Notice that

$$\Omega(p) = \bigcup_k \Omega^{\delta_k}(p).$$

Thus $\Omega(p)$ has the same homotopy type. Q.E.D.

Chapter 18

Vanishing Theorems for Homotopy Groups

In the previous chapter, we discuss the Morse theory of the canonical energy functional on a loop space of a Finsler space. In this chapter, we will show that under certain curvature conditions, the indices of C^∞ geodesic loops are bounded from below. Then we prove some vanishing theorems for the homotopy type of Finsler spaces.

18.1 Intermediate Curvatures

By definition, the Ricci curvature is the trace of the Riemann curvature. In order to reveal more relationship between the topological quantities and the geometric quantities, we introduce the intermediate curvatures between the Riemann curvature and the Ricci curvature.

Let (M, F) be a Finsler space. For a $(k+1)$-dimensional subspace $\mathcal{V} \subset T_xM$, define the Ricci curvature $\mathbf{Ric}_\mathcal{V}$ on $\mathcal{V}$ to be the trace of the Riemann curvature restricted to $\mathcal{V}$. $\mathbf{Ric}_\mathcal{V}$ is given by

$$\mathbf{Ric}_\mathcal{V}(y) = \sum_{i=1}^{k+1} \mathbf{g}_y\Big(\mathbf{R}_y(\mathbf{b}_i), \mathbf{b}_i\Big), \qquad y \in \mathcal{V},$$

where $\{\mathbf{b}_i\}_{i=1}^{k+1}$ is an arbitrary orthonormal basis for $(\mathcal{V}, \mathbf{g}_y)$. Set $\mathbf{Ric}_\mathcal{V}(0) = 0$. We call $\mathbf{Ric}_\mathcal{V}$ the Ricci curvature on $\mathcal{V}$. $\mathbf{Ric}_\mathcal{V}$ is positively homogeneous of degree two on $\mathcal{V}$,

$$\mathbf{Ric}_\mathcal{V}(\lambda y) = \lambda^2 \mathbf{Ric}_\mathcal{V}(y), \qquad \lambda > 0,\ y \in \mathcal{V}.$$

From the definition,

$$\mathbf{Ric}(y) = \mathbf{Ric}_{T_xM}(y), \qquad y \in T_xM.$$

For a tangent plane $\mathcal{V} = P \subset T_xM$, the flag curvature $\mathbf{K}(P, y)$ is related to $\mathbf{Ric}_P$ by

$$\mathbf{K}(P, y) = \frac{\mathbf{Ric}_P(y)}{F^2(y)}, \qquad y \in P.$$

Put

$$\mathbf{Ric}_{(k)} := \inf_{\dim \mathcal{V}=k+1} \inf_{y\in\mathcal{V}} \frac{\mathbf{Ric}_{\mathcal{V}}(y)}{F^2(y)},$$

where the infimum is taken over all $(k+1)$-dimensional subspace $\mathcal{V} \subset T_xM$ and $y \in \mathcal{V} \setminus \{0\}$. We have

$$\mathbf{Ric}_{(1)} \leq \cdots \leq \frac{1}{k}\mathbf{Ric}_{(k)} \leq \cdots \leq \frac{1}{n-1}\mathbf{Ric}_{(n-1)}.$$

Note that

$$\mathbf{Ric}_{(1)} = \inf_{(P,y)} \mathbf{K}(P, y), \qquad \mathbf{Ric}_{(n-1)} = \inf_{F(y)=1} \mathbf{Ric}(y).$$

18.2 Vanishing Theorem for Homotopy Groups

Now we are ready to study the homotopy type of a Finsler space under certain curvature bounds. Let (M, F) be an n-dimensional Finsler space and E denote the canonical energy functional on a loop space $\Omega(p)$ at $p \in M$. We will show that if $\mathbf{Ric}_{(k)} \geq k$ for some $1 \leq k \leq n-1$, then for any geodesic loop $c \in \Omega(p)$ with $L(c) > \pi$ satisfies $\mathrm{ind}(c) \geq n-k$. On the other hand, we will show that for any simply connected closed reversible Finsler space with $1 \leq \mathbf{K} < 4$, every geodesic loop $c \in \Omega(p)$ has length $L(c) > \pi$. Therefore M is a homotopy sphere.

Let (M, F) be a Finsler space and $p \in M$. Consider a geodesic loop $c : [0, 1] \to M$ with $c(0) = p = c(1)$. Fix a non-zero piecewise C^∞ function

$f : [0,1] \to \mathrm{R}$ with $f(0) = f(1) = 0$. Let

$$\mathcal{S}_c(f) := \Big\{ fW;\ W \text{ parallel along } c \text{ and } W \perp \dot{c} \Big\}.$$

Here $W \perp \dot{c}$ means that $W(t)$ is orthogonal to $\dot{c}(t)$ for all $0 \le t \le 1$ with respect to $\mathbf{g}_{\dot{c}(t)}$. Denote by $\tilde{E}$ the restriction of E to $\mathcal{S}_c(f)$. For a vector field $V = fW \in \mathcal{S}_c(f)$,

$$d_c^2 \tilde{E}(V) = \mathcal{I}_c(V,V) = 2C \int_0^1 \Big\{ f'(t)^2 - r^2 \mathbf{K}\Big(P(t), \dot{c}(t)\Big) f(t)^2 \Big\} dt,$$

where $\mathbf{K}(P(t), \dot{c}(t))$ denotes the flag curvature of $P(t) := \mathrm{span}\{\dot{c}(t), W(t)\}$, $C = \mathbf{g}_{\dot{c}}(W(t), W(t))$ and $r = F(\dot{c}(t))$ are constants. We always have

$$\mathrm{ind}(c) = \mathrm{ind}_c E \ge \mathrm{ind}_c \tilde{E}.$$

For further discussion, we need the following elementary lemma.

Lemma 18.2.1 *For any $r > \pi$, there is a piecewise C^∞ function $f : [0,1] \to \mathbb{R}^+$ with $f(0) = f(1) = 0$ such that*

$$\int_0^1 f(t)^2 dt = 1, \qquad \int_0^1 \Big\{ f'(t)^2 - r^2 f(t)^2 \Big\} dt < 0.$$

Proof: Choose a sufficiently small $\varepsilon > 0$. Define a piecewise C^∞ function

$$f(t) = \begin{cases} \sin(rt), & \text{if } 0 \le t \le \frac{\pi-\varepsilon}{r} \\ \sin(\varepsilon)[1 - (rt - \pi + \varepsilon)/2\varepsilon] & \text{if } \frac{\pi-\varepsilon}{r} \le t \le \frac{\pi+\varepsilon}{r} \\ 0 & \text{if } \frac{\pi+\varepsilon}{r} \le t \le 1 \end{cases}$$

For a sufficiently small $\varepsilon > 0$,

$$\int_0^1 \Big\{ f'(t)^2 - r^2 f(t)^2 \Big\} dt = r \sin(\varepsilon) \Big[-\varepsilon \cos(\varepsilon) - \frac{2}{3} \varepsilon \sin(\varepsilon) + \frac{1}{2} \sin(\varepsilon) \Big] < 0.$$

Normalizing f, we obtain the desired function. Q.E.D.

Proposition 18.2.2 *Suppose that $\mathbf{Ric}_{(k)} \ge k$ for some $1 \le k \le n-1$. Then for any geodesic $c \in \Omega(p)$ with length $L(c) > \pi$,*

$$\mathrm{ind}(c) \ge n - k.$$

Proof. Let $r = L(c) > \pi$ and $f : [0,1] \to \mathrm{R}$ be as in Lemma 18.2.1. For $V = fW \in \mathcal{S}_c(f)$ with $\mathbf{g}_{\dot{c}}(W,W) = 1$,

$$d_c^2\tilde{E}(V) = 2\int_0^1 \Big\{f'(t)^2 - r^2\mathbf{K}\Big(P(t), \dot{c}(t)\Big)f(t)^2\Big\}dt$$

where $P(t) = \mathrm{span}\{W(t), \dot{c}(t)\}$ and $r = F(\dot{c}(t))$. Define an inner product $(\ ,\ )_c$ on $\mathcal{S}_c(f)$ by

$$\begin{aligned}(fW_1, fW_2)_c &= \int_0^1 f(t)^2\mathbf{g}_{\dot{c}(t)}\Big(W_1(t), W_2(t)\Big)dt \\ &= \mathbf{g}_{\dot{c}(t)}\Big(W_1(t), W_2(t)\Big). \qquad (18.1)\end{aligned}$$

There is an orthonormal set $\{fW_i\}_{i=1}^{n-1}$ for $\mathcal{S}_c(f)$ with respect to $(\ ,\ )_c$ such that

$$d_c^2E(fW_i) = \lambda_i.$$

By (18.1), $\{W_i(t)\}_{i=1}^{n-1}$ is also an orthonormal set with respect to $\mathbf{g}_{\dot{c}(t)}$. We may assume that $\lambda_1 \le \cdots \le \lambda_{n-1}$. Let

$$\mathcal{V}(t) = \mathrm{span}\Big\{\dot{c}(t), W_{n-k}(t), \cdots, W_{n-1}(t)\Big\} \subset T_{c(t)}M.$$

Observe that

$$\begin{aligned}\sum_{i=n-k}^{n-1}\lambda_i &= \sum_{i=n-k}^{n-1} d_c^2E(fW_i) \\ &= 2\int_0^1 \Big\{kf'(t)^2 - r^2\mathbf{Ric}_{\mathcal{V}(t)}\Big(\dot{c}(t)\Big)f(t)^2\Big\}dt \\ &\le 2k\int_0^1 \Big\{f'(t)^2 - r^2f(t)^2\Big\}dt < 0.\end{aligned}$$

This implies that at least $\lambda_{n-k} < 0$. Thus d_c^2E is negative definite on an $(n-k)$-dimensional subspace of $\mathcal{S}_c(f)$

$$\mathcal{W}(t) = \mathrm{span}\Big\{W_1(t), \cdots, W_{n-k}(t)\Big\}.$$

We conclude that $\mathrm{ind}(c) \ge n - k$. Q.E.D.

According to Proposition 18.2.2, if $\mathbf{Ric}_{(k)} \ge k$, then the index of geodesic loops c with length $L(c) > \pi$ has $\mathrm{ind}(c) \ge n - k$. However, there might be shorter geodesic loops in a Finsler manifold, whose indices are not under control.

For a point $p \in M$, let

$$\mathrm{sys}(M,p) := \inf_c L(c),$$

where the infimum is taken over all geodesic loops c at p. Set $\mathrm{sys}(M) := \inf_{p\in M} \mathrm{sys}(M,p)$. Clearly,

$$\mathrm{sys}(M,p) \geq 2\mathbf{i}_p,$$

where $\mathbf{i}_p$ denotes the injectivity radius at p. By Proposition 18.2.2 and Theorem 17.4.3, we immediately obtain the following

Theorem 18.2.3 *Let (M,F) be a connected Finsler space. Suppose that for some $1 \leq k \leq n-1$,*

$$\mathbf{Ric}_{(k)} \geq k, \qquad \mathrm{sys}(M) > \pi.$$

Then $\Omega(p)$ has the homotopy type of a CW complex with cells each of dimension $\geq n-k$. In particular,

$$\pi_i(M,p) = \pi_{i-1}(\Omega(p),*) = 0$$

for $1 \leq i \leq n-k$.

According to Theorem 13.1.2, an upper bound on the flag curvature implies a lower bound on the conjugate radius. More precisely, if the flag curvature satisfies $\mathbf{K} \leq \lambda$ for some $\lambda > 0$, then $\mathbf{c}_M \geq \pi/\sqrt{\lambda}$ ($= \infty$ if $\lambda = 0$). However, the upper bound on flag curvature implies no lower bound on the injectivity radius. A typical example is the torus $\mathrm{T}^2 = \mathrm{S}^1 \times \mathrm{S}^1$ with the standard flat Riemannian product metric. If we shrink one factor, the Gauss curvature always vanishes, but the injectivity radius approaches zero.

Theorem 18.2.4 *Let (M,F) be a closed simply connected reversible Finsler manifold of dimension $n \geq 3$. Suppose that*

$$\mathbf{Ric}_{(n-2)} \geq n-2, \qquad \mathbf{c}_M > \frac{1}{2}\pi.$$

Then $\mathbf{i}_M = \mathbf{c}_M$.

Proof: We prove it by contradiction. Suppose that $\mathbf{i}_M \neq \mathbf{c}_M$. Then $\mathbf{i}_M < \mathbf{c}_M$ by Lemma 12.2.3. It follows from Lemma 12.3.2 that there is a closed geodesic c with

$$L(c) = 2\mathbf{i}_M < 2\mathbf{c}_M.$$

Assume that $c : [0,1] \to M$ is parametrized proportional to arc-length with $c(0) = p = c(1)$. Since M is simply connected, there is a homotopy c_s such that $c_1 = c$ and $c_0 = p$. We may assume that c_s is piecewise C^∞. Consider the canonical energy functional E on the loop space $\Omega(p)$. We know that the critical points of E are exactly the smooth geodesic loops at p. It follows from Proposition 18.2.2 that if

$$\sqrt{E(\sigma)} = L(\sigma) > \pi,$$

then

$$\mathrm{ind}(\sigma) \geq 2.$$

Fix an arbitrary number δ with

$$\sqrt{\delta} > \sup_{0 \leq s \leq 1} L(c_s).$$

Approximate $\Omega^\delta(p)$ by a finite-dimensional manifold $\Omega^\delta(t_0, \cdots, t_k)$, where $0 = t_0 < \cdots < t_k = 1$ is a partition of $[0,1]$ that is defined as in (17.7). There is a deformation from $\Omega^\delta(p)$ onto $\Omega^\delta(t_0, \cdots, t_k)$ (Lemma 17.4.1). We denote by E^δ the restriction of E to $\Omega^\delta(t_0, \cdots, t_k)$. A geodesic loop $c \in \Omega^\delta(p)$ is a critical point of E if and only if $c \in \Omega^\delta(t_0, \cdots, t_k)$ is a critical point of E^δ. The index of E at $c \in \Omega(p)$ is exactly the index of E^δ at $\sigma \in \Omega^\delta(t_0, \cdots, t_k)$. Take a number δ' with

$$\delta > \delta' > \sup_{0 \leq s \leq 1} L(c_s)$$

such that δ' is not a critical value of E^δ. According to Proposition 17.1.3 there exists a curve c_s^* in $\Omega^\delta(t_0, \cdots, t_k)$ from $c_0 = p$ to $c_1 = c$ such that

$$L(c_s^*) = \sqrt{E^\delta(c_s^*)} \leq \max(2\mathbf{i}_M, \pi) + 2\varepsilon < 2\mathbf{c}_M - 2\varepsilon,$$

where ε is a small positive number. We make the following

Assertation. Each c_s^* can be lifted to a loop in $\mathrm{B}_{\mathbf{c}_M - \varepsilon}(p) \subset T_pM$ by $\exp_p$.

We grant this assertion for a moment. Then $c_1^* = c$ can be lifted to a loop in T_pM by $\exp_p$. This is impossible, because that the pre-image of c under $\exp_p$ in the ball $\mathrm{B}_{\mathbf{c}_M-\varepsilon}(p) \subset T_pM$ is a straight line issuing from the origin.

Proof of Assertation. Let I denote the set of $s \in [0,1]$ for which c_s^* can be lifted to a loop in $\mathrm{B}_{\mathbf{c}_M-\varepsilon}(p)$ by $\exp_p$.

(i) Suppose that $c_{s_o}^*$ can be lifted to a loop $\tilde{c}_{s_o}$ in $\mathrm{B}_{\mathbf{c}_M-\varepsilon}(p)$ by $\exp_p$. Since $\exp_p$ is a local diffeomorphism on $\mathrm{B}_{\mathbf{c}_M}(p)$, for s sufficiently close to s_o, c_s^* can be lifted to a curve $\tilde{c}_s^*$ in $\mathrm{B}_{\mathbf{c}_M-\varepsilon}(p)$. Therefore I is open.

(ii) Let $s_i \to s_o$ be such that $c_{s_i}^*$ can be lifted to a loop $\tilde{c}_{s_i}$ in $\mathrm{B}_{\mathbf{c}_M-\varepsilon}(p)$. Note that $L(c_{s_i}) < 2\mathbf{c}_M - 2\varepsilon$. It follows from the Ascoli theorem that $\tilde{c}_{s_i}$ sub-convergent to a loop $\tilde{c}_{s_o}$ in $\overline{\mathrm{B}_{\mathbf{c}_M-\varepsilon}(p)} \subset \mathrm{B}_{\mathbf{c}_M}(p)$. The limit curve $\tilde{c}_{s_o}$ is the lift of c_{s_o}. Since $L(c_{s_o}) < 2\mathbf{c}_M - 2\varepsilon$, $\tilde{c}_{s_o}$ must be contained in $\mathrm{B}_{\mathbf{c}_M-\varepsilon}(p)$. Thus I is closed.

(i) and (ii) imply that $I = [0,1]$. This proves the assertation. Q.E.D.

Theorem 18.2.5 *Let (M,F) be a complete simply connected reversible Finsler manifold of dimension $n \geq 3$. Suppose that for some $1 \leq k \leq n-2$,*

$$\mathbf{Ric}_{(k)} \geq k, \qquad \mathbf{c}_M > \frac{1}{2}\pi.$$

Then $\Omega(p)$ has the homotopy type of a CW complex with cells each of dimension $\geq n-k$. In particular, $\pi_i(M,p) = 0,\ 1 \leq i \leq n-k$.

Proof. By Theorem 18.2.4, we see that

$$\mathbf{i}_M = \mathbf{c}_M > \frac{1}{2}\pi.$$

Thus every geodesic loop at p has length greater than π. It follows from Proposition 18.2.2. that every geodesic loop c at p has index

$$\mathrm{ind}(c) \geq n-k.$$

By Theorem 18.2.3, we can conclude that $\Omega(p)$ has the homotopy type of a CW complex with cells each of dimension $\geq n-k$. Q.E.D.

According to the Bonnet-Myers theorem (Theorem 13.3.1), if M is a positively complete simply connected Finsler surface with $\mathbf{K} \geq 1$, then $\mathrm{Diam}(M) \leq \pi$ and M is diffeomorphic to S^2. By this observation, Theorems 13.1.2 and 18.2.5, we immediately obtain the following

Theorem 18.2.6 (Homotopy Sphere Theorem). *Let (M, F) be a complete simply connected reversible Finsler manifold of dimension $n \geq 2$. Suppose that one of the following conditions is satisfied*

(a) $\mathbf{Ric}_{(k)} \geq k,\ \mathbf{c}_M > \frac{1}{2}\pi$,
(b) $\mathbf{Ric}_{(k)} \geq k,\ \mathbf{K} < 4$,

where $k = [\frac{n}{2}]$. Then M is a homotopy sphere.

J. Kern proved a diffeomorphism sphere theorem for almost Riemannian Finsler spaces of pinched curvature. See [Ke1][Ke2] for details.

Using Toponogov's comparison theorem [To] for Riemannian spaces, we can prove the following homeomorphism sphere theorem in a much more direct way.

Theorem 18.2.7 *If (M, g) is a simply connected closed Riemannian manifold with $1 \leq \mathbf{K} < 4$, then M is homeomorphic to a sphere.*

This celebrated theorem is the result of efforts by several people: Rauch, Berger [Bg1] [Bg2], Toponogov, and Klingenberg [Kl1][Kl2]. See [ChEb] [Bg3] [Pe3] for further developments.

18.3 Finsler Spaces of Positive Constant Curvature

There are infinitely many non-reversible Finsler metrics on the n-sphere with positive constant curvature. These examples are given in Section 9.2. In this section, we will discuss some basic geometric and topological properties of Finsler spaces of positive constant curvature.

First, we prove the following

Theorem 18.3.1 *Let (M, F) be a positively complete Finsler space with $\mathbf{K} = 1$. Then for any point $p \in M$, there is a point $q \in M$ such that every geodesic $c(t)$, $0 \leq t < \infty$, with $c(0) = p$ passes through q at $t = \pi$, i.e., $c(\pi) = q$.*

Proof: Let $\exp_p : T_pM \to M$ denote the exponential map at p. By Theorem 13.1.2, $\exp_p$ is non-singular on $\mathrm{B}_\pi(p) \subset T_pM$. Take any C^∞ curve $\xi : (-\varepsilon, \varepsilon) \to \mathrm{S}_pM$ and consider the geodesic variation $H : (-\varepsilon, \varepsilon) \times [0, \pi] \to M$ defined by

$$H(s, t) := \exp_x[t\xi(s)].$$

For each fixed s, the variation field

$$J_s(t) := \frac{\partial H}{\partial s}(s,t), \qquad 0 \le t \le \pi,$$

is a Jacobi field along the unit speed geodesic $c_s(t) := \exp_x[t\xi(s)]$, $0 \le t \le \pi$. J_s is C^∞ on $(0,\pi]$. By Lemma 11.2.2, J_s is C^∞ along $c_s(t)$, $t \ge 0$, satisfying

$$J_s(0) = 0, \qquad \mathrm{D}_{\dot{c}_s} J_s(0) = \xi(s).$$

Further, $J_s(t)$ is perpendicular to $\dot{c}_s(t)$ with respect to $\mathbf{g}_{\dot{c}_s(t)}$. Since $\mathbf{K} = 1$, the Jacobi field J_s satisfies

$$\mathrm{D}_{\dot{c}_s}\mathrm{D}_{\dot{c}_s} J_s(t) + J_s(t) = 0.$$

Let E_s denote the parallel vector field along c_s with $E_s(0) = \xi(s)$. We obtain that

$$J_s(t) := (\sin t) E_s(t)$$

In particular, we have

$$\frac{\partial H}{\partial s}(s,\pi) = J_s(\pi) = (\sin \pi) E_s(\pi) = 0.$$

Thus

$$\exp_x(\pi\xi(s)) = H(s,\pi) = \{point\}, \qquad |s| < \varepsilon.$$

Since $\xi(s)$ is an arbitrary curve in S_pM, we conclude that there is a point $q \in M$ such that

$$\exp_p[\pi\xi] = q, \qquad \forall \xi \in \mathrm{S}_pM.$$

This proves the theorem. Q.E.D.

Now let us look at the Bao-Shen metrics $F_k = \alpha_k + \beta_k$ on S^3. See Example 9.3.1 above. F_k has the following properties

(a) $\mathbf{K} = 1$;
(b) $\mathbf{S} = 0$;
(c) $\mathrm{Vol}_{F_k}(\mathrm{S}^3) = \mathrm{Vol}(\mathbb{S}^3)$.

Fix an arbitrary point $p \in \mathrm{S}^3$. By Theorem 18.3.1 and (a), there is a point $q \in \mathrm{S}^3$ such that

$$\exp_p(\pi\xi) = q, \qquad \forall \xi \in \mathrm{S}_p\mathrm{S}^3.$$

By the volume comparison theorem, for any $r > 0$,

$$\mathrm{Vol}_{F_k}(B(p,r)) \le \sigma_3(r), \tag{18.2}$$

where $\sigma_3(r)$ denotes the volume of the metric ball of radius r in $\mathbb{S}^3$. From the proof of Theorem 16.1.1, we see that the equality in (18.2) holds if and only if $B(p,r) \subset \mathcal{D}_p$, that is, the injectivity radius $\mathbf{i}_p \ge r$. By the Bonnet-Myers theorem (Theorem 13.3.1), we know that Diam $\le \pi$. Thus $\overline{B(p,\pi)} = M$. By (c), we obtain that

$$\mathrm{Vol}_{F_k}\Big(B(p,\pi)\Big) = \mathrm{Vol}_{F_k}(\mathrm{S}^3) = \mathrm{Vol}(\mathbb{S}^3) = \sigma_3(\pi).$$

We conclude that $\mathbf{i}_p \ge \pi$. Therefore, $d(p,q) = \pi$; all geodesics issuing from p pass through q; and geodesics from p to q are minimizing.

Now we consider a complete reversible simply connected Finsler space (M, F) with $\mathbf{K} = 1$. By the Bonnet-Myers theorem (Theorem 13.3.1), $\mathrm{Diam}(M) \le \pi$. By the Cartan-Hadamard theorem (Theorem 13.1.2), $\mathbf{c}_M \ge \pi$. By Theorem 18.2.4, $\mathbf{i}_M = \mathbf{c}_M$. We conclude that

$$\mathbf{i}_M = \mathbf{c}_M = \mathrm{Diam}(M) = \pi.$$

Fix a point $p \in M$. By Theorem 18.3.1, there exists a point $q \in M$ such that

$$q = \exp_p(\pi\xi), \qquad \forall \xi \in \mathrm{S}_pM,$$

all geodesics in M are closed of length 2π. We have proved the following

Theorem 18.3.2 ([Sh1])*Let (M, F) be a complete simply connected reversible Finsler space of constant curvature $\mathbf{K} = 1$. Then M is diffeomorphic to S^n and*

$$\mathbf{i}_M = \mathbf{c}_M = \mathrm{Diam}(M) = \pi.$$

Further, all geodesics are closed with perimeter 2π.

So far we have not found any non-Riemannian reversible Finsler metrics on S^n with $\mathbf{K} = 1$. We know that any Landsberg metric defined on an open subset $\mathcal{U} \subset R^n$ must be Riemannian if it has constant flag curvature $\mathbf{K} = \lambda \neq 0$ (Theorem 9.1.1). This is a local result. We conjecture that every reversible Finsler metrics on S^n with $\mathbf{K} = 1$ and $\mathbf{S} = 0$ must be Riemannian.

Chapter 19

Spaces of Finsler Spaces

So far we have considered "individual" Finsler spaces. Even for this purpose it can be very helpful to view the space as a member of a larger collection of Finsler metrics. A simple example of this is the collection of Minkowski spaces. Here we present some of the fundamental work of Gromov [GLP][Gr4] on collections of Finsler spaces. In particular, we will discuss the so-called precompactness and compactness theorems. We expect that there would be further developments in future.

19.1 Gromov-Hausdorff Distance

In the beginning of 80's, M. Gromov [GLP] introduced a generalized Hausdorff distance between two metric spaces. His notion leads to great developments in Riemann geometry in 80's and 90's.

We begin with the classical Hausdorff distance on the collection of subsets in a metric space. Then we generalize it to a "distance function" on the collection of compact metric spaces. For the sake of simplicity, we always assume that metrics are reversible throughout this chapter.

Let (X, d) be a metric space. For subsets $A, B \subset X$, we define the *Hausdorff distance* d_H between A and B in X by

$$d_K^X(A, B) := \inf\Big\{\varepsilon,\ A \subset \bar{U}_\varepsilon(B) \text{ and } B \subset \bar{U}_\varepsilon(A)\Big\}$$

where $U_\varepsilon(A) := \{x \in X,\ d(x, A) < \varepsilon\}$ denotes the ε-neighborhood of A in X.

Let A and B be compact metric spaces. Endow $X = A \amalg B$ with the obvious metric d^X on A and B and

$$d^X(a,b) := \max\Big\{\mathrm{Diam}(A), \mathrm{Diam(B)}\Big\}, \qquad a \in A,\, b \in B. \tag{19.1}$$

The natural embeddings of A and B into $X = A \amalg B$ are isometries. Thus, we can always view A and B as subsets in a metric space.

For arbitrary metric spaces A and B, the *Gromov-Hausdorff distance* d_{GH} between A and B is defined by

$$d_{GH}(A,B) := \inf d^X_H\Big(f(A), f(B)\Big),$$

where the infimum is taken over all metric spaces X and isometric embeddings $f : A \to X$ and $g : B \to X$. By the natural embeddings of A and B into $X = A \amalg B$ with the above metric d^X in (19.1), we obtain that

$$d_{GH}(A,B) \leq \max\Big\{\mathrm{Diam}(A), \mathrm{Diam(B)}\Big\} < \infty.$$

Let $\mathcal{M}$ denote the set of all isometric classes of compact metric spaces.

Proposition 19.1.1 ([Pe2]) *The Gromov-Hausdorff distance d_{GH} is a metric on the isometric classes of compact metric spaces. Moreover, $(\mathcal{M}, d_{GH})$ is complete.*

The proof is quite technical, but elementary. According to Proposition 19.1.1, for compact metric spaces A and B, $d_{GH}(A,B) = 0$ if and only if A is isometric to B. This is not true for non-compact metric spaces. For example, for the set $\mathbb{Q}$ of all rational numbers,

$$d_{GH}\Big([0,1], \mathbb{Q} \cap [0,1]\Big) = 0.$$

To have a better understanding on the Gromov-Hausdorff distance, we introduce the notion of ε-nets. Let (X, d) be a metric space. For $\varepsilon > 0$, a subset A of X is called an *ε-net* of X if

$$d(x, A) := \inf_{a \in A} d(x,a) < \varepsilon, \qquad \forall x \in X.$$

We denote an ε-net in X by N^X_ε. Note that the set $\mathbb{Q} \cap [0,1]$ is an ε-net in $[0,1]$ for any $\varepsilon > 0$. But $\mathbb{Q} \cap [0,1]$ is not isometric to $[0,1]$.

From the definition, we see that $d_{GH}(N^X_\varepsilon, X) < \varepsilon$. Thus we can estimate the Gromov-Hausdorff distance between two metric spaces by their ε-nets. The following lemma is very useful.

Lemma 19.1.2 *Let A, B be compact metric spaces. If there are ε-nets $N_\varepsilon^A = \{x_i\}_{i=1}^k \subset A$, $N_\varepsilon^B = \{z_i\}_{i=1}^k \subset B$ such that*

$$|d^A(x_i, x_i) - d^B(z_i, z_j)| < \varepsilon.$$

Then $d_{GH}(A, B) < 3\varepsilon$.

Proof: We endow the disjoint union $X = A \amalg B$ with the obvious metric d^X on A and B and

$$d^X(x, z) := \min \left\{ d^A(x, x_i) + d^B(z, z_i) + \varepsilon \right\}, \qquad x \in A, z \in B.$$

We see that $A \subset U_{3\varepsilon}(B)$ and $B \subset U_{3\varepsilon}(A)$. Hence $d_{GH}(A, B) < 3\varepsilon$. Q.E.D.

19.2 Precompactness Theorem

The space of all isometric classes of compact metric spaces equipped with the Gromov-Hausdorff distance is a complete metric space. However this metric space is not compact. In applications, for a given class of metric spaces, we want to know if it is (pre-)compact.

Let (X, d) be a compact metric space. For an $\varepsilon > 0$, define $\mathrm{Cov}(X, \varepsilon)$ as the minimal number of closed ε-balls needed to cover X and $\mathrm{Cap}(X, \varepsilon)$ as the maximal number of disjoint ε-balls in X. Clearly,

$$\mathrm{Cov}(X, 2\varepsilon) \leq \mathrm{Cap}(X, \varepsilon).$$

Further, if $d_{GH}(X_1, X_2) < \delta$, then for any $\varepsilon > 0$,

$$\begin{aligned} \mathrm{Cov}(X_1, \varepsilon) &\geq \mathrm{Cov}(X_2, \varepsilon + 2\delta), \\ \mathrm{Cap}(X_1, \varepsilon) &\geq \mathrm{Cap}(X_2, \varepsilon + 2\delta). \end{aligned}$$

Theorem 19.2.1 ([GLP][Gr4]). *Let $\mathcal{C} \subset \mathcal{M}$ be a subset. The following are equivalent:*

(1) $\mathcal{C}$ is precompact.

(2) There is a function $N : (0, a] \to (0, \infty)$ such that

$$\mathrm{Cap}(X, \varepsilon) \leq N(\varepsilon), \qquad \forall X \in \mathcal{C}, \ \varepsilon \in (0, a].$$

(3) There is a function $N : (0, a/2] \to (0, \infty)$ such that

$$\mathrm{Cov}(X, \varepsilon) \leq N(\varepsilon), \qquad \forall X \in \mathcal{C}, \ \varepsilon \in (0, a/2].$$

See [GLP] [Pe2] for the details.

Given $n \in \mathbb{Z}^+, \lambda \in \mathrm{R}, \delta > 0$ and $d > 0$, let $\mathcal{F}(n, \lambda, \delta, d)$ denote the space of all isometric classes of n-dimensional closed reversible Finsler spaces satisfying the following bounds

$$\mathbf{Ric} \geq (n-1)\lambda, \quad |\mathbf{S}| \leq (n-1)\delta, \quad \mathrm{Diam} \leq d.$$

We have the following

Theorem 19.2.2 ([Sh2]) *The subspace $\mathcal{F}(n, \lambda, \delta, d)$ is precompact in $(\mathcal{M}, d_{GH})$.*

Proof. Let $X \in \mathcal{F}(n, \lambda, \delta, d)$. Take a maximal set of disjoint ε-balls $B(x_i, \varepsilon)$, $i = 1, \cdots m$, in X. Assume that $B(x_{i_o}, \varepsilon)$ has the minimal volume among these ε-balls. Then

$$m \, \mathrm{Vol}_F(B(x_{i_o}, \varepsilon)) \ \leq \ \sum_{i=1}^{m} \mathrm{Vol}_F(B(x_i, \varepsilon))$$

$$\begin{aligned} &= \mathrm{Vol}_F\Big(\bigcup_{i=1}^{m} B(x_i,\varepsilon)\Big) \\ &\leq \mathrm{Vol}_F(M) \\ &= \mathrm{Vol}_F(B(x_{i_o}, d)). \end{aligned}$$

Then by the volume comparison theorem,

$$\mathrm{Cap}(X,\varepsilon) = m \leq \frac{\mathrm{Vol}_F(B_{i_o}, d)}{\mathrm{Vol}_F(B(x_{i_o},\varepsilon))} \leq \frac{\int_0^d \left[e^{\delta t}\mathbf{s}_\lambda(t)\right]^{n-1} dt}{\int_0^\varepsilon \left[e^{\delta t}\mathbf{s}_\lambda(t)\right]^{n-1} dt} =: N(\varepsilon).$$

By Theorem 19.2.1, we conclude that $\mathcal{F}(n,\lambda,\delta,d)$ is precompact in $(\mathcal{M}, d_{GH})$.

Q.E.D.

Remark 19.2.3 Let $\mathcal{R}(n,\lambda,d)$ denote the subspace of all isometric classes of n-dimensional closed Riemannian spaces (M,g) satisfying the following bounds

$$\mathbf{Ric} \geq (n-1)\delta, \quad \mathrm{Diam}(M) \leq d.$$

Then $\mathcal{R}(n,\lambda,d)$ is precompact in $(\mathcal{M}, d_{GH})$. This is the so-called Gromov precompactness theorem in Riemann geometry. See [GLP][Gr4].

Example 19.2.1 Let (V, F) be a Minkowski space and $\mathrm{S} = F^{-1}(1)$ denote the indicatrix of F. Let $\dot{g}$ and $\dot{\mathbf{C}}$ denote the induced Riemannian metric and the Cartan torsion on S. See (14.17). From the definition of $\dot{\mathbf{C}}$, we have

$$\dot{g}\Big(\dot{\mathbf{C}}(\dot{\mathbf{C}}(u,v),w),z\Big) = \dot{g}\Big(\dot{\mathbf{C}}(u,v),\dot{\mathbf{C}}(w,z)\Big).$$

Define

$$\|\dot{\mathbf{C}}\| := \sup_{v\in TS} \frac{\sqrt{\dot{g}\Big(\dot{\mathbf{C}}(v,v),\dot{\mathbf{C}}(v,v)\Big)}}{\dot{g}(v,v)}.$$

Assume that for some $\delta < 1$, the Cartan torsion satisfies

$$\|\dot{\mathbf{C}}\| \leq \delta. \tag{19.2}$$

Let $\dot{\mathbf{R}}_v(u)$ denote the Riemann curvature of $\dot{g}$. According to [Kil][Kaw],

$$\dot{\mathbf{R}}_v(u) = \dot{g}(v,v)u - \dot{g}(u,v)v + \dot{\mathbf{C}}(v, \dot{\mathbf{C}}(u,v)) - \dot{\mathbf{C}}(u, \dot{\mathbf{C}}(v,v)). \tag{19.3}$$

For any orthonormal set $\{u, v\}$, the sectional curvature $\dot{\mathbf{K}}$ of $P = \mathrm{span}\{u, v\}$ satisfies

$$\begin{aligned} \dot{\mathbf{K}}(P) &= \dot{g}\Big(\dot{\mathbf{R}}_v(u), u\Big) \\ &= \dot{g}\Big(\dot{\mathbf{C}}(u,v), \dot{\mathbf{C}}(u,v)\Big) - \dot{g}\Big(\dot{\mathbf{C}}(u,u), \dot{\mathbf{C}}(v,v)\Big) + 1 \\ &\geq 1 - \dot{g}\Big(\dot{\mathbf{C}}(u,u), \dot{\mathbf{C}}(v,v)\Big) \\ &\geq 1 - \delta^2 > 0. \end{aligned}$$

This implies the diameter of $(\mathrm{S}, \dot{g})$ is bounded from above.

$$\mathrm{Diam}(\mathrm{S}, \dot{g}) \leq \frac{\pi}{\sqrt{1-\delta^2}}.$$

Let $\mathcal{M}(n, \delta)$ denote the collection of the Riemannian metrics $\dot{g}$ on S induced by a Minkowski norm satisfying (19.2). By the Gromov precompactness theorem, $\mathcal{M}(n, \delta)$ is precompact in the Gromov-Hausdorff topology. ♮

Since the subclass $\mathcal{F}(n, \lambda, \delta, d)$ is precompact in $(\mathcal{M}, d_{GH})$, a natural question arises: how many homotopy types among Finsler spaces in $\mathcal{F}(n, \lambda, \delta, d)$? Unfortunately, there are infinitely many homotopy types among Finsler spaces in $\mathcal{F}(n, \lambda, \delta, d)$. If we pick up only Finsler spaces in $\mathcal{F}(n, \lambda, \delta, d)$, whose metric balls of a fixed radius having simple topology, then the topological type of the space can be controlled.

A *contractibility function* $\rho(\varepsilon) : [0, r) \to [0, \infty)$ is a function which satisfies

(a) $\rho(0) = 0$,
(b) $\rho(\varepsilon) \geq \varepsilon$,
(c) $\rho(\varepsilon) \to 0$ as $\varepsilon \to 0$,
(d) ρ is non-decreasing.

A metric space (X, d) is said to be $LGC(\rho)$ for some contractibility function $\rho(\varepsilon) : [0, r] \to [0, \infty)$ if for every $\varepsilon \in [0, r]$ and $x \in X$, the ε-ball $B(x, \varepsilon)$ is contractible inside $B(x, \rho(\varepsilon))$.

Given a function $N(\varepsilon) : [0, a] \to [0, \infty)$ and a contractibility function $\rho(\varepsilon) : [0, r) \to [0, \infty)$. Let $\mathcal{C}(N(\varepsilon), \rho(\varepsilon))$ denote the subset of all isometric classes of compact metric spaces $X \in \mathcal{M}$ such that X is $LGC(\rho)$ and $\mathrm{Cov}(X, \varepsilon) \leq N(\varepsilon)$ for all $\varepsilon \in [0, \alpha]$. We have the following theorem due to P. Petersen.

Theorem 19.2.4 ([Pe1][Pe2]) *Suppose that* $\lim_{\varepsilon \to 0} \varepsilon^n N(\varepsilon) < \infty$. *Then the subclass* $\mathcal{C}(N(\varepsilon), \rho(\varepsilon))$ *is compact in* $(\mathcal{M}, d_{GH})$ *and contains only finitely many homotopy types.*

Below is a consequence of our volume comparison theorem. Given a contractibility function $\rho(\varepsilon) : [0, r] \to [0, \infty)$. Let $\mathcal{F}(n, \lambda, \delta, d, \rho(\varepsilon))$ denote the subclass of $\mathcal{F}(n, \lambda, \delta, d)$ of Finsler spaces which in addition are also $GLC(\rho)$.

Proposition 19.2.5 $\mathcal{F}(n, \lambda, \delta, d, \rho(\varepsilon))$ *contains only finitely many homotopy types.*

Our next task is to find bounds on the geometric quantities under which a Finsler space is $LGC(\rho)$ for some contractibility function depending only these bounds. In [GrPe], Grove-Petersen proved the following result: given any n, λ, D and v, there exist two numbers $r > 0$ and $R > 1$ such that if a compact Riemann n-manifold (M, g) satisfies the bounds

$$\mathbf{K} \geq \lambda, \qquad \mathrm{Diam} \leq D, \qquad \mathrm{Vol} \geq v,$$

then (M, g) is $LGC(\rho)$ for

$$\rho(\varepsilon) = R\varepsilon, \qquad 0 < \varepsilon < r.$$

The graduate textbook by P. Petersen [Pe3] is an excellent book for related results on Riemannian spaces.

For a Finsler space, the lower sectional curvature bound is replaced by a lower flag curvature bound. However, an additional bound on certain non-Riemannian curvatures is required in order to obtain a similar contractibility function.

Bibliography

M. Abate and G. Patrizio, *Finsler metrics-a global approach, with applications to geometric function theory*, Lecture Notes in Mathematics **1591**, Springer-Verlag, Berlin, 1994.

T. Aikou, *Some remarks on the geometry of tangent bundles of Finsler spaces*, Tensor, N. S. **52**(1993), 234-242.

P. Antonelli, R. Ingarden and M. Matsumoto, *The theory of sprays and Finsler spaces with applications in physics and biology*, Kluwer Academic Publishers, 1993.

J.C. Álvarez Paiva, *Some Problems in Finsler Geometry*, preprint.

J.C. Álvarez Paiva and E. Fernandes, *Fourier Transforms and the Holmes-Thompson volume*, International Mathematical Research Notices **19** (1999),1031-1042.

L. Auslander, *On curvature in Finsler geometry*, Trans. Amer. Math. Soc. **79**(1955), 378-388.

H. Akbar-Zadeh, *Sur les espaces de Finsler á courbures sectionnelles constantes*, Bull. Acad. Roy. Bel. Cl, Sci, 5e Série - Tome LXXXIV (1988) 281-322.

D. Bao and S. S. Chern, *On a notable connection in Finsler geometry*, Houston J. Math. **19**(1)(1993), 135-180.

D. Bao and Z. Shen, *On the volume of unit tangent spheres in a Finsler space*, Results in Math. **26**(1994),1-17.

D. Bao and Z. Shen, *Finsler metrics of constant curvature on the Lie group S^3*, preprint.

D. Bao, S. S. Chern and Z. Shen, *An Introduction to Riemann-Finsler Geometry*, Springer-Verlag, 2000.

D. Bao, S.S. Chern and Z. Shen, *Rigidity issues on Finsler surfaces*, Rev. Roumaine Math. Pures Appl. **42**(1997), 707-735.

D. Bao, S.S. Chern and Z. Shen (ed.), *Finsler Geometry*, Contemporary Math. **196**(1996).

G. Bellettini and M. Paolini, *Anisotropic motion by mean curvature in the context* of *Finsler geometry*, Hokkaido Math. J. **25**(1996), 537-566.

P. Bérard, *Lectures on spectral geometry*, Lecture Notes in Math., vol. **1207**, Springer-Verlag, Berlin and New York, 1986.

M. Berger, *Les variétés Riemanniennes $\frac{1}{4}$-pincées*, Ann. Scuola Norm. Sup. Pisa (III) **14**(2) (1960).

M. Berger, *Sur les variétés à courbure positive de diamétre minimum*, Comment. Math. Helv. **35**(1) (1961).

M. Berger, *Riemannian geometry during the second half of the twentieth century*, IHES preprint, 1997.

R. Bishop and R. Crittenden, *Geometry of manifolds*, Academic Press, 1964.

R. Bryant, *Finsler structures on the 2-sphere satisfying $K = 1$*, Finsler Geometry, Contemporary Mathematics **196**, Amer. Math. Soc., Providence, RI, 1996, 27-42.

R. Bryant, *Projectively flat Finsler 2-spheres of constant curvature*, Selecta Math., New Series, **3**(1997), 161-204.

F. Brickell, *A new proof of Deicke's theorem on homogeneous functions*, Proc. of AMS **16**(1965), 190-191.

H. Busemann and J. P. Kelly, *Projective geometry and projective metrics*, Academic Press, New York, 1953.

H. Busemann, *Intrinsic area*, Ann. of Math., **48**(1947), 234-267.

H. Busemann, *The foundations of Minkowskian geometry*, Comment. Math. Helvet. **24**(1950), 156-187.

H. Busemann, *The geometry of geodesics*, Academic Press, New York, 1955.

H. Busemann and W. Mayer, *On the foundations of the calculus of variations*, Trans. AMS. **49**(1941), 173-198.

L. Berwald, *Untersuchung der Krümmung allgemeiner metrischer Räume auf Grund des in ihnen herrschenden Parallelismus*, Math. Z. **25**(1926), 40-73.

L. Berwald, *Parallelúbertragung in allgemeinen Räumen*, Atti Congr. Intern. Mat. Bologna **4**(1928), 263-270.

E. Cartan, *Les espaces de Finsler*, Actualités 79, Paris, 1934.

J. Cheeger and D. Ebin, *Comparison theorems in Riemannian geometry*, North-Holland Publishing Company, 1975.

J. Cheeger and D. Gromoll, *The splitting theorem for manifolds of nonegative Ricci curvature*, J. Differential Geometry **6**(1971), 119-128.

J. Cheeger and D. Gromoll, *On the structure of complete manifolds of nonnegative curvature*, Ann. of Math. **96**, 413-443.

S. S. Chern, *On the Euclidean connections in a Finsler space*, Proc. National Acad. Soc., **29**(1943), 33-37; or Selected Papers, vol. II, 107-111, Springer 1989.

S. S. Chern, *On Finsler geometry*, C. R. Acad. Sc. Paris **314**(1992), 757-761.

P. Dazord, *Propriétés globales des géodesiques des espaces de Finsler*, Thèse, Universite de Lyon, 1969.

P. Dazord, *Tores Finslériens sans points conjugués*, Bull. Soc. Math. France, **99**(1971), 171-192.

A. Deicke, *Über die Finsler-Räume mit* $A_i = 0$, Arch. Math. **4**(1953), 45-51.

Q. Ding, *A new Laplacian comparison theorem and the estimate of eigenvalues*, Chin. Ann. of Math. **15B**: 1(1994), 35-42.

J. Douglas, *The general geometry of paths*, Ann. of Math. **29**(1927-28), 143-168.

C. Duran, *A volume comparison theorem for Finsler spaces*, Proc. of AMS **126**(1998), 3079-3082.

E.D. Dzhafarov and H. Colonius, *Fechnerian metrics in unidimensional and multidimensional stimulus*, Psychological Bulletin and Review, **6**(1999), 239-268.

D. Egloff, *Some new developments in Finsler geometry*, Ph. D. thesis, Aus dem Institut für Mathematik, Universität Freiburg (Schweiz), 1995.

D. Egloff, *Uniform Finsler Hadamard manifolds*, Ann. Inst. Henri Poincaré, **66**(1997), 323-357.

P. Finsler, *Über Kurven und Flächen in allgemeinen Räumen*, (Dissertation, Göttingan, 1918), Birkhäuser Verlag, Basel, 1951.

P. Foulon, *Estimation de l'entropie des systèmes lagrangiens sans points conjugés*, Ann. Inst. Henri Poincaré **57**(**2**)(1992), 117-146.

P. Foulon, *Entropy rigigity of Anosov flows in dimension 3*, preprint.

P. Funk, *Über Geometrien bei denen die Geraden die Kürzesten sind*, Math. Ann. **101**(1929), 226-237.

P. Funk, *Über zweidimensionale Finslersche Räume, insbesondere über solche mit geradlinigen Extremalen und positiver konstanter Krümmung*, Math. Z. **40**(1936), 86-93.

S. Gallot, *Théorèmes de comparaison entre variétés et entre spectres et applications*, Actes du convegno-studio du C. N. R. (Rome 1986), soumis à Astérisque.

Y. Ge and Z. Shen, *Eigenvalues and eigenfunctions of metric measure manifolds*, Proc. London Math., to appear.

W. Gu and Z. Shen, *Lévy concentration of metric measure manifolds*, Fundamental Theories of Physics, **109**(1999), 169-178.

M. Gromov, J. Lafontaine and P. Pansu, *Structures métriques pour les variétiés Riemanniennes*, Cedic-Fernand Nathan, Paris (1981).

L. W. Green, *Auf Wiedersehensflächen* (in English), Ann. of Math. **78**(1963), 289-299.

M. Gromov, *Filling Riemann manifolds*, J. Diff. Geom. 18(1983), 1-147.

M. Gromov, *Sign and geometric meaning of curvature*, Rend. Sem. Mat. Fis. Milano **61**(1991), 9-123.

M. Gromov, *Volume and bounded cohomology*, IHES Publ. Math. **56**(1983), 213-307.

M. Gromov, *Metric structures for Riemannian and non-Riemannian spaces*, Birkhäuser, 1999.

M. Gromov, *Paul Lévy isoperimetric inequality*, IHES preprint (1980). Published in [Gr4].

M. Gromov, *Dimension, non-linear spectra and width*, Lecture Notes in Mathematics, **1317**(1988), 132-185.

M. Gromov and V. D. Milman, *A topological application of the isoperimetric inequality*, Amer. J. Math. **105**(1983), 843-854.

K. Grove and P. Petersen V, *Bounding homotopy types by geometry*, Ann. of Math. **128**(1988), 195-206.

M. Hashiguchi and Y. Ichijyō, *On some special* (α, β) *metrics*, Rep. Fac. Sci. Kagoshima Univ. **8**(1975), 39-46.

E. Heintze and H. Karcher, *A general comparison theorem with applications to volume estimates for submanifolds*, Ann. Sci. Ec. Norm. Super. **11**(1978), 451-470.

H. Hrimiuc and H. Shimada, *On the L-duality between Finsler and Hamilton manifolds*, Nonlinear World **3**(1996), 613-641.

Y. Ichijyō, *Finsler spaces modeled on a Minkowski space*, J. Math. Kyoto Univ. **16**(1976), 639–652.

M. Ji and Z. Shen, *On strongly convex indicatrices in Minkowski geometry*, Canadian Math. Bull., to appear.

A. B. Katok, *Ergodic properties of degenerate integrable Hamiltonian systems*, Izv. Akad. Nauk SSSR **37**(1973), [Russian]; Math. USSR-Izv. **7**(1973), 535-571.

A. Kawaguchi, *On the theory of non-linear connections II, Theory of Minkowski spaces and of non-linear connections in a Finsler space*, Tensor, N. S. **6**(1956), 165-199.

J. Kern, *Fastriemannsche Finslerschen metriken*, Manuscripta Math. **4**(1971), 285-303.

J. Kern, *Das pinchingproblem in fastriemannschen Finslerschen mannigfaltigkeiten*, Manuscripta Math. **4**(1971), 341-350.

S. Kikuchi, *Theory of Minkowski space and of non-linear connections in Finsler space*, Tensor, N.S. **12**(1962), 47-60.

S. Kikuchi, *On the condition that a space with* (α, β)*-metric be locally Minkowskian*, Tensor, N. S. **33**(1979), 242-246.

W. Klingenberg, *Contributions to Riemannian geometry in the large*, Ann. Math. **69**(1959).

W. Klingenburg, *Über Riemannische Mannigfaltigkeiten mit positiver Krúmmung*, Comment. Math. Helv. **35**(1961).

G. Landsberg, *Über die Totalkrümmung*, Jahresberichte der deut Math. Ver. **16**(1907), 36-46.

G. Landsberg, *Über die Krümmung in der variationsrechung*, Math. Ann. **65**(1908), 313-349.

P. Levy, *Problémes Concretes d'Analyse Fonctionelle*, Gauthier-Villars, Paris, 1951.

M. Matsumoto, *Foundations of Finsler Geometry and special Finsler Spaces*, Kaiseisha Press, Japan 1986.

M. Matsumoto, *Theory of curves in tangent planes of two-dimension Finsler spaces*, Tensor, N. S. **37**(1982), 35-42.

M. Matsumoto, *Randers spaces of constant curvature*, Rep. Math. Phys. **28**(1989),

249-261.

M. Matsumoto, *Projective changes of Finsler metrics and projectively flat Finsler spaces*, Tensor, N. S. **34**(1980), 303-315.

M. Matsumoto, *On Finsler spaces with Randers metric and special forms of important tensors*, J. Math. Kyoto Univ. **14**(1974), 477-498.

M. Matsumoto and X. Wei, *Projective changes of Finsler spaces of constant curvature*, Publ. Math. Debrecen **44**(1994), 175-181.

John Milnor, *Morse Theory*, Princeton University Press, 1963.

M. Meyer and A. Pajer, *On Santaló's inequality*, Geometric aspects of functional analysis, Springer Lecture Notes in Mathematics **1376**, 261-263.

S. Numata, *On Landsberg spaces of scalar curvature*, J. Korea Math. Soc. **12**(1975), 97-100.

K. Okubo, *Some theorems on $S_3(K)$ metric spaces*, Rep. on Math. Phys., **16**(1979), 401-408.

T. Okada, *On models of projectively flat Finsler spaces of constant negative curvature*, Tensor, N. S. **40**(1983), 117-123.

G. Paternain, Finsler structures on surfaces with negative Euler characteristic, Houston J. Math. **23**(1997), 421-426.

P. Petersen, *A finiteness theorem for metric spaces*, J. Differential Geom. **31**(1990), 387-395.

P. Petersen, *Gromov-Hausdorff convergence of metric spaces*, Proc. Symp. Pure Math. **54**, No. 3 (1993), 489-504.

P. Petersen, *Riemannian geometry*, Graduate texts in mathematics, **171**, Springer, 1997.

A. Rapcsák, *Über die bahntreuen Abbildungen metrisher Räume*, Publ. Math. Debrecen, **8**(1961), 285-290.

W. Rinow, *Die Innere Geometrie der Metrischen Räume*, Springer, New York, 1961.

H. Rund, *The Differential Geometry of Finsler Spaces*, Springer, 1959.

Z. Shen, *Finsler spaces of constant positive curvature*, In: Finsler Geometry, Contemporary Math. **196**(1996), 83-92.

Z. Shen, *Volume comparison and its applications in Riemann-Finsler geometry*, Advances in Math. **128**(1997), 306-328.

Z. Shen, *On Finsler geometry of submanifolds*, Math. Ann. **311**(1998), 549-576.

Z. Shen, *Curvature, distance and volume in Finsler Geometry*, preprint (1997).

Z. Shen, *Conjugate radius and positive scalar curvature*, Math. Z. (to appear)

Z. Shen, *On projectively related Einstein metrics in Riemann-Finsler geometry*, Math. Ann. (to appear)

Z. Shen, *On R-quadratic Finsler spaces*, Publicationes Math. Debrecen, (to appear).

Z. Shen, *Funk metrics and R-flat sprays*, preprint (2000).

Z. Shen, *Differential Geometry of Spray and Finsler Spaces*, Kluwer Academic Publishers, 2001.

M. Spivak, *A comprehensive introduction to differential geometry*, Vol III, Publish or Perish, Inc. Berkeley, 1979.

C. Shibata, H. Shimada, M. Azuma and H. Yosuda, *On Finsler spaces with Randers' metric*, Tensor, N. S. 31(1977), 219-226.

Z. Szabó, *Positive definite Berwald spaces (Structure theorems on Berwald spaces)*, Tensor, N. S. **35**(1981), 25–39.

A.C. Thompson, *Minkowski Geometry*, Encyclopedia of Math. and Its Applications, Vol. 63, Cambridge Univ. Press, Cambridge, 1996.

V. Toponogov, *Riemannian spaces with curvature bounded below*, Uspehi Mat. Nauk. **14**(1959).

O. Varga, *Die Krümmung der Eichfläche des Minkowskischen Raumes und die geometrische Deutung des einen Krümmungs-tensors des Finslerschen Raumes*, jour Abh. Mth. Sem. Univ. Hamburg **20**(1955), 41-51.

J. H. C. Whitehead, *Convex regions in the geometry of paths*, Quart. J. Math. Oxford Ser. 3 (1932), 33-42.

Y. L. Xin, *Harmonic maps of bounded symmetric domains*, Math. Ann. **303**(1995), 417-433.

H. Yasuda and H. Shimada, *On Randers spaces of scalar curvature*, Rep. on Math. Phys. **11**(1977), 347-360.

W. Ziller, *Geometry of the Katok examples*, Ergod. Th & Dynam. Sys. **3**(1982), 135-157 .

Index

amenable, 258
angular form, 10
angular metric, 124

Busemann-Hausdorff volume form, 22

Cartan tensor, 81
Cartan torsion, 108
Cheeger constant, 63
Chern connection, 85
Chern curvature, 112
Chern curvature tensor, 127
complete, 139, 175
conjugate radius, 179
conjugate value, 179
connection, 85
contractibility function, 296
convex domain, 241
convex hypersurface, 216
convex value, 241
covariant derivative, 85
critical loop, 272
curvature
  sectional, 132
  Chern, 112
  Gauss, 98
  Landsberg, 112
  mean, 219
  normal, 213
  Ricci, 98, 201
  Riemann, 96, 97
cut-domain, 183
cut-locus, 183

defining function, 8
degenerate geodesic, 269
distance function, 1, 45
distortion, 117
divergence, 21

E-curvature, 118
eigenfunction, 210
eigenvalue, 210
Einstein metric, 151
energy functional, 210, 271
Euclidean metric, 2
Euclidean norm, 2
Euclidean space, 2
Euclidean volume, 20
Euclidean volume form, 20
expansion distance, 64
exponential map, 166

Fchner function, 14
Finsler
  volume form, 31
    induced, 31
Finsler m space, 19
Finsler metric, 12
Finsler space, 12

first eigenvalue, 58
flag curvature, 98
flat metric, 2
flat metric space, 6
fundamental tensor, 81
Funk metric, 3, 15, 115

Gauss curvature, 98
geodesic coefficients, 78
geodesic curvature, 77
geodesic flow, 90
geodesic variation, 95
gradient, 41
Gromov simplicial norm, 256
Gromov simplicial volume, 256
Gromov-Hausdorff distance, 292

Hadamard space, 197
Hausdorff distance, 291
Hausdorff measure, 22
Hessian, 207, 212, 221
Hilbert form, 26, 91
Holmes-Thompson volume form, 27

index form, 180, 269
index of a function, 266
injectivity radius, 183
isoperimetric constant, 62
isoperimetric function, 56
isoperimetric profile, 56

Jacobi equation, 167
Jacobi field, 96

Klein metric, 5, 16, 142

Landsberg tensor, 84
Landsberg curvature, 112
Landsberg metric, 113
Laplacian, 209, 212, 220, 221, 243
Legendre transformation, 37, 51
length structure, 5
Levi-Civita connection, 87, 131
Levi-Civita connection forms, 87
Levi-Civita connection, 86
loop space, 271

mean covariation, 118
mean curvature, 219, 243
mean Landsberg curvature, 116
mean tangent curvature, 118
metric, 1
  Einstein, 151
  Finsler, 12
  Funk, 15
  Klein, 5, 16
  Minkowski, 6
  path, 5
  Riemannian, 13
metric measure space, 19
metric space, 1
Minkowski metric, 6
Minkowski norm, 6
Minkowski space, 6
Morse function, 266

negatively complete, 175
non-degenerate critical point, 266
non-degenerate geodesic, 269
normal curvature, 213

observable diameter, 70

parallel translation, 89
path metric, 5
positively complete, 175
psychometric function, 14

Randers metric, 14
Randers norm, 7
regular metric sphere, 247
reversible Finsler metric, 12
reversible metric, 1
Riccati equation, 223
Ricci curvature, 98, 201
Ricci scalar, 98
Riemann curvature, 96, 97
Riemann curvature tensor, 132

Riemannian metric, 13
Riemannian curvature tensor, 127
Riemannian volume form, 20

S-curvature, 118
sectional curvature, 132
shape operator, 222
Sobolev constant, 58
spray, 79
strictly convex, 8
strictly convex hypersurface, 217
strongly convex, 8

T-curvature, 153
Tangent curvature, 153
tangent cut-domain, 183

Varga equation, 214
volume form, 19

weak Landsberg metric, 116